23 Apr 2013

For Tony,

As always, with love and

The Sea

THE BARD GRADUATE CENTER
CULTURAL HISTORIES OF THE
MATERIAL WORLD

The Bard Graduate Center Cultural Histories of the Material World is a series centered on the exploration of the material turn in the study of culture. Volumes in the series examine the ways human beings have shaped and interpreted the material world from a broad range of scholarly perspectives and show how attention to materiality can contribute to a more precise historical understanding of specific times, places, ways, and means.

Peter N. Miller, Series Editor

Antiquarianism and Intellectual Life in Europe and China, 1500–1800
Peter N. Miller and François Louis, Editors

The Sea: Thalassography and Historiography
Peter N. Miller, Editor

The Sea: Thalassography and Historiography

Peter N. Miller,
Editor

The University of Michigan Press
Ann Arbor

Published in the United States of America by
The University of Michigan Press
Manufactured in the United States of America
♾ Printed on acid-free paper

2016 2015 2014 2013 4 3 2 1

A CIP catalog record for this book is available from the British Library.

Library of Congress Cataloging-in-Publication Data

The sea : thalassography and historiography / edited by Peter N. Miller.
pages cm. — (The Bard Graduate Center cultural histories of the material world)
Includes bibliographical references and index.
ISBN 978-0-472-11867-0 (cloth : acid-free paper) — ISBN 978-0-472-02901-3 (e-book)
1. Oceanography—History. 2. Seas—History. 3. Seas—Historiography. 4. Seas—Social aspects—History. 5. Ocean and civilization—History. 6. World history. 7. Material culture—History.
I. Miller, Peter N., 1964–
GC29.S43 2013
551.46–dc23 2012047391

For Istvan

Das Meer flutet immer
Das Land behält es nimmer

—Goethe, *West-Östlich Divan*

Series Editor's Preface

Why invoke a neologism? While Alexander Agassiz's historic voyages in the Atlantic, Caribbean, and Gulf of Mexico in 1877 and 1880 were published as a "Contribution to American Thalassography," the term has generally lain dormant. Even a recent attempt to encourage more thinking about the sea used the term *Thalassology*. But a term that is not used or known is also a term that carries few erroneous connotations to be cleared away like some luxuriant but obscuring underbrush.

The turn to "Thalassography" explicitly emphasizes the writing about the sea; its coupling to "Historiography" emphasizes the reciprocity we are asserting between historical scholarship in general and historical scholarship about the sea in particular. Thalassography offers a means of exploring global history through local contexts. It represents a microhistorical approach to large-scaled frameworks. In so doing, it focuses on the role of commerce, communications, exchange, and networks. These rubrics are all mediating and so are suited to the movement between local and global that drives this research inquiry.

The essays in this volume explore specific sea-contexts, emphasizing the differences between oceans and seas, and the similarities between different seas around the world. The various contributors also make clear the historiographical stakes in this project: given the state of practice today, *thalassography* offers a model for a future practice that could extend well beyond writing about the "sea."

In this way, *The Sea: Thalassography and Historiography* complements the comparative and material perspective of *Antiquarianianism and Intel-*

lectual Life in Europe and China, 1500–1800 (CHMW 1), the deep relationship between the man-made and the natural announced in the title of this series and the companion volume, *Cultural Histories of the Material World* (CHMW 3), and the practice-based focus of *Ways of Making and Knowing* (CHMW 4). It is no surprise that each of these volumes is deeply engaged with local historiographies. For the project of outlining the possibilities of a cultural history of the material world—the purpose of this series—which has been neglected or simply not noticed for so long, actually requires a historiographical engagement if we are to move beyond a simple game of replacing one older fashionable swing of the pendulum with another newer one.

Acknowledgments

In 2009, celebrating the 400th anniversary of the trans-oceanic voyage that created New Amsterdam, two professors at the Bard Graduate Center, Deborah Krohn and myself, along with a curator at The New-York Historical Society, Marybeth DeFilippis, organized an exhibition about the life of a woman who was born in Amsterdam, came of age in Malacca, and died in Brooklyn. How to study a maritime life was the question that generated the symposium which, in turn, gave birth to this book.

It is a pleasure to thank again all those who made the exhibition and symposium such a success. This book, like all those produced in this series, reflects the careful attention of Daniel Lee, Managing Editor of the Bard Graduate Center's learned publications, Laura Grey, its Art Director, and on the web side, Vanessa Rossi, its Digital Content Developer. At the University of Michigan Press we have benefited from a close working relationship with Thomas Dwyer and Christopher Dreyer. As the volume's editor, I would like again to thank Heather Topcik, Janis Ekdahl, Karyn Hinkle, and Tom Treadway, the librarians whose supply of new books and inter-library loans made it possible to explore this new line of inquiry.

A long time ago, I learned that thinking something new about something familiar was only possible if one held oneself and one's argument to the highest standard of truth. The person who taught me this was Istvan Hont. In the year of his sixty-fifth birthday, after twenty-two years of friendship, I am honored to dedicate this volume to him.

Contents

Introduction: The Sea Is the Land's Edge Also

Peter N. Miller

> There was a depression over the Atlantic. It was travelling eastwards, towards an area of high pressure over Russia, and still showed no tendency to move northwards around it. The isotherms and isotheres were fulfilling their functions. The atmospheric temperature was in proper relation to the average annual temperature, the temperature of the coldest as well as of the hottest month, and the a-periodic monthly variation in temperature. The rising and setting of the sun and of the moon, the phases of the moon, Venus and Saturn's rings, and many other important phenomena, were in accordance with the forecasts in the astronomical yearbooks. The vapour in the air was at its highest tension, and the moisture in the air was at its lowest. In short, to use an expression that describes the facts pretty satisfactorily, even though it is somewhat old-fashioned: it was a fine August day in the year 1913.[1]

The opening of Roberrt Musil's *Man Without Qualities* is as good an introduction to the problematics of thalassography as it is to a long novel about the end of the Habsburg Empire. Musil was trying to locate his story, not without some irony and even perhaps some satire, in space and time, but also to convey by analogy the way in which our microscopic world is affected by changes in the macrocosm. Leo Tolstoy's *War & Peace* does the same. Yet by using weather, rather than war, as the lever, Musil takes something completely ordinary and defamiliarizes it through distance. Its ability to effect change is from far away and is heavily mediated. How weather systems affect other weather systems, let alone humans, is extremely complicated. There is no calculus for the interaction of the weather nor, he proposes, for the interaction of people.

Musil's beginning is apt for ours, too, because in its literal sense it describes a way in which sea affects land, and the weather at sea affects the weather on land. Metaphorically, this stands to remind us that that the histories of what human beings do at sea affects the history they make on land and thus our writing of that history. Translated into historical practice, thalassography affects historiography.

But in just the same way, of course, historiography affects thalassography. Just as the question asked frames the answer given, the history of questions asked shapes the history of answers given. And though, as we shall see, many people have written about the sea, and seas have been studied, especially recently, with great intelligence and enthusiasm, the history of sea studies, and the history of writing about the sea, has been, thus far, no one's concern.

Finally, still in the realm of metaphor but transposed into an even more distant key, Musil models for us the relationship between large-scale narrative and the micro-reality of individual existence. Historians interested in concrete experiences invariably grapple with the additive nature of microhistories: do they ever add up to more than the sum of their little stories? Musil, like Tolstoy before him, believes that they do, but unlike Tolstoy finds in weather a model for how this works. In effect, Musil, like all great writers, was casting about for the starkest contrast between the large and small scale, initiating his readers into one of the great mysteries of existence.

Oceans

Musil models connection: how things concatenate, in ever-changing scale. Meterology is a powerful example of this alchemy, since it operates on a level that we still do not fully comprehend—even though our generation is much much more aware than Musil's of the stakes at play, or risk, in these interactions.

Musil, of course, wrote about an ocean (the "low" was an "Atlantic low"), not a sea. Yet, historicizing the role of oceans did not, of course, begin with him. Adam Smith, in *The Wealth of Nations,* identified ocean-crossing as the greatest event in human history since the establishment of the Roman Empire (the Spanish golden age poet Góngora had gone further: the greatest thing since the advent of Jesus Christ himself).[2]

Ocean-thinking gets sophisticated during the era of European expansion and was theorized then for its importance. But thinking about the oceans goes back to humans' first encounters with it. Pytheas of Massilia wrote about oceans in the 4th-century BCE and had sailed the North Atlantic to Iceland, at least according to Barry Cunliffe. Cunliffe himself has argued for the importance of the oceans surrounding Europe at least as early as the Mesolithic (7000 BCE) where they served as enormous protein stores and, imaginatively, as lures to go beyond—*plus ultra.*[3] Charles Baudelaire, more than two centuries after Sir Francis Bacon saw going beyond as our destiny, and more than five after Dante Alighieri saw in it our glory, saw the black canker of disappointment in the mismatch between our hopes and our reality, as mirrored in the ocean. Sigmund Freud's "oceanic feeling" turned Baudelaire's open wound into a feeling rather of vague helplessness.[4]

As much as the fifteenth century, the twentieth was a century of oceans. The two world wars made the realities of the Atlantic and the Pacific central to the destinies of the world, and in the Cold War that followed the United States projected its power via battle fleets (second in the Atlantic, third in the eastern Pacific, and seventh in the southwestern Pacific). Valuable international commodities, such as petroleum, in turn cut superhighways across the oceans.

And though the rise of the British Empire created its school of "bluewater historians," the ocean itself was always the passive actor in this story: an adversary to be conquered, or a passive space to be tamed and traversed. It was only in the second half of the twentieth century that the Atlantic, on which hinged the West—defined by the North Atlantic Treaty (1949) and NATO—became a subject for study in a way that oceans never had been before. Bernard Bailyn in the United States, and Pierre Chaunu in France, each saw in the early modern Atlantic the beginning of the modern histories of Europe and the Americas. Before long, and continuing with accelerated rhythm up to our own day and the late work of Bailyn and J. H. Elliott, Atlantic history has become one of the key ways in which early modern history has been articulated, with implications for scholarship, teaching, and professional advancement. The birth and flourishing of Atlantic history may well come to be viewed as one of the great historiographical creations of the twentieth century.

Seas

And yet, for those in sympathy with this watery turn, it is something of a puzzlement that the terms *ocean* and *sea* are still used so interchangeably, even thoughtlessly. In 1861, Jules Michelet published *La Mer:* probably the first time a historian had made a body of water his subject. Yet though he uses the word *mer* nearly all the time, Michelet is almost always writing about the ocean. Today, Routledge publishes a series entitled *Seas in History,* with volumes on the Mediterranean, Baltic and North Seas, and the Atlantic, Pacific and Indian Oceans.[5] Even as one can point to an increasing rhythm of publishing about "seascapes," no greater clarity has come with it. Thus, a recent book about perceptions of the ocean in modern Western culture is called *Sea Change: Historicizing the Ocean,* and uses the terms interchangeably. Even John Mack's wonderful cultural history of the sea is as much about oceans as about seas.[6] (Even though, in an earlier essay, Mack himself asked the question: "Yet an enclosed seascape like the Mediterranean is one thing; are open oceans signficantly different?")[7] When a few years ago, the *American Historical Review* convoked a printed forum on "Oceans of History," these highly self-conscious, super-specialized scholars spoke without precision about ocean and sea.[8] And a conference at the Library of Congress on littorals and maritime history led to a volume that, again, made no distinction between how oceans and seas function.[9] Some of this confusion may go all the way back to antiquity: Aristotle talks about *esô thálassa*—inner sea (i.e., the Mediterranean) and *exô thálassa*—outer sea, meaning ocean. But he uses the same word to refer to both.[10]

Our choice here of the term *thalassography* is designed to make clear that our focus is on seas, and that we are distinguishing between seas and oceans. Each offers very different historiographical opportunities. The appeal of oceans—for scholars as for novelists—is fundamentally that they offer us a history that escapes all bounds, just as the ocean encircles the globe. If settled life is left behind at the quay, so too are academic divisions separated by discipline, nation, language, or religion. Instead, everything flows together. Sanjay Subrahmayam, for example, has described his work on the Indian Ocean as essays in "connected history."[11] The very vastness of the Atlantic, and multiplicity of its connectedness, strains at our ability "to gear" its causality.[12] In its mature practice, those ironically inclined might find Atlantic history, too, vulnerable now for the young Bailyn's critique of Braudel's *La Méditerranée,* published in 1951 in *The Journal of Economic History:* dazzling in detail but not a whole.[13] Bailyn's point was that beneath the beautiful prose, *La Méditerranée* was a book that did not do what it purported to do; namely, showing how all the different registers geared together could make the history of the period newly intelligible.

Bailyn's own work—first on London merchants, then on colonial governors, then on political ideology, and then, after a first brilliant career, on the peopling of British North America—now appears as an alternative to what he at the beginning of his career perceived to be Braudel's more rhetorically grounded achievement.[14] Indeed, *Voyagers to the West* (1986) as a piece of craft, with its different registers of argumentation—descriptive; quantitative; structural, "multiple, career-line narration"; and even visual—may be one of the most perfect products of the historian's worktable produced in the twentieth century. And yet, what might be seen now as the triumph of Atlantic historiography—its ability to paint an Atlantic stretching from the American backcountry and Argentinian *pampas* deep into Europe and sub-Saharan Africa, and to conjoin political, economic, social, cultural, and intellectual history—may only have produced a mirror image of the Mediterraneanism that is Braudel's ever-giving gift: all things to all historians in all places at all times.

Then there is *maritime history*—an older term being repurposed to describe the range of complex narratives happening on the water.[15] From our perspective, however, there remain two limitations to this term and practice. On the one hand, by sidestepping the whole question of ocean versus sea, it does not add clarity. And on the other hand, aside from its location, one can legitimately ask whether maritime history is at all different from terrestrial history, or whether it just happens under a different set of local conditions. Indeed, the return of maritime history has coincided with an interest in human mobility, with the sea as background, not subject. To make this point even clearer, let us imagine a "celestial history." If all it did was narrate the daily life of people in space, of their loves, angers, losses, and errors, would we really feel that this was a different kind of history? Or just history elsewhere, perhaps? If we are to talk about thalassography, it must be because putting the sea at the center and the land on the periphery—geographically but also conceptually—somehow makes it a different kind of history. To paraphrase Horden and Purcell, it should be history not *in* the sea, but history *of* the sea, and while this does not preclude a discussion of its constituent material, water, nor does it require it. "Of" the sea remains always about the human experience.

The great irony, of course, noted right from the outset in yet another one of those early reviews of Braudel, is that the single biggest inspiration for maritime history, *La Méditerranée,* is a poor specimen of history *of* the sea. Garret Mattingly noted some of Braudel's limitations as a maritime historian in 1950, and Colin Heywood still others in 2008.[16] It was Braudel's claim to have created a geo-history.[17] And yet, once we isolate out what is original in his work we find that it is not, in fact, geography, and certainly not the sea. Part I of *La Méditerranée,* as we know so well, came to him from Vidal de la Blache by way of Lucien Febvre. Readers of *La Méditerranée* possessing copies of Febvre's *La Terre et l'evolution humaine* (1923) will find there discussions of mountains, plains, plateaux, islands, coasts, and oases, as well as routes and towns—all of which is generally considered the most original part of Braudel's work. Nor can it be part III, with its rather traditional account of the clash between the Eastern and Western Mediterranean empires of Spain and Turkey, as this was first written up by Ranke more than a century earlier in *Die Osmanen und die spanische Monarchie im 16. und 17. Jahrhundert* (1827). Part II, the realm of what Braudel called, following economists, the *conjuncture,* but which, instead following archaeologists, we might call "middle range theory," the interrelationship of politics, economics, and culture—and much better articulated in the second edition than in the first—is fascinating to read, but only because it shows a historiographical revolutionary trying to fit himself into traditional costume.

Today, what seems clear is that what was truly captivating about Braudel's *La Méditerranée* was the sheer audacity of proposing a relationship between so much nearly unrelated data. His rhapsodic, kaleidoscopic, material, and climatological approach, mixing the human, the natural and the divine perspective, puts *La Méditeranée* in a line of direct descent from Michelet's *La Mer.* Readers of Michelet's work today cannot fail to be captivated by the pointillist attention to details, and to aspects of the human-sea interface. At the same time, there is something slightly jarring in the utterly romanticized mode of presentation. Once we see it in Michelet we can recognize it in Braudel. Pirenne, whose *Mohammed and Charlemagne* was published in 1937 (though parts of the argument were in print a decade earlier), surely defined the place of the Mediterranean at the heart of any debate about the "end" of the ancient world, but its style and substance seem to have excited little response until much later—after archaeology could complement or contradict the arguments of its "virtual" material culture.[18]

The impact of Braudel on writing about the oceans and seas (used interchangeably, as is typical) is vast. It has launched journals, conferences, and professorships. Even critics find it an inescapable touchstone.[19] Beyond the details of academic criticism, Braudel still serves up enough force to drive two large grant-driven projects: the VW-Stiftung-backed "East Asian 'Mediterranean," and the Indian Ocean World Centre at

McGill University. Each represents an attempt to derive a comparative project from an idea of Braudel's *La Méditerranée*. Neither acknowledges any value in distinguishing between oceans and seas.

"The East Asian 'Mediterranean,' c.1500–1800: A New Quality in the Development of its Neighbouring Countries" was launched with the support of the Volkswagen Foundation in May 2002, and concluded in May 2009. The idea of the project grew out of a conference held in Paris in March 1997.[20] It was designed to shift attention away from bilateral relations between nation-states and to look at "the supra-regional and international economic (and cultural) exchange within the region of greater East Asia." Illustrating connections, rather than separation or seclusion, was its goal. Roderich Ptak had been exploring the Asian seas for much of the previous decades,[21] but this project represented a huge step towards a staffed and funded venture for which Harrassowitz Verlag undertook publication of fourteen collective volumes mapping out the extensive connections between the different parts of this maritime world (which include three PhD dissertations).[22] At its heart is the fact of trade and how it effects change, through whom, and why and when: all ultimately about the desire for goods, the means by which this desire was satisfied (or not), and the consequences of its satisfaction (or failure to do so).[23]

The root of this project lies in a comment made by José de Acosta in 1590, and made famous by Braudel in 1949: "Until now there has been no discovery in the New World of a Mediterranean sea such as Europe, Asia, and Africa possess, where arms of that great ocean enter and form different seas."[24] Where Braudel turned it back on Europe, David Abulafia has seized upon Acosta's seemingly commonplace treatment of the Mediterranean as a type of body of water, and built out a provocative and extremely stimulating mapping of Mediterranean seas all over the world.[25] Schottenhammer's project, "The East Asian Mediterranean," follows directly from this, and from it emerges also the possibility of "Mini-Mediterraneans"—a status Ptak has himself proposed for the Gulf of Tongking.[26] But there has been some caution about the application of this term to Southeast Asian waters.[27]

At the other extreme, the Indian Ocean World Centre, based at the History Department of McGill University, Montreal (after earlier stints at the Universities of Witwatersrand and Avignon), was established in 2007 to study the history, economy, and cultures of the lands and peoples of the Indian Ocean World, from China to Africa. As in the case of the "The East Asian Mediterranean," this Canadian project also acknowledges that its approach was inspired by Braudel, or at least the Braudel of part I *of La Méditerranée* summarized as: "history as an ongoing interaction between human and natural forces, encompassing geography, environment, climate and disease." The projects of the center focus on the "first global economy"—human migration, slavery, and exchange.[28]

With this "industry" flourishing, what is remarkable is how little attention has been given to the history of histories of the sea. It is, for all those who have worked on this subject, a "given" that writing about seas means being an heir of Braudel. And so before we embark upon our main theme—the impact of thalassography upon historiography—we need to begin at the beginning, with the historiography of thalassography itself. That is the purpose of Miller's opening essay—recovering a lost conversation between Braudel and S. D. Goitein (1900–1985) in the middle of the 1950s, and through it the contours of possibility. These were the years in which Braudel was working on a second edition of *La Méditerranée* while directing the VIth section of the École Pratique des Hautes Études and editing *Annales E.S.C.* Goitein, in turn, was working on a project he entitled "From Spain to India," which soon would split into two projects, the second of which, *A Mediterranean Society*, was published in six volumes between 1967 and 1985, and the first of which, the India book, in four parts (three more are envisioned) beginning only in 2008.[29]

With the appearance, finally, of the India book, the relationship between Braudel and Goitein can be considered for the very first time. What the story of their relationship reveals is the rich possibilities of thalassography and of the Mediterranean, which were visible at the time to the two men, but never fulfilled in the work of either of them. In terms of our accurate understanding of the past, and of Braudel in particular, we see that Braudel in the 1950s was thinking precisely along the lines of an integration of the maritime context (environment, trade, police) with the individuals who drove it (merchants, pirates, government). The works he sponsored through the publication series of the Centre de Recherches Historiques reveal this breadth of vision, and these aspirations were passed on to the younger scholars who carried them out. This Braudel was not the one for whom human life was mere "froth on the waves." He saw in Goitein's Geniza the possibility for a Mediterranean defined not by space but by people. Even the second edition of *La Méditerranée* does not go nearly as far toward revealing to us the breadth of its author's vision as does his work as an academic publisher and administrator.[30] Goitein's, in turn, with an approach to archival artifacts born of the union of philology and ethnography—he was trained in Arabic philology but worked in Israel as an ethnographer of newly arrived Yemenite Jews—connects the 1950s to the antiquarianism of the 1650s.[31]

How Is the Sea Different from the Ocean?

In this context one would ask whether something essential about the coherence of a closed sea is inevitably lost in the wide reaches of the Indian, or the Atlantic, or the Pacific Ocean. If, looking from the perspective of "seas," one sees the limitations of a globalized Mediterraneanism, the best Atlantic history, from the other end of the spectrum, is alert to the constraints of the *histoire du large échelle* that is inevitably the oceanic perspective. Wim Klooster, the noted Atlanticist, acknowledges this in the question which he poses in the very first sentence of his essay: "How has a thalassographic prism enabled new approaches in the historiography of the Atlantic oceanic world?"

In other words, while some still confound oceans and seas, Klooster alerts us to a large difference between them. What is a problem for Atlantic history as a practice—its inevitable centrifugalness—points to the need for a thalassography.

Atlantic history came about, according to Klooster, as a way of overcoming the previous generations' privileging of local, regional, and imperial institutions. "Their Atlantic world was one with clear national divisions, with each colony closely tied to its mother country . . . They displayed scant interest in integration, networks, social history, trans-imperial comparisons, and actors across boundaries. Their Atlantic world was shaped by Europeans, with Native Americans and Africans at best reacting to European initiatives, but not actively creating their own destinies."[32]

Klooster then outlines five paths taken in Atlantic history since the 1980s. In his elegant shorthand, these emphasize agency, adaptation, comparisons, entanglement, and networks. What Atlantic historians thus far have not done, he writes, is to "gravitate to the microecological approach that Purcell and Horden favor for the Mediterranean."[33] The Atlantic Ocean was obviously far too variegated for that. Indeed, Klooster's discerning survey of Atlanticist historiography cannot avoid acknowledging that the scale of the oceanic is inevitably dissociative. All in the end unravels.

The careful pursuit of those five options seems instead to lead toward a "microhistorical," if not always microecological, approach. In this context, microhistory means more than just small scale. For in its origin, as Italian *microstoria,* or German *Alltagsgeschichte,* it was directed against "macro" histories—whether in the first instance of kings, and states, and battles, and in the second—that of the dominant Marxist "social history." Jacques Revel observed wisely that the shift to a microhistorical frame introduces the notion of "scale," and therefore creates the historiographical option of interpretation that moves *between.*[34] Siegfried Kracauer, at the end of the 1960s, but unbeknownst to Carlo Ginzburg, who was simultaneously formulating his own approach to microhistory, wrote of a "law of levels" governing movement between the micro and macro scales. Short of the theodicy that undergirds Leibniz's attempt to legislate this relationship in the *Monadology,* much of this "game" must remain metaphorical or ideological.[35]

Klooster proposes to look closely at the Caribbean—a sea, but also an appendage of the wider ocean. It is here, he argues, that the current tendency to careful, detailed, up-close scholarship can actually be fulfilled, whereas it is likely to be stymied, or certainly attenuated, in the wider wash of the great ocean. Klooster's three avenues for future research—comparatism across imperial boundaries, entanglements of empires, and examination of networks—are all easily, amply, and powerfully on view in the scope of Caribbean history. If "connectivity" is where there are now the greatest gaps in Atlantic history, according to Klooster, it is by looking to the narrower seas that the breakthroughs are to be made.

Klooster's distinction takes us to the heart of the matter, and our jumping off point. For while oceans are big history, seas are small-scale history. Oceans are the grand narrative, seas the microhistories. In other words, the sea can be viewed as the mirror image of the ocean: smaller, with identifiable networks, and full of local meaning. Let us be clear: seas are small *only* in comparison with oceans. This is microhistory only because of the density of available historical texture. But the move to the smaller scale is also a move to a different kind of history, one tuned more closely to questions of the relationships *between.* Klooster, from within the new field of Atlantic history, proposes study of the Caribbean as a way of pursuing these meanings. Indeed, the 2012 Seminar in Atlantic History at Harvard University was devoted to precisely this relationship: "The Caribbean, the Atlantic and the Significance of Regional History." At the same time, in New York City, three museums collaborated on an exhibition entitled "Caribbean: Crossroads of the World." Perhaps the Caribbean will replace the Mediterranean as the "case" on which scholars of all stripes will soon cut their teeth.

Purcell traces for us the implications of thalassography for wider forms of writing history. He begins with the notion of "maritime history." This term, he explains, has two important virtues. First, maritime history allowed for study of the material aspects of the past (ecological, economic, practical) that were often ignored by historians interested in the perceptual or cognitive. Second, maritime histories lent themselves to a regional focus that transcended the narrowness of conventional political categories.

Yet, he argues, both of these gains still stopped short of what could be achieved, and which in fact had now become possible goals precisely because of the success of maritime histories. Thalassography, for Purcell, allows historians to join together cognitive and material lines of inquiry. At the same time, it provides a framework for the comparison of maritime historical systems. Both these dimensions emphasize connectivity, whether between historical tool kits or environmental-economic systems.

Even more crucially for Purcell's argument is that thalassography, but not maritime history, allows for the centrality of the human actor at the point where different registers converge: regional networks with material and cultural historical methods. "It is indeed precisely the nature of the behaviour of individuals," he writes, "which must provide the foundation for the characterisations of the systems of contact between places. Only then can we do history with the physical regions which have patterned those contacts."[36] For him, this human dimension can be articulated through three metaphorical arenas of action—and hence foci for a future thalassography: beach, tide, and backwash. The "beach" is where the sea impacts on people. "Tides" teach us that there is no single disposing human power but rather multiple agencies. "Backwash" reminds us of the recursivity of land and sea, people and nature, past and future.

At its best, the pursuit of "mediterraneans"—or "caribbeans"—around the world could well mean the flourishing of this approach to framing history. As Schottenhammer suggests, the identification of, in her case, an "East Asian 'Mediterranean,'" has a historiographical payoff: "to emphasize the history of exchange relations rather than that of more or less isolated states."[37]

In fact, one could argue that the East Asian Mediterranean, like all possible mediterraneans, is a purely historiographical reality. It is, as Schottenhammer makes clear, a means of bringing Chinese history into the same field of vision as its neighbors.[38] Moreover, criticizing prior attempts to domesticate Braudel to Asian waters, she argues that these have: ignored the human-environment interaction (aside from the Monsoon); imported big, exogenous, and therefore inevitably ill-fitting categories like "nation-state" or "region"; and generally been fixated on issues of progress and thus invariably narrate the rising and falling of China.

Thus if, for Purcell, Braudel had to be made true to himself by emphasizing the micro scale, whether of environment or sociology, for Schottenhammer the essence of Braudel to emerge from a second distillation was "exchange." It is, her work powerfully argues, through the lens of trade in the China seas that things clarify. "Therefore, only by considering China as a part of the East Asian networks, at the same time paying attention to political, socio-economic, and cultural changes both within China and in the East Asian world in general, can we come a bit closer to the historical truth."[39] Historiography is of course never without "real world" implications, however mediated: Subrahmanyam notes its great "subversive potential" in destabilizing the inevitable Sino-centrism of Southeast Asian historiography.[40]

One of Schottenhammer's most interesting observations closely resembles Klooster's proposal for a Caribbean Atlantic. A trade-based vision is essentially one of networks, and network studies can (almost) always be analyzed down to smaller units. Schottenhammer suggests that regions and sub-regions, and even villages, could be studied in terms of their role in the functioning of far-flung trading networks.

The implicit suggestion that thalassographies resolve into microsociological analyses is a project taken up by James Warren in his essay in this volume.[41] The Celebes Sea, or more precisely, "Sulu Zone"—comprising the Sulu Archipelago, the northeast coast of Borneo, the foreland of southern Mindanao, and the western coast of Celebes—offers a local reframing of the insular and nationalist perspectives of Indonesian, Malaysian, and Philippine historiography. Again explicitly inspired by Braudel, Warren uses the zone as a way of exploring "larger scale systemic processes of socio-economic change and a borderless history of a wide-ranging maritime trading network oriented toward China, Singapore, Europe, and the United States."[42]

But true to the thalassographer's microhistorical descent, Warren has the eye of an anthropologist. And so nomadic fishermen, slash-and-burn agriculturalists, and slaves on the one hand, and merchants, sailors, and court consumers on the other, are woven together with practices of consumption, production, and depredation and with materials such as shells, foodstuffs, ceramics, and textiles. To Braudel's many metaphors, Warren adds his own, and uses them to organize his vision of this local sea. The sea is, alternately and simultaneously, a place of abundance, fear, opportunity, change, and of history itself. "It is a central argument of this book," Warren writes of his book on the subject, though it as much true of our own, "that we cannot think of societies and cultures in isolation, as self-maintaining, autonomous, enduring systems."[43]

Thalassography as an Opportunity for Historiography

Reading these essays one begins to perceive that writing about the sea, today, is a way of partaking in the wider historiographical shift towards microhistory, exchange relations, networks, and, above all, materiality—both literally and figuratively. Let us turn to the most beautiful definition of thalassography I have found. Claudio Magris who supplied it, used a different word to describe his project, one that is anchored deeply in the revolution in how the West wrote the past. He calls it the "Philology of the Sea."

> The science of the sea encompasses not only the tracing of currents and routes, the chemical analysis of salinity, the study of stratigraphy and of maps and of the benthic and the pelagic layers of marine life, its euphotic, oliophotic, and aphotic zones, and the measurement of temperatures and winds but also stories of shipwrecks, myths of galleons gone under and ancient leviathans, the amniotic fluid of humanity and the cradle of civilization, Greek beauty, which like Aphrodite arises perfect from the sea, the great temptation of the soul as told by Musil, the clash between symbols of eternity and conviction, that is, life itself, resplendent in a pure present and the plenitude of its meaning.[44]

"Philology" indeed moves us closer to our project. For like it, thalasso*graphy* focuses not on questions of discipline and professionalization, so much as on the writing, and therefore the planning and organizing, of histories of the sea. Revel's explanation that the *jeu d'échelles* is carried by literary style—each scale has its own—explicitly gears explanation to exposition, to the active role of the historian as agent.[45] But this focus on writing immediately raises the question of the connection between thalassographies and the planning and organizing of historical projects more generally. Our argument would be that writing about the sea has been, and remains, an engine of historiographical innovation. Hannah Baader and Gerhard Wolf recently argued for "einer kulturwissenschaftlichen Thalassographie," evoking Aby Warburg's grand vision of a "kunstgeschichtlichen Kulturwissenschaft."[46] And perhaps not only of historiographical innovation. In the realm of fiction-writing as well, an attempt to grasp the sea in literary form gave birth to Peter Matthiessen's extraordinary *Far Tortuga,* an utterly compelling effort at marrying thalassography and the novel. "At sea-ness" permeates the structure as well as the content—fragments, floating, bobbing, coasting merge content and mood, just as air, water, and land melt together in his Caribbean.[47]

Nicholas Purcell's essay reflects on how a self-consciously derived thalassography could be used to push the horizon of historiographical accomplishment. The four essays that follow each explore different implications of thalassography for historiography more broadly. Nicola di Cosmo, as if borrowing from Purcell's notion of connectivity, gives us an example of a story told from these premises, and focuses on the Black Sea. Roxani Margariti suggests that the history of the Indian Ocean could be rewritten through the specifically "connective" lens of *nesiology* (island studies), and gives examples of some of the questions derived from that perspective. David Kirby identifies this set of concerns and possibilities with the concept of the littoral and puts it at the heart of his notion of thalassographic possibility. And, in conclusion, Peter Miller provides an example of this human-based history of the sea, picking up on what Braudel envisioned for the book series he sponsored in the 1950s after a gap of half a century.

Nicola di Cosmo's essay provides a small-scale example of the possibilities inherent in Purcell's proposal. He focuses on connections: "communications, economic exchange, political and administrative control." And these processes, in turn, can be redescribed as networks. Networks then bridge individual and institutional actors. For Di Cosmo, what marks the modernity of the post-1500 world is that "it began to be shaped as a growing mesh of networks of which the maritime routes were the supporting scaffolding."[48] This connection between networks, or systems, is what Di Cosmo is pursuing; not a single maritime history but the point where maritime histories connect (or collide). This, in his terms, is where the possibility of world history occurs; in Purcell's, it is where thalassography starts. For both, modernity can be defined in terms of the converging of networks.

The focus of Di Cosmo's essay, however, is not to illuminate a single maritime historical system—and as a historian of central Asia he writes illuminatingly here about the whole question of whether the steppes constitute a sea or not—but the seam where systems meet. The Black Sea zone in the Mongol century is where the Mediterranean system of the Venetians and the Genose met that of central Asia. That one was maritime and the other terrestrial made things "interesting," according to Di Cosmo. That this was the first time when we "we have a clear, deliberate construction of institutions, social bodies, and political *apparatus* that allowed the linkage between the continent-based overland trade networks and the Mediterranean exchange and transportation system to become operative" makes it a crucial historiographical node.[49]

Roxani Margariti takes islands, perhaps the most obvious instance of disconnected space, and shows, much as Di Cosmo does for the steppe lands, that the received view is upside down. Islands—especially when situated near shore, near other islands, or in sea lanes—functioned as links in a chain. Pitcairn Island, St. Helena, Chateau d'If, Devil's Island, and Alcatraz aside, most islands were not cut off (Chateau d'If and Alcatraz are just offshore, as is Napoleon's first refuge, Elba). She calls, both explicitly and by example, for a new "nesiology"—or island studies.

Nesiology from one perspective, thalassography from another. Because what Margariti's essay in this book documents is the extent to which studying islands means studying their connections to elsewhere, mediated by water. These connections are through people who move around and through objects which move around. No island is an island, Carlo Ginzburg has pronounced.[50] Margariti's study demonstrates this well enough. But Ginzburg's phrase is itself a commentary on Malinowski's Kula Ring, the best piece of nesiology we possess—even if most people think of it as anthropology. One could read Malinowski, too, as a thalassographer, for his subject is the interaction of watery wastes and human intention. In the Trobriand Islands, as in the Dahlak archipelago, the gross categories of nature and culture melt into something much less clear, but much more transparent.

In the Baltic, David Kirby reminds us, the divisions between land and water blur, with its thousands upon thousands of islands and inlets. In this environment, relations between land and sea become complicated, but also, perhaps, paradigmatic. Kirby is author of a wide-ranging survey of the Baltic and North Seas that covers the geology of the region, the sea as imagined by people, the physical realities of travel on the sea from boats to instruments to havens, the economic realities, the fish, the sociology of mariners, life at sea, and the role of women.[51] He is himself somewhat wary of the emphasis on connectivity, noting sagely that as with islands—themselves, in the Baltic, often part of the coastline (85 percent of Finland's coast, for instance, is islands)—littorals can separate as much as they connect. And, more generally still, as the concrete historical example of the Baltic shows, the littoral is often a place where central authority can and needs to demonstrate its ability to act. For Kirby, more microanalysis is required, but at the level of *mentalité,* as well as tonnage. It is worth comparing Kirby's use of "littoral" to frame issues of connectivity with Alain Corbin's use of "beach"—the very same physical place—in terms of the history of human emotions and the search for meaning.[52]

If cultural history and economic history still seem separated even at the level of language, Purcell's ambition to join cultural to material history through the singularity of human action complements Michael McCormick's reenvisioning of economic history in terms of "communications and commerce." In his book, *Origins of the European Economy* (2001), McCormick put the human being in motion, transporting goods, ideas, and, not least, him or herself, at the center of the historian's skein. That both Purcell and McCormick come to discern the horizons of the "cultural history of the material world" through writing about the sea suggests something of the power of thalassography to open up new historiographical terrain.

One of the lingering doubts about our age's pleasure in microhistory must be the concern about whether, additively, one can get beyond anecdotalism as history. Thalassography as a study of the world made by individuals in motion provides an armature that can respect the integrity of the microhistorical without giving up on the ambition that there is (or ought to be) a whole at the end of the tunnel.[53] Moreover, if sea-based historiography was long seen as a way of escaping from nationalist conventions, it also brought to the attention of historians a raft of transnational individuals: people whose life in some way is defined by being on the move, and never belonging. Natalie Zemon Davis's al-Hassan al-Wazam, or Mercedes García-Arenal's and Gerard Wiegers's Samuel Pallache are such examples.[54]

These essays in this volume provide exemplary demonstrations of how a sea-based history-writing that focuses on connectivity, networks, and individuals describes the horizons and the potential of thalassography. The volume's conclusion returns to the Mediterranean, and to the seventeenth century, but also to the Paris of the 1950s. Subrahmanyam, even without knowing of this correspondence, saw very clearly that Goitein's Mediterranean was only half a project, and that until his India Book appeared, our vision of a global Mediterranean could only be impressionistic.[55] Peiresc's correspondence network also stretched from Spain to India, and with it we see not only the details of a global Mediterranean, but we see it in Goitein's own terms: letters, memoranda, fragments of documents, and above all, people. Goitein called his approach "sociography," suggesting that it aimed at a description of society, rather than a theory of society. Looking at the Mediterranean through the lens of Peiresc's archive, we can see that historically this same practice of reconstruction was once called "antiquarianism."[56] But we can also see "The Mediterranean and the Mediterranean World in the Age of Peiresc" as just the kind of project that Braudel would have sought for his series "*Affaires et gens d'affaires*," in which merchant letters were used to reconstruct the daily life and daily thoughts of the merchants who made the Mediterranean home.

This Mediterranean—Braudel plus Goitein—directs us to look backward as well as forward. When Karl Lamprecht, more than a century ago, wrote one of the first significant pieces of economic history while describing it as cultural history he was arguing that how humans transform and interact with the natural world is *cultura* in the old sense and economics in the new. With McCormick, this precise gambit is being revived (the "origins of the European economy" practiced as cultural history). And alongside his work we can now point to a new generation of economic historians who focus on human actors and their interaction with macro-systems such as Francesca Trivellato, David Hancock, and Avner Greif, and who base their work on merchants and their correspondence. Molly Greene even writes of her book on Maltese pirates and their victims, "this study revisits Goitein's argument, but for the early modern rather than the medieval Mediterranean."[57]

The promise of the Goitein-Braudel correspondence, and the aspirations Braudel had for the CRH, are only now being realized, though without those who are doing the work being aware of this lineage. Historians, as Sanjay Subrahmanyam writes in his Afterthoughts, have never quite been able to put the sea at the center. And yet, his image of histories in bottles is apt, for there have always been historians working on these themes, but on their own and disconnected from one another and from wider conversations. One of the goals of this book, then, is to reestablish this link to the past; another is to suggest how it offers ways to connect research programs and scholars in the future.

Thalassography and Imagination

There is another way of thinking about the difference between the connecting seas and the wide, wide ocean. And this has to do with writing. We have already noted that for Revel, it was writing that enabled the microhistorian's movement between scales. Similarly, Jacques Rancière, in his provocative *Names of History* (1992), suggested that if, in contrast with the Mediterranean, the Atlantic, which so dominated the history of the modern era, does not dominate historiography at all, "it is because no writing covered it in advance." The writers who did come to it, he wrote, emphasized its peripheries—whether African slaving, the Caribbean, Cape Horn, or the Pacific isles. We might, in the light of two more decades of scholarly production, want to amend the categorical judgment of the Atlantic's marginality, yet the basic point remains true: there is no great Atlantic epic. Or rather, according to Rancière again, there is such an epic, but it is, rather, an anti-epic: *Moby-Dick.*[58]

Rancière's point, I suppose (he doesn't elaborate), is that *Moby-Dick* is about the Atlantic because the *Pequod* sails out of Nantucket. In fact, though, it is a global quest and thus rather than being one ocean's epic, it is *the* oceanic epic. But Rancière is right where he does not realize it. For *Moby-Dick,* in its very structure, represents the fusion of antiquarian knowledge and narrative "history." Like Braudel's own *La Méditerranée,* Melville's novel balances the tension between detailed description and over-determination (in his case, the romantic quest—in Braudel's, geography). Baader and Wolf's co-optation of Foucault's presentation of the ship as the heterotopia *par excellence* perfectly suits Melville's presentation of thc *Pequod* in his novel.[59]

Without such writing, Rancière claimed, the Atlantic cannot sustain both narrative and structural interpretation and so ends up breaking down quickly into its local parts. In the hands of scholars it can yield only "The *Odyssey* of research in place of the *Odyssey* of the book."[60]

As by now should be clear, thalassography as a kind of microhistory is not about water, but about people. This is the heritage of its Italian and German ancestry (*microstoria, Alltagsgeschichte*). Hence Thalasso*graphy,* not Thalasso*logy.* As a practice that emphasizes the movement between scales, and uses the human experience, at its different scales, as the cogs which gear the story, thalassography also solves what might be called "the Braudel problem." Bailyn's point, long ago, was that the three different layers of time did not add up to a single story; a thalassographic reenvisioning of the Mediterranean along the lines of what is modeled in this volume's essays shows the way forward.

The ocean, infinite and ageless—as in the concluding sentence of *Moby-Dick*—reminds us at all times of the gap between the human and the natural. The ocean only becomes somewhat knowable where it greets us, at the shore. The sea, however, is always in that closer relationship. It beckons us on, like the Sirens, to a sometimes fantastic sense of proximity and possibility. "The earth," Michelet concludes, "supplies you with what you need to live, and of the best; the sea with what you need to rise up."[61] Imagination, whether Melville's, Freud's, or Michelet's, is what we need if we are to understand this dimension of the sea.

For Michelet, we encounter the ocean at the shore, where the human scale meets the infinite but it is only moving outward, away from land, that this water becomes the authentic ocean—and thus it is truly unknowable. Against Michelet we can balance T. S. Eliot, for whom the sea, by contrast, is progressively revealed as it comes towards us. The beach, or strand, communicates this truth. Moreover, its revelation is of the deeper meaning of life on land.

> The sea is the land's edge also, the granite
> Into which it reaches, the beaches where it tosses
> Its hints of earlier and other creation:
> The starfish, the horseshoe crab, the whale's backbone;
> The pools where it offers to our curiosity
> The more delicate algae and the sea anemone.
> It tosses up our loses, the torn seine,
> The shattered lobsterpot, the broken oar
> And the gear of foreign dead men.[62]

In thalassography, the sea's reality is brought home. This dimension might, then, be most evident when exploring seas with the highest proportion of shore to "blue water"—in Purcell's terms, those with more "beach," "tide," and "backwash," in Kirby's, more "littoral." If we were speaking French we could add more *portuaire,* but in English *portuary* is a kind of prayer book.[63]

It is in Italian, however, that we shall end. For the Adriatic, with its complicated currents, islands, geology, and fauna was once described by Braudel as a miniature Mediterranean and "perhaps the most unified of all the regions of the sea" (though it was long denied the name "sea" and called instead the "Gulf of Venice").[64] While writing is not one of the ways by which Braudel saw its unity, the Adriatic offers an example of the way in which, as per Rancière, the writer's imagination creates a space for the historian's writing. (We might wonder whether Klooster's Caribbean could be reconstituted as "another Adriatic," thinking about Derek Walcott's poems or Timothy J. Reiss's scholarship). Pedrag Matvejevic, in his wonderfully pointillist *itinerarium,* perhaps the most authentic heir to Braudel's magnum opus, explains that the word "gulf" itself comes from the Greek *kolpos,* meaning bosom or lap. The gulf invites contact, closeness. It stands between the "bay," which is an even smaller gulf, and the sea, which is a bigger one—but still embracing us, unlike the wide ocean. Even the division wrought by the larger sea between its northern and southern civilizations is mimicked by the role of the Adriatic dividing the religions and worlds, even, of its eastern and western banks.

Before being drawn into Venice's bosom, the Adriatic was already the most domesticated of seas. The Roman *via Appia* stopped at Brindisi, and its continuation, the *via Egnazia,* began directly across the sea, at Durazzo, so that the ships which plied the path could easily be seen as on a road. The arrival at one or another terminus was marked by the assertion of the human. When Hermann Broch's Vergil came home to die he arrived at the roads of Brindisi, where "the sunny yet deathly loneliness of the sea changed with the peaceful stir of friendly human activity, where the channel, softly enhanced by the proximity of human life and human living, was populated by all sorts of craft."[65]

Where water and land meet, the human imagination is always challenged. Goethe's Faust, by the North Sea littoral, watches the waves crash against "the flat, wide shore" and decides "I'll ban the lordly sea, I'll curb its force. / I'll set new limits to that watery plain / And drive it back into itself again."[66] Rilke, sitting at the head of the Adriatic, at Duino, climbed out on to the cliff in a storm out of which he thought he heard a voice calling, "'Who, if I cried out, would hear me among the angels' hierarchies?"[67] The poems he wrote at Duino dwell, like the place itself, on a threshold between worlds: between the visible, everyday world, and the heightened, supra-sensible world of time and love.

If we compare this interpretation of the interaction between nature and the human or, specifically, between large bodies of water and human events, we can discern very clearly a big difference between Musil's oceanic perspective and what we might call Rilke's thalassographic one. With Musil, the huge impersonal forces act on us in ways so complicated and mediated as to register in an almost unfelt, magical way. The scale difference between ocean and individual is too great for us to comprehend its specific causalities. Rilke's experience is the opposite. The sea meets the human in a way that sets in motion a very serious "game of scales": questions about life and death, about memory and forgetfulness, seen from the finite towards the infinite, and the infinite towards the finite. When Burckhardt set out to paint his picture of the Italian Renaissance, the book which remains today a model for cultural history because of the way it tries to connect human lives with the social scale, he noted "On the wide ocean on which we shall venture out, the possible routes and courses are many. . . ."[68]

Revel, from the pulpit of *Annales,* saw the challenge of future historiography as reconciling, somehow, the challenge of the global with the disciplinary.[69] From our perspective, it is clear what kind of historian can most easily do this: the thalassographer. "Connectivity," "backwash"—whatever term one uses—thalassography is all about the movement between small and large scale. Siegfried Kracauer suggested long ago that microhistory was really "just" history because "the higher the level of generality at which a historian operates, the more historical reality thins out."[70] Thalassographies never thin out. This is what distinguishes them from the maritime history of old. In some way, then, this volume does not only define a historiographical practice, but also a human one. All of us, not only the contributors to this volume, play the thalassographer, always searching for connections, always setting off from the land's edge.

NOTES

1. Robert Musil, *The Man Without Qualities,* vol. 1, trans. Eithne Wilkins and Ernst Kaiser 3 vols. (London: 1979), 1.
2. Adam Smith, *An Inquiry into the Nature and Causes of the Wealth of Nations,* eds. R. H. Campbell and A. S. Skinner (Indianapolis, 1981), IV.vii.c.80, 626.
3. Barry Cunliffe, *Facing the Oceans: The Atlantic and its Peoples, 8000 BC–AD 1500* (Oxford University Press, 2001), and *Europe between the Ocean: Themes and Variations, 9000 BC to AD 1000* (New Haven, CT: Yale University Press, 2008).
4. Baudelaire, "L'homme et la mer" begins: "Homme libre, toujours tu chériras la mer!/ La mer est ton miroir; tu contemples ton âme/ Dans le déroulement infini de sa lame,/ Et ton esprit n'est pas un gouffre moins amer." Charles Baudelaire, *Les Fleurs du mal* (Boston: D. R. Godine, 1982), 200.
5. David Kirby and Merja-Liisa Hinkkanen, *The Baltic and North Seas* (London and New York: Routledge, 2000).
6. John Mack, *The Sea: A Cultural History* (London: Routledge, 2011).
7. John Mack, "The Land Viewed from the Sea," *Azania* 42 (2007): 4.

8. See for example Kären Wigen's "AHR Forum Oceans of History: Introduction," *American Historical Review* 111 (2006): 717; this forum contains contributions by Peregrine Horden and Nicholas Purcell, "The Mediterranean and 'the New Thalassology,'" 722–40; Alison Games, "Atlantic History: Definitions, Challenges, and Opportunities," 741–57; Matt K. Matsuda, "The Pacific," 758–80.
9. *Seascapes: Maritime Histories, Littoral Cultures, and Transoceanic Exchanges,* eds. Jerry H. Bentley, Renate Bridenthal, and Kären Wigen (Honolulu: University of Hawaii Press, 2007); Michael N. Pearson, "Littoral Society: The Concept and the Problems," *Journal of World History* 17 (2006): 353–73.
10. Aristotle, *Meterology,* II.1
11. Sanjay Subrahmanyam, *Mughals and Franks: Explorations in Connected History* (New York: Oxford University Press, 2005).
12. See for example, *Soundings in Atlantic History: Latent Structures and Intellectual Currents, 1500–1830,* eds. Bernard Bailyn and Patricia L. Denault (Cambridge, MA: Harvard University Press, 2009).
13. Bernard Bailyn, "Braudel's Geohistory. A Reconsideration," *Journal of Economic History* 11 (1951): 277–82.
14. See Silvia Marzagalli, "Sur les origines de l' 'Atlantic History,'" *Dix-Huitième Siècle* 33 (2001): 17–31; Horst Pietschmann, "Atlantic History—History between European History and Global History," in *Atlantic History: History of the Atlantic System, 1580–1830,* ed. Pietschmann (Göttingen: Vandenhoek and Ruprecht, 2002), 11–43.
15. These recent titles may be taken as in some way representative of the "new" maritime history: *Seascapes: Maritime Histories, Littoral Cultures, and Transoceanic Exchanges; Trade and Cultural Exchange in the Early Modern Mediterranean: Braudel's Maritime Legacy,* eds. Maria Fusaro, Colin Heywood, and Mohamed-Salah Omri (London: IB Tauris, 2010); Molly Greene, *Catholic Pirates and Greek Merchants: A Maritime History of the Early Modern Mediterranean* (Princeton, NJ: Princeton University Press, 2010); *Maritime History as Global History,* eds. Maria Fusaro and Amélia Polónia (St. John's, Newfoundland: International Maritime Economic History Association, 2010).
16. Garret Mattingly, review of *La Méditeranée et le monde méditerranean à l'époque de Philippe II, American Historical Review* 55 (1950): 350–51; Colin Heywood, "Fernand Braudel and the Ottomans: the emergence of an involvement (1928–50)", *Mediterranean Historical Review* 23 (2008): 176.
17. Two valuable and sympathetic recent assessments of Braudel are: *Braudel Revisited: the Mediterranean World, 1600–1800,* eds. Gabriel Piterberg, Teofilo F. Ruiz, and Geoffrey Symcox (Toronto: University of Toronto Press, 2010), and *Early Modern History and the Social Sciences: Testing the Limits of Braudel's 'Mediterranean',* ed. John Mains (Kirksville, MO: Truman State University Press, 2002).
18. Henri Pirenne, *Mohammed & Charlemagne* (New York: Dover, 1954); Pirenne published "Mahomet et Charlemagne" in the *Revue belge de Philologie et d'Histoire* in 1922. A fascinating assessment of this project from later on is Richard Hodges and David Whitehouse, *Mhammed, Charlemagne & the Origins of Europe* (Ithaca, NY: Cornell University Press, 1983). Pirenne figures as

a central character in the majestic works of Michael McCormick (*The Origins of the European Economy: Communications and Commerce 300–900* (Cambridge: Cambridge University Press, 2001), and Chris Wickham (*Framing the Early Middle Ages: Europe and the Mediterranean 400–800* (Oxford: Oxford University Press, 2005), 697.

19. Peregrine Horden and Nicholas Purcell, "AHR Forum: The Mediterranean and "the New Thalassology,'" 722–40.
20. "The Asian Mediterranean Sea/ La Méditerranée Asiatique," organized by Raymond Ptak with the assistance of Claude Guillot, at the Maison des Sciences de l'Homme.
21. See for example his collections of essays containing material dating from 1983: *China and the Asian Seas: Trade, Travel, and Visions of the Other (1400–1750)* (Aldershot, England and Brookfield, VT: Ashgate. Variorum, 1998); *China's Seaborne Trade with South and Southeast Asia* (Aldershot, England and Brookfield, VT: Ashgate. Variorum, 1999); *China, the Portuguese, and the Nanyang: Oceans and Routes, Regions and Trade (c.1000–1600)* (Aldershot, England and Brookfield, VT: Ashgate. Variorum, 2004).
22. http://www.eamh.net/index.html. The website lists the full range of participants and programming sponsored by the project.
23. See Schottenhammer, "Introduction," in *Trade and Transfer Across the East Asian "Mediterranean,"* ed. Schottenhammer (Wiesbaden: Harrassowitz, 2005), 1.
24. José de Acosta, *Natural and Moral History of the Indies,* ed. Jane E. Managan, trans. Frances M. López-Morillas (Durham, NC: Duke University Press, 2003), 123
25. David Abulafia, "Mediterraneans," in *Rethinking the Mediterranean,* ed. W. V. Harris (Oxford: Oxford University Press, 2005), 64–93.
26. Roderich Ptak, "The Gulf of Tongking: A Mini-Mediterranean," in *The East Asian 'Mediterranean': Maritime Crossroads of Culture,Commerce and Human Migration,* ed. Angela Schottenhammer (Wiesbaden: Harrassowitz Verlag, 2008), 53–72.
27. See Heather Sutherland, "Southeast Asian History and the Mediterranean Analogy," *Journal of Southeast Asian Studies* 34 (2003): 1–20.
28. http://www.indianoceanworldcentre.com. For a one-man version of this, also explicitly inspired by Braudel, see Kirti Chaudhuri, *Trade and Civilisation in the Indian Ocean: An Economic History from the Rise of Islam to 1750* (Cambridge: Cambridge University Press, 1985).
29. S. D. Goitein and Mordechai Akiva Friedman, *India traders of the Middle Ages: documents from the Cairo Geniza : ('India book')* (Leiden: Brill, 2008); *Joseph Lebedi. Prominent India Trader.* India Book I. Cairo Geniza Documents [Hebrew] (Jerusalem: Ben-Zvi Institute, 2009); Friedman, *Madmun Nagid of Yemen and the India Trade.* India Book II. Cairo Geniza Documents [Hebrew] (Jerusalem: Ben-Zvi Institute, 2010); Friedman, *Abraham ben Yiju. India Trader and Manufacturer.* India Book III. Cairo Geniza Documents [Hebrew] (Jerusalem: Ben-Zvi Institute, 2010).
30. There is no mention of the *CRH* in Franco Angiolini and Daniel Roche, eds. *Cultures et formations négociantes dans l'Europe moderne* (Paris: EHESS,

1995), nor in Daniel Roche, "Une Expérience de recherche Européenne: Le projet 'Negoce et Culture' (Florence, 1987–1989)," *Revue de Synthese* 111 (1990): 289–91, and only a passing one in Anthony Molho and Diogo Curto, "Les Réseaux Marchands à l'Époque Moderne," *Annales* 58 (2003). There is no memory any longer of the CRH projects and publications.

31. S. D. Goitein, *A Mediterranean Society,* 6 vols. (Berkeley and Los Angeles: University of California Press, 1965–85), "Epilogue," V, 496–502.
32. Klooster, in this volume.
33. Klooster in this volume.
34. Jacques Revel, "Presentation," *Jeux d'échelles. La micro-analyse à l'expérience* (Paris: Le Seuil/Gallimard 1996), 19.
35. Siegfried Kracauer, *History. The Last Things Before the Last.* Completed after the author's death by Paul Oskar Kristeller (Princeton, NJ: Markus Wiener, 1995), 125–28. The examples given by Matti Peltonen (Benjamin, Auerbach, and Geertz) all refer back to Leibniz though he does not reflect on its implications, see "Clues, Margins, and Monads: The Micro-Macro Link in Historical Research," *History and* Theory 40 (2001): 347–59, esp. 355. For Ginzburg's assessment of Kracauer, see Carlo Ginzburg, "Microhistory: Two or Three Things That I Know About It," *Critical Inquiry* 20 (1993), 26–28.
36. Purcell in this volume.
37. Schottenhammer in this volume.
38. Asymmetry is defended as a historical reality by Wang Gungwu, "The China Seas: Becoming an Enlarged Mediterranean," in *The East Asian 'Mediterranean': Maritime Crossroads of Culture, Commerce and Human Migration,* ed. Angela Schottenhammer (Wiesbaden: Harrassowitz Verlag, 2008), 15.
39. Schottenhammer in this volume.
40. Sanjay Subrahmanyam, "Notes on Circulation and Asymmetry in Two Mediterraneans, c.1400–1800," in *From the Mediterranean to the China Sea: Miscellaneous Notes,* eds. Claude Guillot, Denys Lombard, Roderich Ptak, with the assistance of Richard Teschke (Wiesbaden: Harrassowitz Verlag, 1998), 40.
41. This essay can very profitably be read alongside of Jennifer L. Gaynor, "Maritime Ideologies and Ethnic Anomalies: Sea Space and the Structure of Subalternity in the Southeast Asian Littoral," in *Seascapes,* 53–68.
42. Warren in this volume.
43. Warren in this volume.
44. Claudio Magris, "Introduction: A Philology of the Sea," in Pedrag Matvejević, *Mediterranean: A Cultural Landscape,* trans. Michael Henry Heim (Berkeley and Los Angeles: University of California Press, 1999), 1.
45. Revel, "Presentation," *Jeux d'échelles,* 34: "L'invention d'un mode d'exposition n'induit pas seulement des effets de connaissance. Elle contribue explicitement à la production d'un certain type d'intelligibilité dans des conditions expérimentales définies."
46. Hannah Baader and Gerhard Wolf, "Maritime Tableaus, Eine Vorbemerkung," *Das Meer, der Tausch und die Grenzen der Repräsentation,* eds., Baader and Wolf (Zurich-Berlin: Diaphanes, 2010), 7.
47. I am grateful to Stuart Schwartz for bringing this book to my attention.
48. Di Cosmo in this volume.

49. Di Cosmo in this volume.
50. Carlo Ginzburg, *No Island is an Island: Four Glances at English literature in a World Perspective* (New York: Columbia University Press, 2000).
51. Merja-Liisa Hinkkanen and David Kirby, *The Baltic and North Seas* (London and New York: Routledge, 2000).
52. Alain Corbin, *Lure of the Sea: The Discovery of the Seaside in the Western World, 1750–1840* (Cambridge: Polity, 1994).
53. Colin Heywood, "The English in the Mediterranean, 1600–1630: A Post-Braudelian Perspective on the 'Northern Invasion,'" in *Trade and Cultural Exchange in the Early Modern Mediterranean,* 23–44.
54. Natalie Zemon Davis, *Trickster Travels: A Sixteenth-Century Muslim Between Worlds* (New York: Hill and Wang, 2006); Mercedes García-Arenal and Gerard Wiegers, *A Man of Three Worlds: Samuel Pallache, a Moroccan Jew in Catholic and Protestant Europe* (Baltimore, MD: Johns Hopkins University Press, 2003). But there are many more in this genre, such as Eric Dursteler, *Venetians in Constantinople: Nation, Identity and Coexistence in the Early Modern Mediterranean* (Baltimore, MD: Johns Hopkins University Press, 2006), and Nathalie Rothman, *Trans-Imperial Subjects: Boundary-Markers of the Early Modern Mediterranean* (Ithaca, NY: Cornell University Press, 2011).
55. Subrahmanyam, "Notes on Circulation and Asymmetry in Two Mediterraneans, c.1400–1800," in *From the Mediterranean to the China Sea: Miscellaneous Notes,* eds. Claude Guillot, Denys Lombard, and Roderich Ptak (Wiesbaden: Harrassowitz, 1998), 26.
56. Goitein, *A Mediterranean Society,* "Epilogue," V, 496–502.
57. Greene, *Catholic Pirates and Greek Merchants,* 10.
58. Jacques Rancière, *The Names of History,* trans. Hassan Melehy (Minneapolis: University of Minnesota Press, 1994), 86.
59. Baader and Wolf, "Maritime Tableaus," 7–8.
60. Rancière, *The Names of History,* 86.
61. Michelet, *La Mer,* preface by Jean Borie (Paris: Gallimard, 1983), 328: "La Terre vous supplie de vivre; elle vous offre ce qu'elle a de meilleur, la Mer, pour vous relever."
62. T. S. Eliot, "Four Quartets: The Dry Salvages," in *Collected Poems 1919–1962* (New York: Harcourt Brace, 1982), 191–92.
63. And, in French, it has already found its poet laureate: Jean-Luc Le Cleac'h, *Petite Philosphie des Ports Maritimes* (Clermont-Ferrand, France: Pimientos, 2011).
64. Braudel, *The Mediterranean and the Mediterranean World in the Age of Philip II,* 2 vols. (New York: Harper and Row, 1972), vol. I, 124–33.
65. Hermann Broch, *The Death of Vergil,* (New York: Vintage, 1995), 1.
66. Goethe, *Faust. Part II,* trans. David Luke (Oxford: Oxford University Press, 1994), ll.10223ff, 180. For a survey of "The view from the shore" in art and literature see Kirby and Hinkkanen (*The Baltic and North Seas,* 45–52), but note that they do not mention Goethe.
67. "Outside a violent north wind was blowing, but the sun shone and the water gleamed as if covered with silver. Rilke climbed down to the bastions which,

jutting out to the east and west, were connected to the foot of the castle by a narrow path along the cliffs, which abruptly drop off, for about two hundred feet, into the sea. Rilke walked back and forth, completely absorbed in the problem of how to answer the letter [a "troublesome business letter]." Then, all at once, in the midst of his thoughts, he stopped; it seemed that from the raging storm a voice had called to him: 'Who, if I cried out, would hear me among the angels' hierarchies?'" Princess Marie von Thurn und Taxis-Hohenlohe, *Erinnerungen an Rainer Maria Rilke,* 40ff, quoted in *The Selected Poetry of Rainer Maria Rilke,* ed. and trans. Stephen Mitchell (New York: Vintage International, 1989, 315.

68. Burckhardt, *Civilization of the Renaissance in Italy* trans. S.G.C. Middlemore (London and New York: S. Sonnenschein, 1904), 3.
69. Revel, "Histore et sciences sociales. Les paradigmes des *Annales,*" *Annales* 34 (1979): 1374.
70. Kracauer, *History. The Last Things Before the Last,* 118. He adds, baldly, that "we do not learn enough about the past if we concentrate on the macro units."

ONE

Two Men in a Boat: The Braudel-Goitein "Correspondence" and the Beginning of Thalassography

Peter N. Miller

The concluding section of Abraham ibn Daud's *Sefer ha-Qabbalah* (c. 1261 CE) tells the story of four men—rabbis—in a boat sailing from Bari. The boat is captured by Muslim pirates and three of the rabbis are distributed, or ransomed away, to the cities of Fustat (Old Cairo), Qairawan, and Cordoba. The story, plainly, is a parable of the establishment of autonomous, post-Babylonian, Jewish communities in the West, even down to the silent evocation of the medieval Jewish parody "From Bari shall go forth Torah, and the Word of the Lord from Otranto."[1]

Near the beginning of their magnum opus, *The Corrupting Sea,* Peregrine Horden and Nicholas Purcell present the historiography of the Mediterranean upon which their work rests in terms of "four men in a boat." Without reference to ibn Daud, or perhaps even knowing of his tale, they proceed straightforwardly to survey the contributions made by their chosen ancestors—Mikhail Rostovtzeff, Henri Pirenne, Shlomo Dov Goitein, and Fernand Braudel.[2]

It is with the last of these men that this essay is concerned. And though real boats play a major role in their lives, as well as work—Goitein, famously, travelled from Germany to Palestine in 1923 on the same vessel as Gershom Scholem, and Braudel, even more famously, drew close to Lucien Febvre on the long passage from Brazil back to France in 1937—it is to a metaphorical boat that I point. For though they seem very different scholars, and though no one has thus far found any reason to link them directly—Horden and Purcell, for example, make no effort to show that any of their four figures were personally connected—the fact is that for about ten years Braudel and Goitein were "in the same boat," as we would say: engaged upon a common project. It is this story that I wish to tell for the first time.

And it is a story of no small significance. For these first thalassographers were two of the greatest historians of the twentieth century, one known to all of today's practitioners, the other known to far too few. The one defined the study of the sea in terms of the slow-moving facts against which the lives of individuals vanished, as in the long exposure times of early photography. The other defined the study of the sea in terms of the fast-changing events and thoughts and actions of the individuals whose lives gave meaning to the inanimate world around them. Together Fernand Braudel and Shlomo Dov Goitein defined—and define still—the parameters of a possible thalassography. As scholars, Braudel and Goitein seem to represent two distinct paths in twentieth-century historiography. It is fitting, therefore, to begin this volume with a study not of their work—others have, and can, do this much better—but of their unknown relationship. For historians, even great ones, are human beings and their interactions can have a shaping impact even on their intellectual creations.

On the first page of *A Mediterranean Society,* borrowing a Gibbonian trope, Goitein (1900–1985) writes that the "idea of this book was first conceived by me on September 17, 1954, while in Oxford, searching the treasures of the Bodleian Library for Geniza documents relating to the India trade for a book still in preparation on that subject." In the very next sentence, Goitein explains that "On June 15, 1955, I wrote about my intention to publish a volume of this character to Clemens Heller of the École Pratique des Hautes Études (VIe section), Paris, who had done much to further the preparation of the book about the India trade."[3]

Heller appears nowhere else in Goitein's published oeuvre. He remains, surely, an unknown name to nearly every reader of *A Mediterranean Society.* But those with an eye for historiography and an ear for dates would note that the head of the VIe section in 1954 was none other than Fernand Braudel.

This veiled reference is the only indication either ever gave of the other. For Braudel never cites Goitein in *La Méditerranée,* and Goitein himself explains that he did not read Braudel until after the publication of its English translation—*The Mediterranean*—in paperback in 1972 (he writes 1966). Of course, Braudel's history would have been more compelling if he could have figured out how to accommodate the reality of human agency, and Goitein's more usable if he had been able to tell a story rather than curate individual documents.

On the surface, this one sentence on the first page of Goitein's book seems to point to one of the great near-misses in the history of twentieth-century practice. Yet it actually points to a relationship between Goitein and the VIe section of the École Pratique des Hautes Études (EPHE) that extended over 10 years and 124 letters to and from Heller and his friend and colleague, the head of the section, Fernand Braudel.[4] This connection to the VIe section is not mentioned by Goitein anywhere, not even in the India Book, published in 2008, but initially solicited by Heller and Braudel in the fall of 1954 for the VIe section's publication series.[5]

The one sentence mentioning Heller masked all of that. It also cast a retrospective sheen on the past. For Heller did not "further the preparation of the book on the India trade," he—and through him Braudel—actually commissioned it and paid for much of the research that went into it, and into what became *A Mediterranean Society*.[6] Goitein, in that book's preface, explained his leaving the Indian Ocean for the Mediterranean Sea in a second sentence that also conceals more than it reveals. Goitein mentions that when, on August 18, 1958, he received a letter from G. E. von Grunebaum of the Near East Center at the University of California inviting him to publish in their new series he was "off India and on the Mediterranean." It was in trying to satisfy Grunebaum that "there ripened the bold resolution first to make a survey of the documentary Geniza in toto."[7] But it was Braudel, in fact, who first proposed publishing this kind of book.

The story of these letters provides the "backstory" to one of the great scholarly projects of the twentieth century, Goitein's reconstruction of a medieval trading society from fragments of documents, mostly letters. It also suggests the breadth of Braudel's vision and a sense of his priorities very different from what one might deduce from his printed scholarship alone. The whole story of the pre-history of Goitein's *Mediterranean Society* cannot be told here, but his relationship with Braudel and the VIe section can. And, finally, this story outlines a historiographical orientation which, if sadly impossible fifty years ago, may now finally be coming into view, in which human mobility and connectivity provide the historian's armature.

Fernand Braudel is famous as a scholar. We evaluate his achievement, as we do all scholars, based on what he wrote. Sometimes, of course, we celebrate a scholar by the celebrity of their students. But Braudel should also be famous as an administrator. It was under his guidance of the VIe section of the École Pratique des Hautes Études (EPHE) that a range of institutional projects were launched which culminated in the refounding of the VIe section as the École des Hautes Études en Sciences Sociales (EHESS) in 1975 and the creation of the Maison des Sciences de l'Homme on the Boulevard Raspail as its home. And, in particular, at the EPHE Braudel launched a series of publishing projects of extreme originality. When we evaluate his intellectual orientation we must include these publications, which he often solicited, encouraged, organized, and subsidized, alongside his own.

In short, I am suggesting that the academic administrator or, more precisely, the administrative initiatives undertaken by an academic can and should, under certain circumstances, be understood as more or less mediated reflections of an intellectual vision. But once we broaden out our scope for the history of scholarship to include initiatives launched and managed but implemented through others, then we are also broadening out our range of actors from the professor-administrator to the administrator-academic.

If we look at Braudel's tenure at the VIe section, and then the EHESS, we see powerful administrator-academics. But one name stands out as of exceptional importance for our story, and exceptional interest over all: Clemens Heller (1917–2002). He is ignored in Lutz Raphael's essential treatments of the *Annales* empire, but he does have a substantial role in F. Gemelli's study of how Braudel created the EHESS.[8] In this story, Heller and Braudel worked as a team, with Heller leading the way on international diplomacy, especially with American foundations. Born in Austria, and later a refugee in the United States, Heller went to Paris in 1949 after completing his doctorate at Harvard University (and founding the Salzburg Seminar) in order to work on account books from sixteenth- and seventeenth-century Europe. He became a confidant of Braudel's and moved through a series of positions in the VIe section—(Chargé de conférences (1952–1956), Chargé de la mise en œuvre du programme des Aires culturelles (1955), Chargé des fonctions de sous-directeur d'études (EPHE-VIe) (1957), Chargé du secrétariat et de la coordination de la Division des Aires culturelles (1957–1972), Directeur d'études associé (1972)—and after his official retirement in 1985 followed Braudel as Administrator of the Maison des Sciences de l'Homme (1986–1992). Yet the working relationship between Braudel and Heller was much closer than any title might indicate. Indeed, we might want to consider their connection as something close to that of Warburg and Saxl when they remade the KBW into more than a *privatgelehrter*'s library in 1920s Hamburg.[9]

Heller has not been studied much on his own, reflecting the general tendency of historians of scholarship to focus on the scholars who write the scholarship, rather than the administrators who make it possible for that scholarship to be written. What follows, therefore, could be as much a demonstration piece for taking seriously the academic dimension of institutional histories, as it is a piece focusing on Heller. And, I should add, because of his position and ideas, the Fonds Heller in the archive of the EHESS is a fantastic trove for anyone interested in the global shaping of historical scholarship in the second half of the twentieth century.

At the center of this institution-building project was the Centre de Recherches Historiques (CRH), created by Febvre and Braudel, born with the VIe Section itself in 1949 but only really active from 1951. Lutz Raphael has written about it within his study of the *Annales* project and also in a separate monograph published in the *Cahiers* of the École des Hautes Études en Sciences Sociales.[10] And at the core of the CRH lies a group of publication series which reflected Braudel's vision.

Indeed, the publication series launched by the CRH in the early 1950s may represent Braudel's postwar thinking even better than his *Mediterranean* which, while only published in 1949 was substantially finished by 1938 or 1939. The Braudel of the early 1950s was a different person and the second edition of *The Mediterranean* grew directly out of the research foci of the CRH, as represented in its monograph series.[11]

The two most significant collections established by the CRH were "Affaires et gens d'affaires" and "Ports-Routes-Trafics." These series are anchored in a perception that economic activity is *cultura* in the strong sense, and that studying merchants is therefore central to any future cultural history. The first volume in "Affaires et gens d'affaires" was Armando Sarpori's *Le Marchand Italien au Moyen Age*, with its four chapters divided into "Physionomie du marchand," "Les marchands au travail," "Les italiens dans le monde," and "Les sources," which included statutes, notarial acts, books, merchant letters—described as absolutely essential but hard to find—"manuels de commerce, mémoires et journaux, et chroniques."

Right from the start there was equal emphasis on the role of the historian as a key figure in this story, not as someone hidden behind a veil of pseudo-objectivity. As Lucien Febvre wrote in his preface to Sapori's book, "the author of this very fine book . . . this author is a man . . . concerned with yesterday, preoccupied with tomorrow—because tomorrow and yesterday find a common link within the heart of man."[12]

In his preface to an exactly contemporary publication in the parallel series *Ports-Routes-Trafics* (Étienne Trocmé and Marcel Delafosse, *Le Commerce Rochelais de la fin du XVe siècle au début du XVIIIe*), Fernand Braudel made an impassioned plea to treat ostensibly economic sources as windows into lived life.

> It is another, and rarer thing, which would fill many of our colleagues in history with joy, not to consider economic life as a reality in itself, a closed world of numbers, but as a sector of life among others, and one that must be replaced in a set of circumstances and varied links to restore its full value. For Etienne Trocmé, his merchants are men, his city is not only streets, a harbor, some ramparts, a 'suburb', local foods, distant commerce—but also institutions in the richest sense of the word. La Rochelle gradually reveals itself to us in the fullness of its life, in its dignity, and in its urban *force d'état.*[13]

For Braudel and the CRH, letters were the type of source that offered the most. The key statement of purpose was made by Braudel in his preface to a volume in *Affaires et gens d'affaires* on the letters of a Marseille merchant family at the turn of the seventeenth century.

"[Their interest,] if I am not mistaken, is to insert us into the heart of the practices and realities of merchants' everyday lives: these realities, on their own or because of their repetition, often go beyond the merely anecdotal detail." Then, taking a single letter as an example, he notes that it sheds light on "the city and the position of Marseille, the role of the Genoese silk buyers, the importance of exchange in Lyon, the attraction of Seville and Cadiz, the transport of 'American' cochineal eastwards, the price of silver reals in Marseille, the journey of ships towards Alexandria and Tripoli."[14]

Braudel, whose famous book, *The Mediterranean,* was famously depopulated, in these prefaces emphasized the value of letters precisely because of their individualized, "granular," view of reality. And he identified this with the aspirations of the VIe section as a whole.

> The VIe section of the École des Hautes-Études was even more interested in these letters as it had in the works two analogous publications [*Lettres d'un marchand vénitien Andrea Berengo (1553–1556)* (Affaires et gens d'affaires 10, 1957); *Simon Ruiz et les 'asientos' de Philippe II* (Affaires et gens d'affaires 6, 1953)] . . . But these letters from Marseillais merchants have their value and will acquire importance by being thus compared to similar documents and plunged back into an

> economic history of the XVIth century as seen day by day, explained by its actors, a history which, in general, has but distant relationships with the images presented by, when they are presented, the great books of synthesis.[15]

One of Braudel's younger followers, J. Gentil da Silva, who published his edition of merchant letters as *Affaires et gens d'affaires 9* in 1956, described the way letters had to be read. On the face of it, they often might appear of little value. "For their correspondence does not say all about their activities or their calculations. The hardest, and we are going to try in a moment, is to understand their language, to follow their procedures, their tactics, their art of diagnosing and taking action."[16]

In short, in the early 1950s, there was no place in the world in which merchant letters were as central, as valued, and as "theorized," as the Centre de Recherches Historiques in Paris.[17]

Our thalassographic close encounter begins on October 10, 1954, when Heller, at the heart of this project, wrote a letter out of the blue to Goitein, sparked by the appearance in *Speculum* of Goitein's essay, "From the Mediterranean to India: documents on the trade to India, South Arabia and East Africa from the eleventh and twelfth centuries."[18] Heller had read it, and was impressed. Right off the bat, Heller puts the collaborative project on the table, linking the Centre's interests to those of Goitein (or, so interpreting them as to link): "I have just read with great interest your article in SPECULUM. For me the article has been one of the most exciting and rewarding ones I have seen in a long time, especially since the Geniza papers promise to be of great value for the studies on commercial credit in antiquity & the middle ages, which I am now undertaking."

Heller then went on to write:

> But I should like to write you today not so much on my own behalf, but as a member of the 6th Section of the École des Hautes Études at the Sorbonne and its Centre de Recherches Historiques. Under the leadership of Prof. Fernand Braudel of the College de France we are just developing a programme for a comparative study of Asiatic & Occidental economic & social history & of economic & social movements which involve as well Asia as the Occident.
>
> It is intended to build up several research projects, & to invite international cooperation for them, to expand the publications of

> the 6th Section in the direction of these investigation [sic], and to start, beginning in October 1955, a new Journal (in French & English) devoted to the study of the problems of comparative economic history.
>
> Obviously Jewish Commerce will have to constitute a major problem of investigation, and you can well see how the Geniza papers, and the programme you outline for the future, are of immense interest to us. I should therefore like to ask you, whether it would be possible for you, not only to keep the Centre abreast of your studies, but to cooperate in its efforts and perhaps to you use our journal for other articles by yourself or by colleagues of yours which deal with the economic and legal implications of the Geniza papers.

Heller put this into very concrete terms, opening the door to what would be a nearly sixty-year-long project.

> I note that you speak only of a "possible" publication of the Geniza documents. I take it that the publication is by no means assured, but if this should not be the case, it might be possible for the Centre to assume responsibility for the publication, unless costs turn out to be entirely prohibitive.

In conclusion, invoking Braudel's personal research project again, Heller concludes:

> Prof. Braudel has asked me to inquire whether the Geniza papers contain important material on the Indian and on Mediterranean trade in the 15th & 16th century, and whether you think it worthwhile to set some young scholars in Paris to work on these papers.[19]

Goitein immediately replied, writing on October 27 from England, "I am enthusiastic about the prospect of cooperating with you." But then he quickly qualified, identifying himself almost reflexively, and defensively, against what he might have perceived as too-high expectations. This is a passage of extraordinary value for its autobiographical hue.

> I am a philologist and historian of Mediaeval Islam and Judaism, but will never write economic history. I regard it my duty to edit, to translate and explain as exactly and completely as possible those texts of the Cairo Geniza. The conclusions to be drawn from them for the

> history of economy will be made by the scholars specializing in this field. In my article "The Cairo Geniza as a source for the History of Muslim Civilization" in the forthcoming issue of Islamic Studies (Paris) you will find a number of important conclusions drawn from my collection of papers dealing with the trade to India. But these are summaries of facts rather than abstractions made by a specialist.

Nevertheless, at this stage he presented the gap between his own philologist's approach and that of the historian, perhaps with Braudel as the model, as fruitful. "Thus, I believe that cooperation with the Centre de Recherches Historiques will be very useful." Heller marked this in orange pencil in the margin.

Turning to details, Goitein explained that he was "delighted that you consider the possibility of publishing my book *on the trade in the Indian Ocean.*" Heller underscored this passage. "So far," Goitein continued, "only very informal talks have been made with other scientific bodies." Then Goitein turned to an issue that would run through his Parisian correspondence, and perhaps be of decisive importance. "Of course the book *has to appear in English* (Heller's emphasis). For although French is not unknown to me, I could not take upon myself the responsibility to translate these difficult texts into a language which I never use actively."

The shadow of Braudel looms always over this correspondence. In concluding this first letter, Goitein turns directly to Heller's transmitted question about the possible presence in the Geniza of material bearing on the later, early modern, history of the Mediterranean. "Please inform Professor Braudel," Goitein writes, "that the Geniza contains a lot of material from later times, especially, I believe, from the 18. century, from the 16. too, while—if I am not mistaken—there is little from the 15." Braudel had raised the possibility of setting a young person on to this material, a very forward-thinking proposal. Goitein agreed, but with an illuminating caveat. "I think, it is an excellent idea to set a young scholar to work on this material, but only one, for the very secret of the study of these fragments is that one person has to concentrate on a whole field."[20] This insistence on one person reflects his own view that the Geniza could only be understood if one had a view of it as a whole, and therefore the need for a person to have mastered everything in order to say anything. Whether this is in fact true—could it not be said of any documentary trove? And yet it is not standard practice in other fields—or not, there is no doubt that this premise had more than anything to do with the rate at which Goitein produced his own books on the Geniza.

This first exchange of letters flags many of the main themes that will recur in the later correspondence of these two men. From a letter written on December 1, 1954, from Jerusalem, Goitein noted that Heller's letter reached him on November 8 in Cambridge, just before he left England. He had been out of the country for sixteen months and was drowning in old business. It makes no reference to his earlier letter, however.[21] He thought his book on India would nicely complement that of a Mr. Kirpas on the India trade in the sixteenth century.[22] Goitein explained his time frame, and his dependence on foundation financial support. What he really hoped, eventually was "in collaboration with my colleagues and graduates, to cover the whole field of the documentary Geniza."[23]

Based on Goitein's one letter, Heller offered him a subsidy, which Goitein later acknowledged as 60,000 francs per month. One can only marvel at the remarkable sure-footedness and speed with which Heller—and behind him, of course, Braudel—worked.[24] Goitein again brought up the language question, asking if he could publish in English. He was already thinking in terms of a complex project. "This volume I would be exclusively in French, containing a very detailed Introduction and the Translations of the texts, while a second, smaller, volume would bring the texts (of course in Hebrew letters) with a philological Introduction and detailed Arabic (in Latin or/and Arabic letters)—French Indices. Do you think this is feasible?"[25]

Heller must have tried to paint a picture of Braudel's project by sending him some of the volumes produced in the CRH's series, for in his letter of February 6, 1955, Goitein thanked Heller for the two packets of publications of "your institute." By this date, the CRH would have published at least fifteen titles. "I believe now," Goitein replies, that my *Indiabook* fits better into the series 'Affaires et Gens d'Affaires.'" This was, as we have noted, the flagship series in which Braudel promoted his argument about the centrality of merchants in the making of the modern world through the writings of merchants themselves. "In English," Goitein concludes, "I would call it 'The Jewish India merchants of the Middle Ages'. In German, there exists even a word corresponding exactly to the Arabic 'Musafirun al-Hind,' namely 'Indienfahrer'; I doubt, however, whether there exists such a word in French."

Goitein offered that the texts were already typed and ready for translation, and that only time in a major library was needed for him to complete the introduction.[26] Goitein's formal request for funding was made to the CRH in a second letter of February 6, 1955. This was addressed to

Heller but Goitein explains, as if to a committee, that he had collected "about 160 items referring to the trade between the Mediterranean and the countries of the Indian Ocean" and wanted to remain in Europe to complete the study. He wrote that he believed the Centre would recognize the value of this work because it has done so much for "research in the history of economy."[27]

Goitein's first letter to Braudel was written on February 20 in order to thank him personally for the grant which he received of the 60,000 francs. "Dr. Clemens Heller informs me that you most kindly consented to grant me a scholarship of 60.000 francs per month for a period of two months, in order to enable me to work on my book on the India merchants of the Middle Ages.

> I was indeed deeply moved by this generous allocation, and in particular by the spirit in which it was granted. The connections between the Centre de Recherches Historiques and myself were brought about entirely by the initiative of the Centre. This I call a real policy in planning and executing of scientific research. The stupendous literary output of the Centre shows that this farsighted policy is bearing fruit.

But in the midst of this very important letter Goitein's seeming insecurity about his lacking command of the French language surfaced. "I have to apologize for writing to you in English instead of French. I am a great admirer of France and in particular of her literature and science. However, as I have to use constantly four different languages for my work of research, teaching and writing, I found it was not conducive to the health of the soul to try using French as a means of expression."[28]

In his letter to Heller of February 20, which accompanied that to Braudel, Goitein explained that "After having looked through the publications of the Centre, I think we better shelve the idea of printing the Judaeo-Arabic originals. This is a big Orientalist undertaking and is better left to a project of *bringing out the Geniza in toto*" (Heller's underscore). "Thus," he concludes, "the book will be in French from the beginning to the end." This distinction between the India Book and the *Geniza in toto,* or "documentary Geniza," is fundamental, for in it lies the seed of the ultimate bifurcation of Goitein's project.[29]

Heller's response of March 30, 1955, stressed the capabilities of the Imprimerie Nationale in the area of oriental-language printing. He also explained that the CRH had made a series of appointments in "Oriental problems."[30] Turning to the question of Goitein's team, Heller asked about Eliyahu Ashtor and Murad Mikh'ail and their readiness to publish, and emphasized the value the CRH put on collaboration. A broader collaboration was for a broader project. It was Braudel himself to whom he assigned the vision for this.

> And this brings me to a very large question with [which?] M. Braudel and I have discussed without coming to any definite conclusions: that is whether we should propose to cooperate with you in bringing out the *Geniza* in toto if you do not find another institution to support such publication. But this may well be beyond our possibilities and economic means. I wonder if you would care to make any comments?[31]

Goitein's response directly took up Braudel's challenge.

> You are perfectly right in the assumption that the India book should be planned within the frame of the whole Geniza projects; and it is a testimony to the broadness of your and Braudel's conceptions of History that you consider to bring out the Geniza as a whole. No doubt, this treasure house of information on social and economic history should be made accessible, if not in a total edition, at least in the form of Regesta.

This was the kernel of *A Mediterranean Society*.

Goitein was also willing to collaborate with the CRH, but he suggested broadening it out to include also the Hebrew University and "some American organisation or University." In Goitein's view "this would solve also the question of the language. As Hebrew and German, the languages with which we are familiar, are ruled out for a publication of the character of the Geniza project, we should use English, the language in which our publications appear, as far as they are not destined for the Hebrew reading public." Again, we see that language was an ongoing issue for Goitein. He explained that his notes were all in English and that translating from it—which was already translated from the medieval Hebrew or Arabic—into a fourth language, French, "would be a source of many mistakes or inaccuracies." Therefore, he concluded, "I thought to bring out the India book solely in French and to leave the philological aspects to a later publication, which should be a part of the Geniza project as a whole." And this, presumably, would have been in English.

In conclusion, Goitein invited Heller to Jerusalem to meet the possible participants in the big Geniza project. He specifically mentioned Ephraim Urbach, F. Baer and H. H. Ben Sasson. One of the fascinating what-ifs of our story is the possibility, had Heller gone to Jerusalem, and had this collaboration blossomed, of a very different Jerusalem School arising in the 1950s, one linked much more closely to Braudel, the CRH, and the *Annales*.[32]

Writing back immediately (six days after Goitein put pen to paper), Heller reported that a discussion of the Geniza papers had been held, but that more information about scope was needed. He had, however, been able to secure Braudel's agreement to Goitein's proposed institutional collaboration. "In principle, an arrangement could be considered, in which the University of Jerusalem would assume responsibility for the scholarly preparation of the volumes, in which we would take care of the printing by the *Imprimerie Nationale,* and an American institution would assume responsibility for additional expenses."[33]

Again, turning the correspondence around quickly, Goitein wrote back from Jerusalem on the 25th. "My final decision is that I offer the Centre my India book for publication and suggest that you put the costs for translation and printing into your budget of 1956." Goitein planned to be in Europe in the summer writing the book's Introduction and would know the "exact extent" of the book by October. He proposed meeting Heller in Paris at the end of October or early November and going over the details. He also suggests auditioning Paul Klein, once of Paris and the Bibliothèque Nationale and now of Jerusalem, to see how well he translated three or four extensive documents.[34]

Since Goitein's definitive-sounding pronouncement had made no mention of language of publication, or of the relationship of the India Book to the wider Geniza project, Heller queried precisely this in his next letter, written after a lapse of four weeks. "Please do let me know whether the idea of joined publication of the Geniza papers is dead or not, what should be done about the translation of your book and whether it is going to include the original text or not."[35]

Heller also put his money where his mouth was, offering Goitein 120,000 francs for the summer.[36] In exchange, Goitein made clear his relationship to the CRH. "I offer to the Centre," he wrote to Heller, "the publication of my book 'The Geniza Archives of the Jewish India Merchants.' You will have to put into your budget for 1956 . . . the cost of translation from Hebrew, as well as the printing [of] the French text and of the originals, which will be in Hebrew characters with some additional dots (denoting the Arabic). . . . As to the Geniza project as a whole, you need not worry about it *now,* as no other volume will be ready, before the end of . . . ."[37]

Heller, going back to the U.S. for the summer, slightly lamented Goitein's failure to rise to the bait about the Geniza project *in toto.* "With more information about the great Geniza project, I might have had an occasion to discuss it over there." He left the Rockefeller Foundation as his forwarding address.[38]

This caught Goitein's attention. Five days later he sent Heller a plan for what a published Geniza might look like. "Our provisional plans for the Geniza: 8 volumes to be brought out in about twelve years.

> 1. The Archives of the India merchants. To be ready in 1956 (Goitein). [In the margin, Heller penciled "To be ready in 1956."]
> 2. The Archive of Nahrai ben Nissim: letters from Egypt, North Africa, Sicily, Syria, and Palestine. To be ready in 1957 (Murad Mikha'el).
> 3. Letters from other archives related to the Mediterranean . . .
> 4. A general volume illustrating social life. I am collecting material for this volume all the time. [In the margin, Heller noted "Goitein ed."]
> 5. Letters from the Egyptian country side or inner Egyptian correspondence.
> 6. Marriage, Divorce and other documents related to family life.
> 7. Other legal deeds.
> These seven volumes will contain material in the main from eleventh thru [*sic*] the thirteenth centuries.
> 8. A selection from later documents (16 to 18 centuries).

Goitein concluded: "I personally intend to undertake, beside the volume on the trade to India, only volume 4. I see, however, suitable candidates for the other volumes, although some of them will have to grow." Heller commented in the margin: "Edition of Geniza Papers relating to Social & Economic History. Edition S.D. GOITEIN."[39]

This is the letter Goitein referred to on the opening page of *Mediterranean Society.* It clearly is written as a response to Heller's challenge to produce something for him to show the Rockefeller Foundation.

Heller's response to the big plan was immediate and welcoming. But he also perceived a certain ambiguity, if not downright ambivalence, in Goitein's reply. Was it one volume they were discussing, or several? And were they all the same, and if not, how different? Nevertheless, Heller's—

and the CRH's—position was clear: "I do not know whether such a permanent relationship with Paris would be of interest to you, but we would certainly welcome it."[40]

Heller, in turn, in his next letter, asked if Goitein, who was then in England, could come to Paris by the end of October in order to meet Braudel.[41] Goitein was "awfully sorry," but deferred the invitation. Instead, he enclosed "a table showing the final arrangements of the texts in the India volume (which has again a new name). I should very much appreciate if Prof. Braudel had time to have a look on it, for then he would get a more detailed idea about the content of the book."[42] This "table" seems not to have survived.

A long letter from Heller that may have crossed with Goitein's shows how deeply Goitein's project had gotten bound up with those of the CRH. Heller explained that "M. Braudel would like to bring out the 1st Volume of the Geniza Papers in an exemplary fashion, and therefore, considerations of money should not bother you in deciding whether or not to publish documents in full." This was a pretty extraordinary extension of carte blanche. Heller went on. "I presume you are in agreement that we should publish all the eight vol. together; naturally, we could add related volumes, which you may care to suggest." He asked what Goitein had done about translation because budget decisions needed to be made before Braudel's departure. Heller also noted that the VIe section was undertaking a major expansion in its Asian and Islamic studies. He asked for Goitein's feeling about the "Geniza Papers" being published in a series containing volumes on India, China, and the world of Islam. Heller ended by asking for a quick reply "so that I can still discuss it with M. Braudel."[43]

In his reply, dashed off only two days later, and just after the post arrived at 8:30 a.m.—"when one is burning to run off to the Library"—Goitein rejected the Islamic series as a home for his India Book. He suggests instead "Intermediate civilization or something similar." He also noted that while the "The India volume—as well, I hope that about social life—is my literary property, [but] the Geniza project as a whole is an affair of the Hebrew University & we have to await its decision. It will certainly be decided before the India volume is ready for print."[44] This is somewhat puzzling as the Geniza documents were themselves scattered around the world; the claim that the Hebrew University had a right to the project could only stem from the fact of his being salaried by the University. But why Goitein thought that his curatorial or interpretative work and that of his students somehow belonged to the University in a way that differed from the India Book is not self-evident. In the light of what came later, we cannot rule out the possibility that this was a stalling tactic.

In a letter written on that very same day (September 23) Heller proposed splitting the India Book into two parts. "In talking over the publication of the Indian book"—presumably with Braudel, but perhaps with others as well—"the suggestion was made to publish your introduction, the translation of the texts, perhaps in French and English, (and, if you think it valuable, the transcriptions) and the indexes, etc., in one volume, and to publish a complete set of the original documents, in the form of microfilm or microcards, available for general distribution."[45]

Goitein's trip to Paris brought him in on the 13th and took him out on the 20th.[46] A month later, after thanking Heller for hospitality during his visit, Goitein announced that under separate cover he was sending the French translation of the preface to the India Book, made by Klein. He apologizes for the matter of fact style, and says the book itself "will be more colourful in general,"[47] The preface was sent the following day, accompanied by the question "When will Professor Braudel be back from USA?"[48] Intriguingly, although Goitein seems to have been in no hurry to accommodate his schedule to Braudel's, he did seem acutely attentive to his comments.

Goitein's next letter to Heller, on December 12, begins where the previous one had ended: "I wonder whether Prof. Braudel has already come back from his visit to the States." He had spoken to Prof. Mazar, President of the Hebrew University, who was enthusiastic about the project. He and Goitein had agreed that

> 1) The publication of the Judaeo-Arabic Geniza documents will be a joint undertaking of the HU and the CRH. 2) the HU will provide the scholars and graduates doing the collection, deciphering, translation and preparation for printing, while the CRH will put at the disposal of the undertaking its wide experience in scientific planning, especially in the fields of economic and social history. Likewise, the Centre will bear the financial burden of the undertaking.

In the margin, Heller scrawled "!!"

The first volume in this series was to be Goitein's essay, "From Spain to India," and it was to appear in French. Goitein, in a marginal note, waived the idea of an English summary. He also looked to a formal collaboration between the Hebrew University and the CRH which included the opportunity for "HU scholars and graduates to work in Paris, while visits of relevant French scholars will be welcome in Jerusalem."[49]

The next letter from Heller, sent by his secretary, Alice Vidal, on his behalf, was dated January 16, 1956. It apologized for the delay but explained that "Mr Braudel was absent for some time." And since the remainder of the letter discusses the "joint edition of the Geniza papers," the explanation seems to imply that on the Centre's part it was Braudel who developed the publication strategy presented now in this letter. "You say," it begins, that "the Centre will bear the financial burden of the undertaking. This needs more precisions to avoid later misunderstanding. Offhand, we are ready to pay for all costs of publication, including translation, and I presume that is what you mean." Offhand or not, this is no small offer on Heller's part. But he said that the volumes ought to appear in French, and that no English edition should appear until at least two years had passed after the French edition's publication. This, again, seems very open minded on the part of Heller and Braudel. And, Braudel agreed that Goitein's choice of translator, Mr. Catane, né Klein, could be entrusted with the India Book.

Heller asks if Goitein wanted a formal letter "written by Mr. Febvre or Mr. Braudel to the President of the Hebrew University, proposing in a formal manner this joint undertaking?"[50]

On January 22, Goitein replied, correcting the trivia—how "Catane" should be spelled—and aiming at the truly important—"Is Professor Braudel satisfied with M. Catane's French style?" His approval was necessary in order for the translator to be he hired. Summarily, Goitein reiterated his approval to points a, b, and c of Heller's letter of January 16 and proposed that no official communication was necessary above and beyond his letter of December 12, 1955—the one in which Goitein approved doing the India book in French.

Goitein was also now in a position to inform Heller that the India book would not be completed in the fall of 1955. No comment about the delay, and certainly no apology for the delay, is offered up by Goitein. Instead, turning to the offer made by Heller "that a young man who had worked on the Geniza, should come out to Paris, to study under Braudel and thus form a kind of link between the Centre and the Geniza project," Goitein offered up his current assistant, Joseph Eliash.[51] Having a Geniza person in Paris, Goitein added, would help a great deal with proofreading "and if Prof. Braudel too would be satisfied with him, we could confide to him, later on, on his own account, e.g. the volume on the Mediterranean trade, as represented in the Geniza papers." Here, we see Goitein thinking in terms of Braudel's project as first articulated to him by Heller a year earlier.[52]

As for the question of Goitein's man in Paris, Heller thought that he might be able to use Rockefeller Foundation money to cover the costs if his time was split between medieval and modern research matters. "But, I shall try in any case," Heller concluded, "to use French funds for him so that he could concentrate on the Geniza."[53]

Goitein, having proposed Eliash and interested Heller, now reported in his next letter that he wanted to put off the visit until 1957. Having proposed Catane and gotten Heller's approval, Goitein now backed away, saying that before he submitted a formal proposal to translate the India Book "I would like to make sure that Prof. Braudel is really satisfied with his French style." Amidst Goitein's puzzling but now increasingly typical hesitation, one detects the ghostly presence of Braudel in their epistolary exchange.[54] Heller immediately picked up on this. Catane's French was fine, he wrote back. Just "go ahead and get him to do the work," he concluded.[55]

The India Book seemed almost within grasp in late March 1956—but then slipped away. "I hope to be able to send a typewritten copy of the volume of the Texts (about 400 pages) in about three weeks time," Goitein wrote. The book as a whole, he thought, could not be printed until the French part was ready. "I do not believe to be able to finish it in 1956."[56] Heller procured funds to begin the translation.[57] Yet by May, the India book, which only a few months earlier was three weeks away from completion, was now slipping out to sea. "This India-book will be a magnificent collection of documents, like one of those big publications which were the pride of the nineteenth century. However, although I am working on it daily, it will take quite a time, before the whole will be ready for print."[58] In this clearest presentation of the content, Goitein was gesturing at grand collections of documents, such as the *Monumenta Germaniae Historicae*, rather than at a connected monograph. Does that mean that Goitein had always envisioned it taking this form, or that he no longer felt able to reduce the material to a single narrative line? We do not know.

A substantial gap in the correspondence ensues. Where the record picks up again in January 1957, it is Heller complaining that he received no answer to a letter he wrote in early autumn.[59] In reply, puzzled, Goitein writes that this letter of January 16 was the first he had received since Heller and Vidal's of April 20.[60] The remainder of 1957 is very poorly attested in the archive. Yet we do know that Heller offered Goitein 350,000 francs for the summer months of 1957.[61]

These were months of crisis. In the fall of 1957, the long silence of the archive was broken by a letter from the great French orientalist Claude Cahen to the medievalist Maurice Lombard. It sheds a wholly different light on the correspondence. Not planning to come to Paris for a fortnight, and fearing that Braudel was away, Cahen wrote to Lombard rather than to Heller whom he did not know. That summer, at the Orientalists Congress in Munich he had met Goitein

> and we spoke of his plans about the Geniza. He said he had intended to publish a book under the auspices of the Centre de Recherches Historiques and in accordance to the offer that Braudel made to him, on the documents of the Geniza in relation to commerce in the Indian Ocean. But later, because of language questions, and especially because the letters he had written to the Center remained without a response, or with late or insufficient responses, he changed his mind and because he will spend some time at the University of Philadelphia, he would figure out how to publish it there.[62]

Thus we see that Heller's silence from the second half of 1956 through the end of 1957 seemed to have soured Goitein on the project. There is the question, of course, as to whether this only served as a more publically acceptable pretense than the language issue, which Cahen mentioned and which clearly did bother Goitein.

Cahen did not know Goitein but had tremendous respect for him as a scholar and for his work. Though having himself no connection with the CRH, he thought it would be a black eye for France if a project of importance promised to the Centre wound up being published overseas. "I am not able to judge if Goitein has any particular susceptibilities; what I know is that he is a great scholar, that his book will be major, and, although I have no relationship with the Centre, that it will be an unfortunate affair that a book promised to a French collection ends up being published abroad."[63]

And while unable to speak in the name of the Centre, Cahen nevertheless expressed what he believed was Braudel's commitment to the project. However, he did not believe that he had succeeded in persuading Goitein. "In Munich I could not naturally speak in the name of the Centre; however I believed I could affirm to him that Braudel, whom I recently heard say that which yourself, when we talked in the spring, had confirmed. And I tried to plead the case of the Centre, without much apparent success."[64]

The latest turn in the story, Cahen related, was that he had heard from Goitein, who had come to treat him as something of an official French representative, that he had met Heller in New York and that they had smoothed things over, "recovering" the project for the CRH. Goitein promised an article for *Annales* and Heller promised to arrange the translation. Cahen made a point, again, of how uncomfortable Goitein was with a language over which he had no command, which apparently extended even to checking another's translation (this is implied, not stated). "Now Goitein, who seems to treat me as an agent for France, writes to me that he met Clemens Heller in New York who seems to have been able to arrange things, to recover the project for the Centre, and also to obtain the promise from Goitein of an article for the Annales; the book could be translated in English instead of French, which would suit Goitein better since he doesn't feel capable of writing good French and distrusts translations that he is not able to review closely enough."[65]

In conclusion, Cahen reported that Goitein remained suspicious ("sur le qui-vive et méfiant") and that some official communication from Braudel was now needed to assuage him. Heller's lack of accessibility was scored for pouring fuel on the fire. Goitein remained wary, so much so that he even said that if he did not hear from the Centre within six weeks of his sending his article to the *Annales*, he would pull it.

> It is nevertheless obvious that he remains on-guard and suspicious; he probably wishes a detailed and official letter or a contract in due form to clarify and guarantee the conversation. However, he told me that since the interview with Heller, 'with the exception of a very flimsy short letter, I have not heard from them since then.' As for the Annales article, which he ought to have sent a week after the letter he sent me, this is what he writes: 'In case I shall not hear from the Centre for, say, six weeks after having sent the article there, I shall offer it somewhere else.'[66]

Therefore, Cahen concluded, someone at the Centre—Braudel if he were there, but certainly Heller—should immediately write to Goitein "so that there be no new rupture, which this time would be irremediable." While some in Paris might be inclined to think him "especially difficult," he was convinced that publishing this material was worth bearing these sorts of "little inconveniences."[67]

Lombard immediately passed along Cahen's letter to Heller. He asked after the state of things and offered to write to Goitein himself in the absence of Braudel if Heller thought it useful.[68]

Lombard set off alarm bells on the rue Varenne. The very next day, Heller drafted a long letter to Goitein. He began, knowing his correspondent, with money matters. Travel monies were being put into the 1958 budget for Goitein. The Agence Friedland, which the CRH used in Paris, would send a *bon d'échange* to Globe Travel Service in Philadelphia. Moreover, dangling the opportunity to see new Geniza materials before Goitein's eyes, Heller allowed that the cost of a trip from Paris to Russia could be paid for by the CRH.

Heller then turned to the publication schedule.

> If I understood our conversation in New York correctly you proposed that in 1958 the first four volumes of the *Geniza Papers* should be ready for publication:
>
> -Vol. I: Goitein and Baneth, Introduction to the Geniza Papers;
> -Vol. II to IV: Indian Merchants.
>
> Several other volumes were discussed for later date, especially those by Mr Murat Michaely and Mr Golb. We also discussed the possibility of two articles for the *Annales,* a shorter one which you would send before the end of the year and another one which you would send later next year. I presume it would be best if you would be kind enough to draw up a detailed program as you envisaged it so that there would be a formal document serving as a base for all future action.

Turning back to money, Heller demonstrated the depth of the Centre's commitment to Goitein.

> As far as financial agreements are concerned I understand that the *Centre de Recherches Historiques* will cover the cost of your trip to Europe next year (plane tourist class or the equivalent on a boat) and will cover your *frais de séjour* up to 100,000 francs a month during the months June, July and August. Please let me know whether any additional financial requirements exist and I shall let you know whether or not they can be assumed by the Centre. Naturally we are completely responsible for the cost of the publication of the books.[69]

And the day after that, it was Braudel himself who wrote to Goitein. This was the formal action that Cahen had thought necessary.

> Mr. Heller told me about the conversation you had with him during his stay in New York. In principle, I agree with everything you discussed and in particular that which Mr. Heller suggested to you in his letter of November 7th, 1957. In other words, I agree with the printing of the 'Documents de la Genisa' and agree to your trip to Europe in 1958. I must unfortunately leave Paris once again for a month and hope that Mr. Heller can resolve all the details with you. However, if I must sign official documents upon my return, I will do what is necessary as soon as possible.[70]

Amidst the official-sounding language, Braudel added a concluding paragraph making clear how important Goitein was for him, as well as for the VIe section. "I am taking this opportunity to reiterate that the École Pratique des Hautes Études is proud to be of use to you in the publication of your work and I look forward to the articles that you kindly put forward to the *Annales*."[71]

Goitein wrote back immediately, still a bit on edge, accompanying the submission of an article on "The Socio-Economic Background of the Muslim Friday Worship." In case it was "too Islamistic," he wrote, "you need not worry; I shall have no difficulty publishing the article in an orientalistic journal. Please inform me of the decision of the editor of the *Annales* and, in case it is negative, I shall advise to which journal in Paris to send it."[72]

Still feeling Cahen's prod, Heller wrote back three days later—which is to say, immediately after the letter arrived—thanking Goitein for submitting the article and informing him that a formal letter from Braudel as editor accepting the article would come soon.[73] Undated, in the dossier, is a note from Braudel to Heller saying that he was going to write to Goitein telling him that he had accepted the article.[74]

It is at this point that one realizes that there had been no direct communication between Braudel and Goitein between February 1955 and November 1957. All Goitein's queries about Braudel's involvement might in fact be read as a desire for direct contact. And yet there was none.

Braudel's letter seems, at least on the surface, to have assuaged Goitein's doubts. In any event it elicited from him a response, dated November 21, 1957, reflecting on the history of their relationship.

> Please accept my sincere thanks for your kind letter of November 8. I feel it is a great honor for me that my proposed publication of the Geniza papers has been included in the magnificent program of your Centre.
>
> I appreciated in particular that it was the initiative of your section which created the connection between your school and my work on the Geniza, and I hope that the publication will live up to the high standard set by the École Pratique des Hautes Études.[75]

On the same day, November 21, Goitein addressed a longer letter to Heller, responding to the plans laid out in Heller's letter of November 7, which itself resumed, or purported to resume, the conclusions reached during their conversations in New York. The main point Goitein wished to clarify was delivery date. Heller had said all four volumes would be ready in 1958. This was incorrect. "Concerning the publications discussed by us in New York, I would like to make the following qualifications to your letter: The material for volume I, 'An Introduction to the Study of the Documentary Geniza,' is ready." It would be organized into a book during his visit to Europe in summer of 1958. But as for the India book, Goitein wrote, "I told you that the last volume, which contains the texts (Arabic and Hebrew letters), is more or less ready for print. Concerning volume I (the general introduction) and volume II (summaries, translations, commentaries), I can only say that they will certainly not be ready next summer, for, as you remember, I wrote the whole thing first in Hebrew and *in extenso.*"There is some confusion here, because volume I refers to what elsewhere is called volume II, and volume II to what elsewhere is called volume III."However, we have decided, at our meeting in New York, as you remember, to bring out this book in English for the benefit of many scholars in India and elsewhere who are not fluent in French . . . In addition, only a selection of texts will be translated in full, while many others will be given in summaries. All this implies a complete re-shuffle of all I have done . . . I do ~~not~~ [*sic*] hope to complete a considerable part of it until next May."

Finally, Goitein promised to send to *Annales* an article dealing with some aspect of the India project. "In connection with my work on the India book, I shall send a comprehensive article on one of its aspects to the *Annales.*" This never happened.[76] From then on, things fell apart. The correspondence focuses on the details of printing, and by the middle of 1958, Goitein had already become an author for the University of

California Press. Heller carried on his correspondence, trying to keep some contact for the CRH, but it was clear by then that Goitein's heart was not any longer in this relationship.[77]

So, what are the stakes in this story?

At one immediate level, it gives some concreteness to the mystery of the India Book, one of the great historical mirages of the twentieth century. It is said by Goitein to have been ready in 1955, 1956, 1958, 1959, and 1963, and yet it was never ready, not even in 1985 when Goitein died. Why Goitein was unable to pronounce it as finished probably offers us a very fine insight into his function as a scholar, and his vision of the Geniza. Goitein viewed perfect scholarship as complete compilation. We might long have abandoned this, but working with a defined corpus, even one of astounding difficulty, gave Goitein the illusion that this was possible. And thus, the ever-receding perfection, and ever-receding publication target, was an inevitable function of this illusion.

Without the India Book, scholars in cognate fields have of course suffered. Even an incomplete India Book would have been better for scholarship than none. But Goitein, the philologist, obviously did not think this way. He wanted it perfect, and trusted only himself to deliver that perfection to posterity. The bitter irony of the outcome is inevitable: the work was not finished under his supervision, appeared half a century later, under the care of another, in a form different from how Goitein might have done it, and with editorial and technical decisions that were another's. And yet, the first, introductory volume has been followed by three others, yielding the four volumes that Goitein had imagined, though configured somewhat differently.[78] And still to be published but already in some state are three additional works—a biography of Halfon ha-Levi, an edition of the documents related to Halfon ha-Levi, and a Judeo-Arabic lexicon compiled from the Geniza documents.[79]

Without the India Book we have looked with only one of Goitein's eyes. The full impact of Goitein therefore, will only now be dawning. The very recent work of Avner Greif, Roxani Margariti, Marina Rustow, Jessica Goldberg, Philip Lieberman, and others demonstrates that the usefulness of what Goitein assembled is finally now being harvested by those interested in the dynamics of social and economic life in the Middle Ages.[80] And wider still, scholars with interests as diverse as Chris Wickham and Sanjay Subrahmanyam have incorporated Goitein's work into their own and others, such as David Hancock, Francesca Trivellato, and Sebouh David Aslanian are carrying Goitein's approach, knowingly or unknowingly, to domains far from his own.[81] In a way, then, only now is the promise held out by the CRH, of putting Goitein into the *Affaires et gens d'affaires,* finally being realized: a vision of economic history as cultural history, with individual experiences at the center, but so densely documented as to endow the whole with more than anecdotal significance.

More profoundly, however, the absence of the India book has obscured the greatness of Goitein's vision. Long before there was transnational or global history, long before transmission, translation, networks, communication, correspondence, or exchange were the *Schlagwörte* of the day, Goitein saw a historical project driven by and exemplifying all of these. Had he sent the India Book off into the world at half the size of its 1964 state, and then finished the five volumes of *Mediterranean Society,* Goitein's would justly have already been hailed as one of the greatest feats of historical scholarship of the twentieth century, and even of modern historical research tout court.

It is very difficult to assess the impact on Goitein of his decade-long association with the CRH. On the surface, there was obviously very little: no references to the kind of projects sponsored by the CRH filtered into Goitein's published oeuvre, and no real sense of conversation with work being done at, or sponsored by, the Centre. Nor was there any acknowledgement for the years of subsidy and encouragement provided from the highest levels.

And yet, perhaps the "miss" was not complete. For we have the epilogue Goitein prepared for the concluding volume of *A Mediterranean Society.* In it, he offered as an autobiographical account the description of his method as "interpretative historical sociography." This category he supports through a reading of Clifford Geertz's *The Interpretation of Cultures* (1973). Geertz, whom he would have met during his retirement to the Institute for Advanced Study at Princeton, might have resonated so powerfully with Goitein because of his own experience as an ethnographer. For we know that Goitein worked for many years on and with Yemenite Jews. This led him to read anthropological theory and, on his visits to England, to consult with Evans-Pritchard and Fortes about the Yemenites. From his fieldwork, and for the benefit of future fieldworkers, Goitein put together a "questionnaire for the study of the Jews of Yemen, especially those from villages and small towns." This is published.[82] Sociography, then, might gesture at the encounter between Goitein the philologist and Goitein the ethnographer.[83]

And yet, readers of *A Mediterranean Society*'s epilogue recognize in its tone not merely the valedictory, but also the desire to situate his work in a social scientific context. And if it was not that of his Parisian patrons, that still does not mean he was not in some more mediated way shaped by the encounter with them. We know that he referred to his "Geniza Lab," just as the CRH was often, especially in its early years, referred to as a *Laboratoire de recherches historiques*.[84]

And this brings us to Braudel, who always is included in lists of the greatest historians. The Heller-Goitein correspondence sheds further light on the administrator whose vision drove the CRH, some of which can be deduced from prefatory materials to the volumes in *Affaires et Gens d'Affaires* and *Port-Routes-Trafics*. Braudel the head of the VIe section and the CRH was much more open to human-centered history than Braudel the historian of *The Mediterranean*. The Braudel who valued Goitein was the Braudel who wrote "For to challenge the enormous role that has sometimes been assigned to certain outstanding men in the genesis of history is by no means to deny the stature of the individual as individual and the fascination that there is for one man in poring over the fate of another."[85]

And yet, as great an administrator as Braudel was, one wonders whether in every case he maintained at arm's length a relationship with those he was trying to woo. While delegation to Heller might have made sense from an administrative point of view, when dealing with someone like Goitein it clearly would have helped had the great man himself taken more of a role in this pursuit. Some share of the blame for the collapse of this collaboration must then be laid at his door.

For whatever reason, it is the failed meeting of Braudel and Goitein that pains. For had they met, had Goitein published with the CRH, had he, as might be expected, have gone and lectured as Braudel's guest at the École and, conversely, had Heller or even Braudel come to Jerusalem, one could envision a very different shape to both the Jerusalem School and to the *Annales* School.

The impersonal social science history of the 1950s and 1960s would have had a hard time digesting Goitein's human-scaled project. But the force of his vision could have helped hasten the age of microhistory, mentalités, merchants, and the Mediterranean and commerce might have featured then, as they do now. But one could go even further. Goitein's project was about letters, and had *Affaires et gens d'affaires* become a dominant historical vision, as well it might have, the possibility of joining economic to cultural history, a vision that went back to Karl Lamprecht at the end of the nineteenth century, might well have come to pass. A cultural history cognizant of material reality could have emerged in the 1950s—it is only happening now. And who knows how this in turn might have helped sharpen up our vision of other republics of letters, both in and around the Mediterranean. In the possible collaboration between Braudel and Goitein, as in its failure, we can discern historical seascapes of past and future.[86]

NOTES

Earlier versions of this chapter were presented to the annual meeting of the Medieval Academy of America, the Maison Méditeranéenne des Sciences de l'Homme in Aix, the Mediterranean Seminar at the École Normale Supérieure in Paris, and at the Scholion Center at the Hebrew University of Jerusalem. I wish to thank especially Youval Rotman, Brigit Marin, Wolfgang Kaiser, Claudia Moatti, Bernard Vincent, and Daniel Schwartz for these invitations. I am extremely grateful to Brigitte Mazon at the Archive of the École des Hautes Études en Sciences Sociales for facilitating my work in the Fonds Heller and for permission to quote from its contents.

1. I am indebted to the brilliant essay by Gerson D. Cohen, "The Story of the Four Captives," in *Studies in the Variety of Rabbinic Cultures* (Philadelphia and New York: The Jewish Publication Society, 1991), 157–208.
2. Peregrine Horden and Nicholas Purcell, *The Corrupting Sea: A Study of Mediterranean History* (Oxford: Wiley-Blackwell, 2000), 31–39.
3. S. D. Goitein, *A Mediterranean Society* 6 vols. (Berkeley and Los Angeles: University of California Press, 1965–1985), vol. I, vii.
4. There are sixty-nine letters from Goitein to Heller; forty-two letters from Heller to Goitein; six letters from Goitein to others (two to Braudel, one to the University of California Press, one to Mouton & Co.); three letters from Heller to others (one to the University of California Press, one to Paul Lemerle of CNRS, one to a travel agent); two from Braudel (one to Goitein, one to Heller); one from Cahen to Lombard; one from Lombard to Heller; 1 from Lemerle to Heller. In addition to these letters, the Goitein dossier in the Fonds Heller in the EHESS in Paris contains the annual research reports Goitein drafted for Heller in exchange for funding for the years 1958, 1959, 1960, 1964, as well as a preface to the India book and a list of its documents drawn up for Heller in 1955.
5. S. D. Goitein and Mordechai Friedman, *India traders of the Middle Ages: documents from the Cairo Geniza, the 'India Book,'* Études sur le Judaïsme Médiéval 31 (Leiden: Brill Publishers, 2008). Publication of the India Book does seem to have turned attention to Goitein's contribution, as in a recent forum in the *Jewish Quarterly Review* that devotes two articles to Goitein and twentieth-century historiography, though without reference to the contents of either the Heller or Goitein archives: Elliott Horowitz, "Scholars of the Mediterranean and the Mediterranean of Scholars," and Fred Astren, "Goitein, Medieval Jews, and the 'New Mediterranean Studies'," *Jewish Quarterly Review* 102 (2012): 477–90 and 513–31.
6. S. D. Goitein, "Preface," in *A Mediterranean Society,* vol. 1 (Berkeley and Los Angeles: University of California Press, 1965), vii.
7. Goitein, "Preface," in *A Mediterranean Society,* viii.

8. Giuliana Gemelli, *Fernand Braudel, e l'Europa universale* (Venezia: Marsilio, 1990). No biography of Heller exists yet. Basic facts can be found in Paul Lewis, "Clemens Heller, 85, Founder of Postwar Salzburg Seminar," *New York Times,* September 6, 2002, http://www.nytimes.com/2002/09/06/world/clemens-heller-85-founder-of-postwar-salzburg-seminar.html?scp=1&sq=clemens%20heller&st=cse. He is currently the subject of the *Habilitationsprojekt* of Anne Kwaschik, Von Harvard nach Paris, Clemens Heller, und die Internationsalisierung der europäischen Sozialwissenschaften nach 1945, http://www.geisteswissenschaften.fu-berlin.de/frankreichzentrum/forschung/kwaschik.html.
9. This association between the CRH and Warburg Institute is not adventitious. See A. Tenenti, "Quindici anni di attività del Centre de recherches historiques di Parigi," *Studi storici,* (1967): 203–11, esp. pages 204, 211. "Il Centre ha assolto a più riprese la funzione di laboratorio e di campo d'esperienza: . . . ora il Centre, come in certo qual modo anche il Warburg Institute ad esempio, ha voluto imboccare una strada ben diversa : e ci sembra che sia riuscito sufficientemente a dimostrare—non solo con la rapidità esteriore del suo sviluppo—quale diversa capacità di accelerazione acquistava la ricerca una volta impostata su di un piano collettivo o di équipe."
10. Lutz Raphael, *Die Erben von Bloch und Febvre. 'Annales'-Geschichtsschreibung und 'nouvelle histoire' in Frankreich 1945–1980* (Stuttgart, Germany: Klett-Cotta, 1994), 169–99; idem, "Le Centre de recherches historiques de 1949 à 1975," *Les Cahiers du Centre de Recherches Historiques* 10 (1993), http://ccrh.revues.org/index2783.html. Gemelli emphasizes 1951 because that was when Frederic Lane arrived with the Rockefeller Foundation's money, and the Centre's activities could accelerate (*Fernand Braudel e l'Europa universale,* 256).
11. Gemelli, *Fernand Braudel,* 256–57, citing an observation by David S. Landes.
12. "l'auteur de ce très beau livre . . . cet auteur est un homme. . . . Soucieux d'hier, préoccupé de demain—parce que demain et hier trouvent dans le coeur de l'homme un lien commun." Armado Sapori, *Le Marchand Italien au Moyen Age. Conférences et bibliographie.* Afffaires et Gens d'Affaires I, Introduction de Lucien Febvre (Paris: Armand Colin, 1952), vii.
13. "C'en est un autre, et des plus rares, qui remplira d'aise beaucoup de nos collègues en histoire, que de ne pas considérer la vie économique comme une réalité en soi, un monde fermé de chiffres, mais bien comme un secteur de la vie entre quelques autres, et qu'il faut remplacer dans un ensemble de circonstances et de liaisons diverses pour lui rendre sa pleine valeur. Pour Etienne Trocmé, ses marchands sont des hommes, sa ville non seulement des rues, un havre, des remparts, une 'banlieu', des nourritures proches, des commerces lointains—mais aussi des institutions au sens le plus riche du mot. La Rochelle peu à peu se révèle à nous dans la plénitude de sa vie, dans sa dignité et sa force d'état urbain." Fernand Braudel, "Avant-propos," Étienne Trocmé et Marcel Delafosse. *Le Commerce Rochelais de la fin du XVe siècle au début du XVIIIe.* Ports-Routes-Trafics V (Paris: Armand Colin, 1952), i.
14. "Leur intérêt, si je ne me trompe, est de nous introduire au coeur des pratiques et réalités de la vie quotidienne des marchands: ces réalités, par elles-

mêmes ou du fait de leur répétition, dépassement souvent le détail simplement anecdotique." Then, taking a single letter as an example, he notes that it sheds light on "la ville et la place de Marseille, le role des Génois acheteurs de soie, l'importance des changes de Lyon, l'attraction de Seville et de Cadix, les transpsorts de cochenille 'américane' vers le levant, le cours de réaux d'argent à Marseille, les voyages de navires vers Alexandrie et Tripoli." Fernand Braudel, "Avant propos," *Lettres de négociants marseillais: Les Frères Hermite (1570–1612).* Affaires et gens d'affaires 3, ed. Micheline Baulant, (Paris: Armand Colin, 1953), v.

15. "La VIe Section de l'École des Hautes-Études s'est d'autant plus intéressée à ces lettres qu'elle avait en chantier deux publications analogues [*Lettres d'un marchand vénitien Andrea Berengo (1553–1556)* (Affaires et gens d'affaires 10, 1957); *Simon Ruiz et les 'asientos' de Philippe II* (Affaires et gens d'affaires 6, 1953)] . . . Mais ces lettres de marchands marseillais ont leur valeur et gagneront en importance à être ainsi rapprochées de documents analogues et replongées dans une histoire économique du XVIe siècle vue au jour le jour, expliquée par ses acteurs, histoire qui, en général, n'a que de lointains rapports avec les images qu'en offrent, quand ils les offrent, les grands livres de synthèse." *Lettres de négociants marseillais,* v.
16. "Car leur correspondance ne dit pas tout de leurs activitiés ou de leurs calculs. Le plus difficile, et nous allons nous y essayer dans un instant, c'est encore de bien comprendre leur langage, de suivre leurs procédés, leur tactique, leur art de diagnostiquer et d'agir." José Gentil da Silva, *Stratégie des Affaires à Lisbonne entre 1595 et 1607. Lettres marchandes des Rodrigues d'Evora et Veiga,* Affaires et gens d'affaires IX (Paris: Armand Colin, 1956), 9.
17. The best known demonstration of this practice was, however, not produced in Paris (nor, even, by an academic). Iris Origo's *The Merchant of Prato* (1957) made use of Francesco Maria Datini's personal correspondence; Federigo Melis, who had begun publishing on Datini in 1954 and in May 1955 opened an international exhibition on him, strictly separated the business history from the personal. Melis, for all his later inclination to Braudel and Paris (he made Braudel President of the newly founded Istituto Datini in 1967) seems to have resented Origo's non-quantitative approach (Giampiero Nigro, "Introduction," in *Francesco di Marco Datini. The Man, the Merchant,* ed. Nigro (Florence: Firenze University Press, 2010), xiii–xiv). The generally favorable review of Origo's book in *Annales*—by none other than Gentil da Silva—acknowledged that it was *au courant* with the latest work being done in and around the CRH, but was critical of its narrowly biographical focus (José Gentil da Silva, "Un capitaliste toscan du XIVe siècle," *Annales. ESC* 13 (1958): 398–402.
18. *Speculum* 29 (1954): 181–97.
19. Heller to Goitein, 10 October 1954, Archives de l'École des Hautes Études en Sciences Sociales, Fonds Heller, Dossier "Israël", unpaginated [all subsequent references are to this archive and will include only the date of correspondence].
20. Goitein to Heller, 27 October 1954.
21. Goitein to Heller, 1 December 1954. This seems like a direct answer to part of Heller's first letter, but Goitein's next letter, written from Jerusalem, makes no mention of having answered earlier—unless there is a letter

that is missing which might make sense, since the letter of December 1, below, does not directly respond to Heller's first, but as if to a different, more advanced, round of questions.

22. I have been unable thus far to identify this work.
23. Goitein to Heller, 1 December 1954.
24. Goitein to Heller, 18 December 1954. From its beginnings the CRH subsidized research done by scholars elsewhere. These included the research of P. Jeannin on the traffic of Rouen (1955–1956) and M. Mollat in the archives of Lille (also 1955–1956). "Le démarrage improvisé : 1949–1956," *Les Cahiers du Centre de Recherches Historiques,* 10 (1993), http://ccrh.revues.org/index2788.html.
25. Goitein to Heller, 18 December 1954.
26. Goitein to Heller, 6 February 1955.
27. Goitein to Heller et al., 6 February 1955.
28. Goitein to Braudel, 20 February 1955.
29. Goitein to Heller, 20 February 1955.
30. "As you know, your book will be printed by the *Imprimerie Nationale* and they have a very good department for oriental characters (better than Brill). Technically, there should be no difficulty and economically, I think we could support the cost of including the printing of the originals. Besides, the *Centre de Recherches Historiques* has undergone some important transformations since I wrote you first and [the] character of its publications will expand considerably towards Oriental problems. This week, four sinologists have been appointed to the Faculty, at the *VIème Section,* one indologist and one specialist on Northern Africa. This is only the beginning of further developments in a programme of social and economic studies regarding Asia, the Islamic world and Russia. A young French rabbin, Mr. Schwarzfuchs, has already joined our effort and I am enclosing an outline of the programme he set himself." Heller to Goitein, 30 March 1955.
31. Heller to Goitein, 30 March 1955.
32. Goitein to Heller, 14 April 1955. Goitein allowed himself a bit of humor here: "On that occasion, we could discuss the Geniza business. I shall be glad to put you up. Like most Israelis, during weekdays we are mostly ichtyophagists, but you will not starve." On the Jerusalem School, see David N. Myers, *Re-Inventing the Jewish Past: European Jewish Intellectuals and the Zionist Return to History* (New York: Oxford University Press, 1995).
33. Heller also politely declined Goitein's invitation to Jerusalem, saying that there was no money available for this. Heller to Goitein, April 20, 1955.
34. Goitein to Heller, 25 April 1955.
35. Heller to Goitein, 25 May 1955.
36. Goitein to Heller, 28 May 1955, referring to Heller to Goitein of May 3.
37. Goitein to Heller, 31 May 1955.
38. Heller to Goitein, 10 June 1955.
39. Goitein to Heller, 15 June 1955.
40. Heller to Goitein, 29 June 1955.
41. Heller to Goitein, 14 September 1955.
42. Goitein to Heller, 16 September 1955.
43. Heller to Goitein, 21 September 1955.

44. Goitein to Heller, 23 September 1955.
45. Heller to Goitein, 23 September 1955.
46. Goitein to Heller, 11 October 1955.
47. Goitein to Heller, 20 November 1955.
48. Goitein to Heller, 21 November 1955.
49. Goitein to Heller, 12 December 1955.
50. Vidal (for Heller) to Goitein, 16 January 1956.
51. "His curriculum vitae is interesting and makes him an ideal candidate for the position offered. His name: Joseph Eliash. He is 24, of German Jewish origin, but educated in Baghdad, where he first visited a French school (St Joseph of the Peres Carmes), then a Jewish higher Secondary school with Arabic and English as languages of instruction, he made the London matric Honours, then came here and studied at the Hebrew University "Middle East in modern Times" and Economics (B.A., 1954) and is going to get his M.A. in Arabic language and Literature and Islamic studies this summer. He is married to a girl from Rumania, who knows French well."

    "Mr. Eliash has worked on the Geniza since January 1955 and I am able now to state that I am very satisfied with his work. Besides Arabic and Hebrew, he speaks French, English and German. He has a very amiable personality and it is pleasant to work with him." Goitein to Heller, 22 January 1956.
52. Goitein to Heller, 22 January 1956.
53. Heller to Goitein, 30 January 1956.
54. Goitein to Heller, 15 February 1956.
55. Heller to Goitein, 20 February 1956.
56. Goitein to Heller, 25 March 1956.
57. Heller to Goitein, 10 April 1956.
58. Goitein to Heller, 18 May 1956
59. Heller to Goitein, 16 January 1957.
60. Goitein to Heller, 24 January 1957.
61. Goitein to Heller, 9 October 1957. This is Goitein's first letter on University of Pennsylvania stationery.
62. "et nous nous sommes entretenus de ses projets sur la Geniza. Il m'a dit qu'il avait eu l'intention de publier sous les auspices du Centre de Recherches Historiques un ouvrage sur les documents de la Geniza relatifs au commerce dans l'Océan Indien, conformément à l'offre que Braudel lui en avait faite. Mais ultérieurement, un peu pour des questions de langue, mais surtout parce que les lettres qu'il avait écrites au Centre étaient restées sans réponses, ou avec des réponses tardives et insuffisantes, il y avait renoncé, et, puisqu'il va passer quelque temps à l'Université de Philadelphia, il allait voir comment le publier là-bas." Cahen to Lombard, 4 November 1957.
63. "Je ne suis pas en état de juger si Goitein a des susceptibilités particulières; ce que je sais est que c'est un grand savant, que son ouvrage sera capital, et, bien que je n'aie aucun rapport avec le Centre, qu'il serait de facheux exemple qu'un ouvrage promis à une collection française finisse par paraitre à l'étranger." Cahen to Lombard, 4 November 1957.
64. "Je ne pouvais naturellement pas à Munich parler au nom du Centre; toutefois j'ai cru pouvoir lui affirmer que Braudel, auquel je l'avais récemment

entendu dire, ce que toi-même, lors de notre entretien printanier, m'avais confirmé. Et j'ai essayé de plaider la cause du Centre, sans succès apparent." Cahen to Lombard, 4 November 1957.

65. "Maintenant Goitein, qui parait me traiter en fondé de pouvoir pour la France, m'écrit qu'il a rencontré a New-York Clemens Heller, et celui-ci parait avoir réussi à arranger les choses, à recouvrer le projet pour le Centre, et aussi à obtenir de Goitein la promesse d'un article pour les Annales; le livre pourrait etre traduit en anglais au lieu de français ce qui arrangerait mieux Goitein, qui ne s'estime pas capable de rédiger en bon français et se méfie des traductions qu'il ne peut revoir d'assez près." Cahen to Lombard, 4 November 1957.

66. "Néanmoins il est évident qu'il reste sur le qui-vive et méfiant; il souhaite certainement qu'une lettre officielle et circonstanciée ou un contrat en bonne et dûe forme vienne préciser et garantir la conversation. Or, me dit-il, depuis l'entrevue avec Heller, 'with the exception of a very flimsy short letter, I have not heard from them since then.' Et, pour ce qui est de l'article pour les Annales, qu'il devait envoyer une semaine après l'envoi de la lettre qu'il m'écrit, voici ce qu'il m'écrit: 'In case I shall not hear from the Centre for, say, six weeks after having sent the article there, I shall offer it somewhere else.'" Cahen to Lombard, 4 November 1957.

67. "J'ai cru bon de te mettre au courant pour que tu puisses tout de suite, avec Braudel s'il est là, et avec Heller en tous cas, aviser à ce qu'il n'y ait pas une nouvelle rupture, qui serait cette fois irrémédiable. Peut-etre seront-ils enclins à le trouver bien difficile; mais il me semble que la chose vaut la peine de passer sur de petits inconvénients comme ceux-là." Cahen to Lombard, 4 November 1957.

68. "Où en est exactement la question de la publication des documents de Geniza que Goitein doit faire dans la collection du Centre et celle de la parution de son article pour les *Annales* ? Et serait-il bon—en l'absence de Fernand Braudel—que j'entre en contact avec Goitein pour arranger les choses et calmer ses appréhensions? Ou bien vous chargez-vous de lui écrire vous-meme, vous qui le connaissez bien?" Lombard to Heller, 6 November 1957.

69. Heller to Goitein, 7 November 1957. "Baneth," sometimes spelled here "Paneth" is David H. Baneth, a specialist of Maimonides, Judah Halevi, etc. See *Studia Orientalia: Memoriae D.H. Baneth Dedicata,* ed. J. Blau et al. (Jerusalem: The Magnes Press, 1979), which includes an essay by Goitein. I thank Sanjay Subrahmanyam for bringing this to my attention.

70. "Monsieur Heller m'a mis au courant de l'entretien que vous aves eu avec lui pendant son séjour à New York. Je suis en principe d'accord pour tout ce dont vous avez discuté et en particulier pour ce que Monseiur Heller vous a proposé dans sa lettre du 7 novembre 1957. Autrement dit, je suis d'accord pour l'impression des 'Documents de la Genisa' et pour votre voyage en Europe en 1958. Je dois malheureusement m'absenter encore une fois de Paris pour un mois et j'espère que Monseiur Heller pourra régler tous les détails avec vous. Toutefois si je dois signer des documents officiels après mon retour, je ferai le nécessaire aussi vite que possible." Braudel to Goitein, 8 November 1957.

71. "Je profite de cette occasion pour vous redire combien l'École Pratique des

Hautes Études est fière de pouvoir vous être utile dans la publication de vos travaux et j'attends avec impatience les articles que vous avez bien voulu proposer pour les *Annales.*" Braudel to Goitein, 8 November 1957.

72. Goitein to Heller, 12 November 1957.
73. Heller to Goitein, 15 November 1957.
74. Braudel to Heller, undated, (after November 1957).
75. Goitein to Braudel, 21 November 1957.
76. Goitein to Heller, 21 November 1957.
77. I plan to discuss Goitein and this latter part of the story in a separate publication.
78. The Ben Zvi Institute in Jerusalem published in 2009–2010 Mordechai Friedman's three volumes of Goitein's work on Joseph Lebedi, Madmun Nagid of Yemen, and Abraham ben Yiju.
79. All these, as well, will be edited by Friedman.
80. Avner Greif, *Institutions and the Path to the Modern Economy: Lessons from Medieval Trade* (Cambridge and New York: Cambridge University Press. 2006); Jessica Goldberg, *Trade and Institutions in the Medieval Mediterranean: The Geniza Merchants and Their Business World* (Cambridge: Cambridge University Press, 2012); Marina Rustow, *Heresy and the Politics of Community: The Jews of the Fatimid Caliphate* (Ithaca, NY: Cornell University Press, 2008); Roxani Margariti, *Aden and the Indian Ocean Trade: 150 Years in the Life of a Medieval Arabian Port* (Chapel Hill: University of North Carolina Press, 2007); Phillip Lieberman-Ackerman, "A Partnership Culture: Jewish Economic and Social Life Viewed through the Legal Documents of the Cairo Geniza" (PhD diss., Princeton University, 2007).
81. Sanjay Subrahmanyam, "Notes on Circulation and Asymmetry in Two Mediterraneans, c.1400–1800," in *From the Mediterranean to the China Sea: Miscellaneous Notes,* eds. Glaude Guillot, Denys Lombard, and Roderich Ptak (Wiesbaden: Harrassowitz Verlag, 1998), 32; David Hancock, *Oceans of Wine: Madeira and the Organization of the Atlantic Market, 1640–1815* (New Haven, CT, Yale University Press, 2009); Francesca Trivellato, *The Familiarity of Strangers: The Sephardic Diaspora, Livorno, and Cross-Cultural Trade in the Early Modern Period* (New Haven, CT: Yale University Press, 2009); Sebouh David Aslanian, *From the Indian Ocean to the Mediterranean: The Global Trade Networks of Armenian Merchants from New Julfa* (Berkeley and Los Angeles, London: University of California Press, 2010); Chris Wickham, *Framing the Early Middle Ages: Europe and the Mediterranean, 400–800* (Oxford: University of Oxford Press, 2005), 718n48.
82. Goitein, *The Yemenites: History, Communal Organization, Spiritual Life* (Jerusalem, 1983), 345–55.
83. Goitein, *A Mediterranean Society,* vol. V, 496–502.
84. Raphael, *Die Erben von Bloch und Febvre,* 162n22. I owe the reference to Goitein's habitual use of the term *Geniza Lab* to Adina Hoffman and Peter Cole. See their discussion of Goitein in *Sacred Trash: The Lost and Found World of the Cairo Geniza* (New York: Schocken Books, 2011), ch.10.
85. Fernand Braudel, "The Situation of History in 1950," in *On History,* trans. Sarah Matthews (Chicago: University of Chicago Press, 1980), 20.
86. The second part of this study is published in the *Mitteilungen des Kunsthistorischen Institutes* 55 (2013).

TWO

Atlantic and Caribbean Perspectives: Analyzing a Hybrid and Entangled World

Wim Klooster

How has a thalassographic prism enabled new approaches in the historiography of the Atlantic world? In order to answer this question, I delve into the writing of Atlantic history since its coming of age in the last two decades. I distinguish five different ways to write Atlantic history: one school stresses agency, another adaptation, a third privileges comparisons, a fourth entanglement, and a fifth studies networks. But, for the sake of clarity, let me start with the question: What is Atlantic history? The most convenient definition is that of Sir John Elliott: Atlantic historians study "the creation, destruction, and re-creation of communities as a result of the movement, across and around the Atlantic basin, of people, commodities, cultural practices, and values."[1] This definition is perhaps erring a little on the cultural side, but it will do for now. What should be clear from the outset is that Atlantic historians do not gravitate to the microecological approach that Purcell and Horden favor for the Mediterranean.[2] The Atlantic Ocean was obviously far too variegated for that. In his book, *La Mediterrannée,* Fernand Braudel noted the contrast between the "exhausting process of colonizing the New World carried out by the Iberians," and "the facility with which the Mediterranean dweller traveled from port to port." Mediterranean migration, he added, was not true transplantation, but merely removal, and the new occupant would feel quite at home in his new habitat."[3] In the French Atlantic alone, one historian writes, "the range in climate and topography must have been daunting to any eighteenth-century administrator, ranging from ice floes to equatorial rain forests, from rocky islets to humid swamps and tundra, across two major oceanic wind and current systems and another half-dozen major regional subsystems."[4] The Atlantic world does have limits, however, not only geographically but temporally. The general (albeit not universal) agreement is that Atlanticists tackle the period between 1492 and circa 1830—these dates marking the start of Europe's exploration and conquest of the Americas, and the completion of the process that dissolved the chief Atlantic empires.

Prior to the recent popularity of Atlantic history, there had always been historians trying to come to terms with the Atlantic world. Yet this was historiography of another kind, not necessarily inferior, but with different foci. In times past, historians emphasized local, regional, and imperial institutions. Their Atlantic world was one with clear national divisions, with each colony closely tied to its mother country, even if autonomous developments in the colonies received increasing attention as the decolonization of Africa and Asia gained pace after World War II. They displayed scant interest in integration, networks, social history, trans-imperial comparisons, and actors across boundaries. Their Atlantic world was shaped by Europeans, with Native Americans and Africans at best reacting to European initiatives, but not actively creating their own destinies. The pioneers of Atlantic history, those who in the decades after World War II began to envisage a unified Atlantic world as a perspective, although not yet as a research field,[5] saw it invariably as white and projected its location north of the equator. Their New World was usually synonymous with British North America. Nor did students of the South Atlantic offer an alternative. Specialists of colonial Latin American history were equally parochial—a comparative perspective was almost completely missing from the *Cambridge History of Latin America*, published in 1984.[6]

Much has changed in the last two decades. As it has spawned journals, book series, conferences, and PhD programs, Atlantic history has become ubiquitous in the process. Why do we study Atlantic history? One reason why Atlantic history has excited so much interest is that it breaks with precedent. It moves away from the study of nation-states and continents, replacing it with a broader horizon and a more integrated view of history. The makeover is not nearly complete. Much of what passes for Atlantic history is hardly different from colonial American history—a relabeling of older, currently less fashionable historiographies. The explanation, it would seem, is that the field was launched in the United States, at least in part, in an attempt to align the country's history with the reality of a post-civil-rights-era nation. Privileging Pilgrims and Puritans in the opening stages of colonial history had to make way for a narrative that incorporates the history of blacks and Hispanics from the start. An Atlantic approach to North America must therefore introduce elements of African and Latin American history. So far, so good, but the distinctly North American context also prescribes the boundaries of practicing Atlantic history. Virtually all who teach it are actually professors of American History who teach an occasional class on the Atlantic world. And since most Americanists are not linguistically adventurous, their Atlantic world is bound to remain confined. Like those of the typical historians of the Mediterranean, their publications only deal with relatively local history, without wider significance for the Atlantic world.[7]

What crops have Atlantic studies yielded? In an article published a dozen years ago, Jerry Bentley provided an overview of the types of processes that have come into focus by adopting oceanic perspectives, and the Atlantic world figured prominently: "the Columbian exchange, the Atlantic slave trade, the organization of plantation societies in the Americas and Oceania, the formation of diaspora communities, migrations of European peoples to the Americas and other temperate regions, indentured-labor migrations of mostly Asian peoples to the Americas and to tropical regions around the world, and the construction of global networks of trade, communication, and exchange."[8]

Migration studies have, indeed, blossomed. Alison Games used a London port register of 1635, containing information about almost 4,878 people moving across the Atlantic, and found information for an additional 1,360 to shed light on early migration to British America, concluding that migration was central to colonial viability. She followed in the footsteps of Bernard Bailyn, who compiled and analyzed a large variety of data on the migrants to British North America in the closing years of the colonial period.[9] Combining insights about sending and receiving regions has led to sometimes surprising conclusions. It turns out, for example, that both English and French migrants to North America forsook mobile and rapidly developing countries for fairly traditional lives across the ocean. Still, North America was not a static world. The processes of settlement and development, Bailyn argued, were varied and constantly shifting.[10] Immigration was also a business, and a new type of economic endeavor at that, as Marianne Wokeck has revealed. Entrepreneurs "trading" Germans increased their profit margins at the expense of the voyagers, who had to put up with inferior accommodations and poor food.[11]

Like the historians detailing European migration, those specializing in the Atlantic slave trade are connecting knowledge about both sides of the ocean. Their subject matter, even more dynamic in nature, involving 12.5 million Africans, has been charted in painstaking detail in the Transatlantic Slave Trade Database under the supervision of David Eltis and others.[12] Using these data to make estimates for the overall slave trade, Eltis and Richardson have broken down the Middle Passage into multiple slave trades, organized by a great many European states which each had their preferences for African origins and American outlets. They have also reconstructed regional patterns of emigration in Africa and immigration in the Americas, explaining demographic trends and making sense of ebbs and flows. Their work has brought to light that in one area of Africa (Congo-Angola), the effect of the growing demand for slaves, and their rising prices was to extend catchment areas further into the interior, whereas in other parts of Atlantic Africa the result was for enslavement to rise dramatically in coastal regions, after which other littorals were drawn in.[13]

Apart from their research on the transatlantic leg of the Africans' odysseys, historians have detailed the voyages of slaves from the interior of Africa to the coast and from their disembarkation in America to their final destination.[14] They have assessed the impact for Africa of the continent's participation in the Atlantic world, not only in demographic but also economic and cultural terms.[15] Some have pointed out that involvement in the transatlantic slave trade has led to "the ongoing creation of new collective identities in Africa."[16] A student of Yoruba culture has raised the following questions about cowry shells, the currency widely used in the Atlantic slave trade: "Why and how were cowries stripped of their external meanings and reconstituted within the frameworks of the Yoruba cultural traditions? What types and forms of knowledge, beliefs, and ideas did cowries engender? How were cowries used to construct, shape, and coordinate new forms of social, political, and economic relationships following the ruptures associated with the Atlantic economy?"[17]

The African slave trade was, of course, a form of migration, but to understand the way it was organized as a business, scholars have to integrate information not about two, but three continents. In other words, only an integrated method will provide the answers. In his classic book, *Way of Death,* Joseph Miller dissected the Portuguese slave trade in a subtle way, distinguishing between the major players, the politically and economically weak, and the slaves themselves. The powerful interests included metropolitan Portuguese merchants, African merchant princes, authorities in Lisbon, and their financial backers from Britain.

From the perspective of these princes and the Portuguese merchants, who exported their woolens and gunpowder to obtain Brazilian plantation crops or precious metals, slaves entered the picture only to settle accounts. African middlemen on the coast and Brazilian planters were weaker positioned. Perennially in debt to the European financiers who supplied the credit that was the slave trade's lifeblood, their aim was to survive and avoid bankruptcy. The ready availability of this commercial credit wreaked havoc on Angola, which was flooded with trade goods, for whose payment the slaving frontier moved back, which led to the shipment of less healthy Africans across the sea. Given the resulting high mortality of Brazilian slaves, constant new slave imports were needed.[18]

Like Miller's book, the work of Stephen Behrendt exemplifies the potential of Atlantic history, combining information on three continents to elucidate developments in each one.[19] In fact, such an approach is the only proper one for students of the slave trade, if they want to understand the way slave merchants used transaction cycles to reduce risk and costs and maximize profit. The African slave trade required a monumental coordinating effort that affected for example, the Baltic, where British ships arrived in the early spring to buy timber needed for carpenters to transform regular ships into slave carriers. Planning provisions was a key ingredient of the slave trade. Ships left their European home ports with vast amounts of provisions—especially those vessels that left for drought-plagued Angola. Ship captains generally "organized their slave trades around the production schedules of local crops."[20] Buying African produce to sustain the enslaved cargoes was so important that it outweighed the loss of life among European crews, who tended to die at high rates during the rainy season. The time of departure from Africa depended in part on Caribbean conditions, as captains timed their arrival in the West Indies to coincide with the availability of provisions that could be used to buy slaves in Africa. The slave trade was thus shaped by transatlantic supply and demand.

Agency

The cultural dimension is also increasingly studied transatlantically. There has been an outpouring of studies on the retention and transformation of African beliefs and cultural practices. This body of work explicitly reacts against the old theory that Africans arrived in the Americas as amorphous crowds whose diverse backgrounds hampered cultural continuity and encouraged cultural amalgamation. For example, Paul Lovejoy examined the proposition that Islam was one African tradition that resisted such creolization. His investigative wish list contains questions such as: "To what extent did deportation and exile heighten commitment to Islam? In what ways were enslaved Muslims able to avoid detection under slavery? How were Islamic theories of behaviour for Muslims in a land of Unbelief interpreted and applied in the Americas? . . . How did their experiences compare with those of enslaved Muslims in non-Muslim countries in West Africa and the Islamic lands of the Sahara and North Africa?"[21]

Another way to establish the transmission of African culture is by positing the existence of certain African cultural zones, as John Thornton has done. He divides sub-Saharan Africa into three culturally distinct zones as well as seven distinct subcultures—the main three zones being Upper Guinea, Lower Guinea, and the Angolan coast. These cultural "groupings" were not scattered in the slave trade, he claims. If that occurred, it was when they were sold and put to work in the Americas, but even those processes did not prevent them from finding others who knew their language and shared their values.[22] In line with this approach, some publications aim to show that distinct African ethnicities survived and consolidated in the Americas. One example is the Igbo peoples of the Nigerian hinterland, who, according to one historian, were "first-comers" in some American colonies. "In such cases, they set the basic patterns of material, social and ideological culture of enslaved communities to which succeeding waves of . . . slaves acculturated. In other colonies . . . Igbo were 'second-comers' who 'Igboized' existing institutions and cultural patterns as people drew on ancestral material, social and ideological resources in order to adapt both to slavery and to the culture of slaves already there."[23] Not everyone is convinced. Some fail to see the presumed existence of today's ethnolinguistic "tribal" identities. Others claim that whatever African identities did exist in the Americas "were often convenient reconstitutions or inventions."[24] African ethnicities probably did not survive intact, but their affinities—for instance, in the spiritual realm—were close enough for a process of "Africanization" to occur, as an essentially African core of religious beliefs was melded.[25]

The quantitative data that have been amassed on the Atlantic slave trade have left some scholars to complain that individual voices were muffled and that Africans are merely represented as victims, not active agents. As an antidote, biographies have been published on a number of colorful individuals exported to the Americas and their descendants.

Spelling out their protagonists' contributions to the Atlantic world, these studies have agency written all over them. It cannot be a coincidence that the blacks thus spotlighted were all instrumental in obtaining their own freedom.[26] In assigning agency to Africans and Native Americans, Atlanticists have broken with past practice. It would be unthinkable for a historian to announce, as K. G. Davies did in the preface to *The North Atlantic World in the Seventeenth Century* (1974), that "this is a book about Europeans," adding, "Africans and Indians enter the story as the human material on which the Europeans worked, usually clumsily, often disgracefully, but seldom without effect."[27] We have come a long way since then. Nonwhites have secured positions in historical arenas where they used to be totally absent. Their participation in wars and revolutions may not have been entirely overlooked by older generations of historians, but has been studied in much more depth. In particular, the Haitian Revolution has established itself as one of the great revolts during the Age of Revolutions. The contributions of blacks and Native Americans to the American Revolution are also getting the recognition they deserve.[28]

The Enlightenment provides another example. African slaves in the Americas, it is said, "were often the origin point in the Enlightenment enterprise of the universal collection and systematization of nature even if that enterprise was otherwise managed by well-connected white naturalists." "Colonials used Africans as discoverers of curative specimens and procedures, as collectors of such specimens, as magical manipulators of their fellow slaves' environments, as river navigators and forest pathfinders, and, finally, as importers and adapters of agricultural methods."[29] But Africans were not merely instrumental. Laurent Dubois has encouraged historians to develop a truly Atlantic approach to the history of ideas during the Age of Revolutions, thus avoiding the mistaken assumption that "Europeans and European colonists were the exclusive agents of democratic theorizing." After all, he argues, the process of abolition of slavery in Saint-Domingue took place as slaves had won both territorial and ideological control over the French colony during the revolution.[30] The study of the black reformulation of enlightened and revolutionary ideas seems, indeed, to hold great promise.

Adaptation

In the 1970s and 1980s, scholars widely believed that slaves in the Americas had been creolized—assimilated in ways that left no room for initiatives of their own. Today, these slaves are seen first and foremost as self-conscious men and women armed with cultural strategies. Terminology has changed accordingly, with many specialists now preferring *enslaved Africans* over *African slaves.* Yet no consensus exists about the modes of adaptation to life in the New World. The growing emphasis on African retention has not silenced those who adhere to creolization theory. On the contrary, they have come back roaring, albeit with a twist or two. In an influential article, Ira Berlin stressed the key role played by "Atlantic creoles"—slaves with wide experience in the Atlantic world who were born in America or along the Atlantic littorals of Africa or Europe. Usually of mixed parentage and often linguistically versatile, these Atlantic creoles served as intermediaries between Africans and Europeans, starting in the sixteenth century.[31] Linda Heywood and John Thornton have built upon this concept, asserting that Atlantic creoles hailing from Congo and Angola were often insiders rather than outsiders in the early Dutch and English colonies in the Americas. Creolization on their native soil had started at least as early as 1491, when Congo's king converted to Catholicism. Their ability to show that they were Christians frequently helped these creoles obtain their freedom and become property owners.[32]

In any colony in the Atlantic world, neither the African, nor the Native American, nor the European background triumphed completely—unless, of course, the native population had completely died out.[33] It was rather a mixture of elements that shaped the new worlds. Although in most places, the predominant character of the emergent culture was shaped by white European settlers, the distinctive regional nuances were formed by the interaction of all ethnic groups. In many parts of the New World, this process of cultural synthesis began when settlers came directly from Europe. Totally unequipped, they relied on Native provisions for their survival.[34] If natives rapidly thinned out in some parts of the Caribbean, African slaves, maroons, European pirates, privateers, soldiers, and other adventurers and fortune-seekers helped create new communities afloat and ashore that contributed to a process of mutual accommodation. Where natives survived, they came to inhabit borderlands, middle grounds.[35] In these encounters, the element of surprise or even bafflement was paramount, leading each group to respond with innovation.[36] In his discussion of this process as it unfolded in the English colonies, T. H. Breen found eight elements that shaped cultural creativity: "1) the backgrounds of migrants and native Americans before colonization; 2) the perceptions that members of the three racial groups formed not only of themselves but also of representatives of other races; 3) the motives of different white settlers for moving to the New World; 4) the timing of initial contact; 5) the physical environment in which contact occurred; 6) the shifting demographic configurations; 7) the character of local economies; 8) the force of individual personality."[37]

The ongoing creation and reshaping of communities in colonial New England among natives and Europeans has prompted Faren Siminoff to note that we should not conceive of a clash between large collectives—Europeans versus natives—"but also as an interaction of smaller, emerging communities of interest that split off from these larger groups. . . . With this in mind, there are no clear-cut winners or losers in this Atlantic cauldron but simply a vast matrix in which all groups were immersed and from which they emerged altered."[38]

One important point of contact was religion. Wherever the Europeans entered, they brought with them their Christian beliefs, which they usually tried to spread among the nonwhite populations. Such efforts were not necessarily crowned with success. In the Spanish province of Texas on the northern frontier of New Spain, contact between Spaniards and Hasinai natives was facilitated by the joint use of gender to find a common ground, "to make the other intelligible and knowable, to find points of access to one another's alien culture." The outcome of this process was not the increase of knowledge about the other side. The Spanish had no idea, for example, that the iconography featuring the Virgin Mary was interpreted by the Hasinais in a distinctive way. These Indians were convinced that the Spanish sought "to compensate for their lack of real women and thereby offset any appearance of hostility."[39] Indians throughout the Americas did not give up their own mental frameworks, but adjusted them to their changing needs as contacts with the Europeans intensified.[40] But converting to Christianity they usually did not; at the most, they converted to colonialism.[41] What was forged among North America's Eastern Algonkians during the French missionary process was a confluence of native and Jesuit cosmologies. The Algonkians perceived baptism as "a means of establishing, securing, and processing a relationship with French allies." Baptism was also understood as a kind of medicine, a rite that invoked the healing intervention of a figure the Jesuits called 'He-Who-Made-All.'"[42] Andean natives also combined elements of their own religion with Christian concepts, but the original Spanish and Andean ingredients intermingled in such a way that they were no longer recognizable. Hybridity was the norm. Religious elements were mixed, not substituted or replaced, writes Kenneth Mills.[43]

Nor did preexisting trade networks vanish overnight. In western Amazonia, native groups struggling to survive the brutal Spanish expeditions and the introduction of a demanding labor system (the *encomienda*) restructured their commercial links. When Jesuit missionaries came in and used long-distance riverine and interfluvial networks for their own purposes, both interethnic relations and the nature of overall commerce changed.[44] The post-contact history of western Amazonia reminds us of the all-important role of violence in interactions between natives and strangers, whites and nonwhites. The emphasis on hybridity should not obscure the fact that to the Indians, the middle grounds and borderlands that appear as cradles of a new shared civilization were in reality homelands. And if it was a middle ground, it soon developed into a divided ground, not only between Indians and whites, but also between various groups of Europeans. Europeans were initially drawn into native conflicts, but ended up using these divisions for their own purposes, which meant that Indian struggles became more a product of native affiliation with Europeans. Alan Taylor has shown how Iroquoia, the homeland of the Iroquois Indians, developed into such a divided ground, as Great Britain and the emerging United States won over different groups of Iroquois in the American Revolution.[45] In general, American borderlands offered Europeans states opportunities to find native commercial partners and military allies as much as they allowed Indians to exploit European imperial rivalry for their own ends.[46] On the African Gold Coast, Europeans also interfered in local conflicts, as they attempted to play petty native states against each other. But at the same time, the abundance of forts belonging to various European powers also gave African rulers leverage vis-à-vis their European partners.[47]

Englishmen legitimized the use of violence against Indians by reference to the need to preserve the social order. The relationship with Indians was promoted as a marriage, but with a twist: the ultimate goal was for the natives to fear the Europeans.[48] The hegemonic underpinnings of European colonial projects have also been underscored for the French. Casting aside conventional scholarly wisdom about the absence of French prejudice against natives, Guillaume Aubert has shown that sixteenth- and seventeenth-century French discourse was rife with notions about innate differences between classes and races. The belief that both physical traits and moral virtues could be transmitted from one generation to the next left its imprint on New France and the French Caribbean.[49]

Unfulfilled Promises

It is thus in the realm of culture that Atlantic history has seen its most fruitful application, especially in the analysis of nonwhite agency and the hybridity of new communities, conventions, and cosmologies. In pursuing these themes, Atlanticists have answered the call from postcolonial critics of the 1980s for a revision of the place of colonies and colonial subjects in the imperial world. Politics, economics, and political economy are less well endowed than cultural studies. The same goes for certain methodologies, in part because they require the mastery of different historiographies, in part because of the lack of linguistic facility—and if by its very nature the field transcends borders, historians must of course pass linguistic frontiers. Underdeveloped remain, therefore,

a. comparative perspectives both within and across imperial boundaries,
b. the study of the entanglement of neighboring empires, and
c. the examination of networks.

Comparison

Some Atlantic historians have already shown the way in comparative studies. In a recent monograph, Filipa da Silva raises the question why the Dutch and Portuguese Atlantic empires took different shapes and followed different paths in the seventeenth century. Her case study of West Africa shatters the myth of Dutch commercial superiority and Portuguese economic inefficiency. She shows that the successes of the Dutch derived primarily from naval and military power, while Portuguese colonial survival depended on settler strength.[50] David Eltis also stresses that the Dutch were not the cutting-edge entrepreneurs they are sometimes made out to be. Comparing the most advanced English colonies of Barbados and Jamaica to their Dutch counterparts and Portuguese Brazil, Eltis argues that the English Atlantic was the most modern, based it was on the production of cash crops and the introduction of a type of slavery founded specifically on gang labor with tight control of the slaves.[51]

Comparison can obviously lay bare similarities as well as contrasts, as exemplified in two separate studies comparing Baltimore, Maryland, with towns in South America. Camilla Townsend concludes that in many respects the "economic cultures" of Baltimore and Guayaquil, Ecuador, were very different in the early nineteenth century. What set them apart were the elite preference for coerced versus free labor, the salaries employers were willing to pay, the access poor people had to education, and the expectation that few or many townsmen would become part of the informal or criminal circuits. Mariana Dantas, on the other hand, emphasizes analogy. The urban environments of Baltimore and Sabará in Minas Gerais, Brazil, ensured more opportunities for blacks to make the transition to freedom and to climb the social ladder. The increase of the black population and their expanding involvement in the urban economy contributed significantly to the growth of these towns.[52]

The most comprehensive comparative study to date is Sir John Elliott's widely acclaimed *Empires of the Atlantic World.* Each of the twelve chapters provides an in-depth and interwoven comparison of arrival, settlement, the confrontation with natives, the exploitation of natural resources, the relationship between crown and colonists, the ordering of society, America as sacred space, empire and identity, societies on the move, war and reform, empires in crisis, and decolonization. The (rather underwhelming) conclusion is that despite the many differences—some temporary, others structural—both empires were remarkably similar. Comparing national Atlantics can also have the opposite effect of warning us against homogenizing. Patricia Seed has shown that for all their common technological and ecological backgrounds, European nationals went about establishing political legitimacy in the Americas differently. The rituals, ceremonies, and other acts of possession reflected distinct cultures—Spanish, Portuguese, English, French, and Dutch—fully intelligible only to those who had grown up in them.[53] As such, Seed's study does for the Europeans what students of the African diaspora have done for Africans: connect American cultures to their Old World backgrounds.

Since the Atlantic world was not only predicated on the stream of people, goods, and ideas, but also the flow of germs, the comparison of New World disease environments is another relevant theme in early American history. In *Mosquito Empires,* J. R. McNeill recently moved beyond the traditional emphasis on the devastating impact of common Eurasian diseases on the indigenous Americans. McNeill is interested in the ecological changes wrought by European settlers. These changes created landscapes that were particularly suitable to mosquitoes which transmitted yellow fever and malaria. Time and again, these maladies prevented foreign armies from capturing Spanish American provinces.

During the Age of Revolutions, the same diseases thwarted the attempts of European forces to subdue colonial revolts across the sea.[54]

Connectivity

As argued earlier, much of Atlantic history is still old imperial wine in new bottles, obscuring the structural entanglement between the various national Atlantics.[55] Colonists did not live behind impermeable walls nor did imperial authorities only occupy themselves with their own realms. Discrete Atlantics never existed, since mutual influence, border disputes, and wars were the norm. Several studies have highlighted the interpenetration of colonial empires in border areas. We know much more today about the close collaboration between Englishmen and Dutchmen in their respective colonial projects and about the countless disputes between Spanish and English diplomats, officials, missionaries, soldiers, slaves, and settlers that were part and parcel of the settlement process.[56]

To do justice to the past, it is not enough to document the close connections between neighboring colonies. Ports, towns, and colonies were tied to each other across vast distances. A.J.R. Russell-Wood has observed that the human, political, commercial, and cultural ties between Salvador (Brazil's capital until 1763), and Dahomey and the Bight of Benin were so close that West Africa might be regarded as part of the hinterland of Salvador.[57] Horden and Purcell make the same point for the Mediterranean when they argue that distance is inverted in that places linked by sea are close, while neighbors on land may be distant in terms of interaction.[58]

Reflecting on developments in research on the African diaspora and Atlantic history, Kristin Mann has observed that the "reciprocal influences of the Americas on Africa during the era of slavery are well known, if not yet fully understood. Less commonly appreciated is the fact that the diaspora was not bilateral but multilateral. As a consequence, we need a model of it that is not two-dimensional but three dimensional. Influences moved not only back and forth between specific regions of Africa and the Americas but also between different parts of Africa and of the Americas. Indeed, they circulated in flows of differing reach and proportion all around the Atlantic basin."[59]

Also circulating around the Atlantic basin were a great variety of revolutionary ideas in the late eighteenth and early nineteenth centuries. These ideas had no class or imperial boundaries, were introduced, transmitted, transformed, and implemented in very diverse theaters, following as well as shaping the rhythm of imperial decline. All of this is beginning to be charted, for ideas and values, but also for trade flows.[60] As historians discover that simple imperial exchange of goods was only part of the picture, they abandon the mercantilist model of the Atlantic world. The Madeira wine trade, David Hancock maintains, did not correspond to a hub-and-spokes model, looking more like a spider-web trade with lines of correspondence going in every direction.[61] Portuguese New Christian merchants, for their part, set up trading networks that were organized multilaterally, with a high degree of interconnection, and crossing many imperial boundaries. Their activities spanned the entire Atlantic, connecting Lisbon, Madrid, Seville, Amsterdam, Lima, and Potosí. Each merchant house "that participated in the network was a quasi-autonomous unit: it controlled its own pool of capital and it was responsible for the success or failure of its commercial ventures."[62]

Numerous communication networks sprang up throughout the region and across imperial frontiers. News about trade and privateering was passed on by fishing boats, canoes, or cargo ships on their way from one American port to another. Besides this, information was obtained on desert islands, which served as meeting places for traders from different empires. Other communication networks linked the Mennonite, Quaker, and Pietist communities that were involved in recruiting immigrants to North America.[63] Analyzing communication is also a rewarding way to reconstruct how each of the various national Atlantic realms functioned. One study has shown that efforts by the metropolitan state to establish a monopoly of information were bound to fail in the French Atlantic. The constant reliance on mercantile networks to obtain news about events across the sea led to the demise of the hierarchical reporting system typical of absolutism. Correspondence between the state and the colonies also grew dramatically in the English Atlantic, as not only governors of colonies and colonial agents in London, but increasingly other interest groups—new crown officials, local political leaders, merchants, and London agents—communicated with each other. In the process, the English Atlantic became closely politically integrated at the same time that its political culture became more complex.[64]

An Atlantic Microcosm: The Caribbean

The dearth of studies on Atlantic connectivity can be explained in part by the difficulty of locating connective tissue across oceanic space. Changing the perspective from ocean to sea allows us to lay bare inter-imperial connectedness with much greater clarity. The sea of choice is the Caribbean. Nowhere in the world did such a concentrated set of colonies exist with multiple imperial allegiances, enabling widespread hybridity.[65] Examples abound: residents of British Jamaica founded Saint-Domingue's first Masonic lodge; the architecture of Bridgetown, Barbados, acquired a distinctly Dutch flavor; and the black Catholic population on the Dutch island of Curaçao often listened to a Spanish priest in church on Sundays.[66]

Trade—in particular, illicit trade—formed the foundation of inter-imperial exchanges in the Greater Caribbean. The ubiquity of smuggling surprised Europeans such as Gregorio Robles, a young man from the diocese of Toledo, Spain, who set off on a voyage to the New World in 1687. Prompted by curiosity and in search of adventure, he embarked on a fifteen-year trip which would lead him to many parts of Spanish America. Robles did not take any notes during his travels, but his fabulous memory compensated for his inability to read and write. Most striking in the account that he gave upon his return to a minister of the Council of the Indies is how routinely the prevailing laws were ignored in the Spanish Caribbean. Robles was particularly mindful of contraband trade—a many-headed monster that reared its head almost everywhere, especially in the colonies bordering the Caribbean Sea. In Santo Domingo, Robles saw merchants from Dutch and English colonies come ashore and move about as if they were at home, while he witnessed several foreigners keeping shops in New Granada and publically weighing the hides and dyewood they received in payment for their European merchandise.[67]

What Robles witnessed was in fact free international trade taking place in spite of metropolitan rules and regulations.[68] This contraband trade in the Spanish colonies grew in scope and frequency after the rise of the regional *entrepôts* of Curaçao and Jamaica in the third quarter of the seventeenth century. Smugglers based in northern European colonies did not only target the Spanish possessions. Starting in the early eighteenth century, the Danes of St. Thomas, the Dutch of St. Eustatius, and the English of Jamaica also found customers for their slaves in Saint-Domingue, as they did in the other French colonies of St. Christopher, Guadeloupe, and Grenada. By the mid-eighteenth century, Caribbean smuggling became multidirectional, with settlers from all Caribbean colonies taking initiatives to engage in trade across imperial boundaries. In many parts of the Caribbean, smuggling was by then no longer merely a means to make a living, but had become a lifestyle founded on inter-imperial connections.

The intensity of legal and illegal intra-Caribbean trade guaranteed a constant diffusion of news and rumors across the islands and the mainland bordering the Caribbean Sea. All social groups availed themselves of incoming information to make sense of the world around them. Slaves were no exception. One rumor in particular circulated in slave circles in the Caribbean: many slaves believed that emancipation had been decreed by a distant king, but that local authorities and/or slaveholders withheld the freedom. Varieties of this rumor surfaced from at least 1669 (Bermuda) through 1848 (St. Croix).[69] In some versions, slaves identified an African king as the great liberator. In 1768, slaves in Martinique spread the rumor that a powerful king had arrived from Africa, purchased all the slaves from the colonial government, and would soon organize their return to Africa.[70] In Cuba, half a century later, slaves of the Congo nation believed that the king of Congo had sent letters to the island ordering the slaves' freedom.[71] Usually, however, the emancipation decree was attributed to a European monarch.

The misplaced belief in an emancipation decree was the catalyst for numerous slave uprisings in the Americas. Remarkably, virtually all these revolts occurred in the Greater Caribbean—and not in Brazil, the interior of Spanish America, or on the Eastern Seaboard of North America—suggesting that the area had its own rumor mills whose reach was regional. Often, slaves took up arms in the belief that they had the backing of the imperial ruler. Enslaved Africans in early nineteenth-century British Caribbean, for example, mistakenly believed that their king was an abolitionist. The royal emancipation rumor motivated slaves in revolts in Barbados, Jamaica, and Demerara (Guiana) in the two last decades of British slavery.[72] These particular rumors seems to have been based on the presentation of a recent motion in the House of Commons, which provided for the emancipation of all slave children after a certain date, the restriction of slave punishments, and the eventual abolition of slavery altogether.[73] In the Spanish colonies, a slave code issued by King Charles IV in 1789, which somewhat ameliorated working and living conditions for slaves, had similar consequences.[74] The activities in France of the Society of the Friends of the Blacks, an abolitionist society, combined with news about the French Revolution, also inspired slave rebellions motivated—at least in part—by the belief that an emancipatory document had been introduced, but was hidden. It even contributed to the outbreak of the massive revolt in St. Domingue in 1791.[75]

As several versions of this rumor spread across the Caribbean in the 1790s, they resuscitated older variants which had hibernated in the wake of past revolts or conspiracies. At the same time, news about the Spanish slave code was probably interpreted in French colonies as confirming the belief of a freedom decree signed by the French king, while the French Revolution may have confirmed lingering convictions among slaves in the Spanish Caribbean that the king of Spain had ordered their emancipation. In later decades, actual emancipation in neighboring empires also affected slave rebelliousness. The best example is the decree of general slave emancipation issued by the provisional government of the French Second Republic in April 1848. The measure fomented the rumor in Danish St. Croix that emancipation had officially been declared, but was withheld. On July 3, eight thousand slaves refused to start their working day and gathered in front of a government fort in the town of Frederiksted demanding their freedom. Ignoring the advice of other officials, the Danish governor told the assembled slaves "Now you are free, you are hereby emancipated."[76]

As Julius Scott has argued, there was a "Caribbean reality of a regional community where geographic proximity was often more important than national boundaries."[77] Crisscrossed more intensively and thus more intimately than the vast Atlantic basin, the Caribbean can be fruitfully studied as an Atlantic microcosm. What emerges from a scrutiny of Caribbean life is the daily exchange of crops and commodities, as well as cultural elements across imperial borders, which fomented economic entanglement and a development along culturally hybrid lines.

Studies of Caribbean patterns should not take the place of Atlantic history as a field. There is still much work to be done in other corners of the Atlantic world, especially on the eastern side. Thus far, their choice of topics manifests that Atlantic historians are primarily interested in the construction and evolution of American worlds. The full impact of Europe's engagement with the Atlantic world still has to be assessed, and this may well depend on the development of Atlantic history as a field in that continent. It also seems odd that the Barbary corsairs are conspicuously absent from Atlantic histories, in spite of the effect they had on transatlantic shipping, the fear they struck into the hearts of merchants and migrants, and the firsthand knowledge of Islamic lands that was the legacy of the Europeans' sojourns in Algiers, Tripoli, Tunis, and Saleh.[78] The European experiences in Barbary also complicate our understanding of what slavery meant to Europeans at home and abroad.[79] Atlanticists should not, of course, try to monopolize the debate on the Barbary corsairs, but perhaps we can share them with the Mediterraneanists. After all, as every thalassographer knows, one may distinguish between oceanic histories and their protagonists, but it is not possible to isolate them.

NOTES

1. John Elliott, "Afterword. Atlantic History: A Circumnavigation," in David Armitage and Michael J. Braddick, eds., *The British Atlantic World, 1500–1800* (New York: Palgrave, 2002), 233–49, 239.
2. Peregrine Horden and Nicholas Purcell, *The Corrupting Sea: A Study of Mediterranean History* (Oxford, England and Malden, MA.: Blackwell Publishers, 2000).
3. Fernand Braudel, *The Mediterranean and the Mediterranean World in the Age of Philip II* (New York: Harper Colophon Books, 1972), 237.
4. Kenneth J. Banks, *Chasing Empire across the Sea: Communications and the State in the French Atlantic, 1713–1763* (Montreal and Kingston: McGill-Queen's University Press, 2002), 8.
5. See Philip D. Morgan and Jack P. Greene, "Introduction: The Present State of Atlantic History," in Jack P. Greene and Philip D. Morgan, eds., *Atlantic History: A Critical Appraisal* (Oxford: Oxford University Press, 2009), 3–33, 4.
6. Kenneth Maxwell, "The Atlantic in the eighteenth century: a southern perspective on the need to return to the 'big picture,'" *Transactions of the Royal Historical Society,* 6th series, 3 (1993): 209–36, 211.
7. Peregrine Horden and Nicholas Purcell, "The Mediterranean and 'the New Thalassology,'" *American Historical Review* 111, no. 3 (June 2006): 722–40, 730.
8. Jerry H. Bentley, "Sea and Ocean Basins as Frameworks of Historical Analysis," *Geographical Review* 89, no. 2 (April 1999), 215–24, 221.
9. Alison Games, *Migration and the Origins of the English Atlantic World* (Cambridge, MA.: Harvard University Press, 1999). Bernard Bailyn, *Voyagers to the West: A Passage to the Peopling of America on the Eve of the Revolution* (New York: Knopf, 1987).
10. Meaghan N. Duff, "Adventurers across the Atlantic: English Migration to the New World, 1580–1780," in Wim Klooster and Alfred Padula, eds., *The Atlantic World: Essays on Slavery, Migration, and Imagination* (Upper Saddle River, NJ: Pearson Prentice Hall, 2005), 77–90, 77. Leslie Choquette, *Frenchmen into Peasants: Modernity and Tradition in the Peopling of French Canada* (Cambridge, MA.: Harvard University Press, 1997). Bernard Bailyn, *The Peopling of British North America: An Introduction* (New York: Vintage Books, 1988), 49–60.
11. Marianne S. Wokeck, *Trade in Strangers: The Beginnings of Mass Migration to North America* (University Park.: Pennsylvania State University Press, 1999).
12. The expanded and online database (http://www.slavevoyages.org), launched in 2007, was coordinated by David Eltis, David Richardson, Stephen D. Behrendt, and Manolo Florentino.
13. David Eltis, "The Volume and Structure of the Transatlantic Slave Trade: A Reassessment," *William and Mary Quarterly* 58, no. 1 (January 2001), 17–46,

34–35. David Eltis and David Richardson, eds., *Extending the Frontiers: Essays on the New Transatlantic Slave Trade Database* (New Haven, CT: Yale University Press, 2008).

14. Joseph C. Miller, *Way of Death: Merchant Capitalism and the Angolan Slave Trade, 1730–1830* (Madison: University of Wisconsin Press, 1988). Linda A. Newson and Susie Minchin, *From Capture to Sale: The Portuguese Slave Trade to Spanish South America in the Early Seventeenth Century* (Leiden: Brill, 2007).
15. Robin Law, *The Slave Coast of West Africa, 1550–1750: The Impact of the Atlantic Slave Trade on an African Region* (Oxford: Clarendon Press, 1991), Martin Klein, "The impact of the Atlantic slave trade on the societies of the Western Sudan," in Joseph E. Inikori and Stanley L. Engerman, eds., *The Atlantic Slave Trade: Effects on Economies, Societies, and Peoples in Africa, the Americas, and Europe* (Durham, NC: Duke University Press, 1992), 25–47. James F. Searing, *West African Slavery and Atlantic Commerce: The Senegal River Valley, 1700–1860* (Cambridge: Cambridge University Press, 1993).
16. Joseph C. Miller, "Central Africa during the Era of the Slave Trade, c. 1490s–1850s," in Linda M. Heywood, ed., *Central Africans and Cultural Transformations in the American Diaspora* (Cambridge: Cambridge University Press, 2002), 21–69, 22.
17. Akinwumi Ogundiran, "Of Small Things Remembered: Beads, Cowries, and Cultural Translations of the Atlantic Experience in Yorubaland," *International Journal of African Historical Studies* 35 (2002): 427–57, 429.
18. Miller, *Way of Death,* 662–66.
19. Stephen D. Behrendt, "Markets, Transaction Cycles, and Profits: Merchant Decision Making in the British Slave Trade," *William and Mary Quarterly* 58, no. 1 (January 2001), 171–204, 179, 184. Stephen D. Behrendt, "Ecology, Seasonality, and the Transatlantic Slave Trade," in Bernard Bailyn and Patricia L. Denault, eds., *Soundings in Atlantic History: Latent Structures and Intellectual Currents, 1500–1830* (Cambridge, MA: Harvard University Press, 2009), 44–85, 59–60, 83.
20. Behrendt, "Ecology, Seasonality, and the Transatlantic Slave Trade," 68.
21. Paul E. Lovejoy, "Slavery, the Bilād al-Sūdān and the Frontiers of the African Diaspora," in idem, ed., *Slavery on the Frontiers of Islam* (Princeton, NJ: Markus Wiener Publishers, 2004), 1–29, 9, 18. See also Paul E. Lovejoy, *Identity in the Shadow of Slavery* (London: Continuum, 2000).
22. John Thornton, *Africa and Africans in the Making of the Atlantic World, 1400–1680* (Cambridge: Cambridge University Press, 1992), 186–205.
23. Douglas B. Chambers, "'My own nation': Igbo Exiles in the Diaspora," *Slavery and Abolition* 18, no. 1 (1997): 72–97, 84. See also Michael Gomez, *Exchanging Our Country Marks: The Transformation of African Identities in the Colonial and Antebellum South* (Chapel Hill: University of North Carolina Press, 1998); Gwendolyn Midlo Hall, *Africans in the Americas: Continuities of Ethnicities and Regions* (Chapel Hill: University of North Carolina Press, 2001); Toyin Falola and Matt D. Childs, eds., *The Yoruba Diaspora in the Atlantic World* (Bloomington: Indiana University Press, 2004); Matt Schafer, "The Mandinka Legacy in the New World," *History in Africa* 32 (2005): 321–69.
24. David Northrup, "Igbo and Myth Igbo: Culture and Ethnicity in the Atlantic

World, 1600–1850," *Slavery and Abolition* 21, no. 3 (2000): 1–20, 3. Philip D. Morgan, "The cultural implications of the Atlantic slave trade: African regional origins, American destinations and New World developments," *Slavery & Abolition* 18, no. 1 (1997): 122–45, 136.

25. James H. Sweet, *Recreating Africa: Culture, Kinship, and Religion in the Afro-Portuguese World, 1441–1770* (Chapel Hill: University of North Carolina Press, 2003), 132.
26. Robin Law and Paul Lovejoy, eds., *The Biography of Mahommah Gardo Baquaqua: His Passage from Slavery to Freedom in Africa and America* (Princeton, N.: Markus Wiener, 2001); Randy J. Sparks, *The Two Princes of Calabar: An Eighteenth-Century Atlantic Odyssey* (Cambridge, MA: Harvard University Press, 2004); Vincent Carretta, *Equiano the African: Biography of a Self-Made Man* (Athens: University of Georgia Press, 2005). See also Jon F. Sensbach, *Rebecca's Revival: Creating Black Christianity in the Atlantic World* (Cambridge, MA: Harvard University Press, 2005); Beatriz G. Mamigonian and Karen Racine, eds., *The Human Tradition in the Black Atlantic, 1500–2000* (Lanham, England: Rowman & Littlefield, 2010).
27. K. G. Davies, *The North Atlantic World in the Seventeenth Century* (Minneapolis: University of Minnesota Press, 1974), xxi–xxii.
28. Sylvia R. Frey, *Water from the Rock: Black Resistance in a Revolutionary Age* (Princeton, NJ: Princeton University Press, 1991). Colin G. Calloway, *The American Revolution in Indian Country: Crisis and Diversity in Native American Communities* (Cambridge: Cambridge University Press, 1995). Alan Taylor, *The Divided Ground: Indians, Settlers, and the Northern Borderlands of the American Revolution* (New York: Alfred A. Knopf, 2006). Simon Schama, *Rough Crossings: Britain, the Slaves and the American Revolution* (New York: HarperCollins, 2006). Patrick Griffin, *American Leviathan: Empire, Nation, and Revolutionary Frontier* (New York: Hill & Wang, 2007). Jim Piecuch, *Three Peoples, One King: Loyalists, Indians, and Slaves in the Revolutionary South, 1775–1782* (Columbia: University of South Carolina Press, 2008).
29. Susan Scott Parrish, "Diasporic African Sources of Enlightenment Knowledge," in James Delbourgo and Nicholas Dew, eds., *Science and Empire in the Atlantic World* (New York: Routledge, 2008), 281–310, 291, 305. For the view that enslaved Africans played a key role in North American rice culture, see Judith A. Carney, *Black Rice: The African Origins of Rice Cultivation in the Americas* (Cambridge, MA: Harvard University Press, 2001). A debate on this topic has appeared in the *American Historical Review:* David Eltis, Philip Morgan, and David Richardson, "Agency and Diaspora in Atlantic History: Reassessing the African Contribution to Rice Cultivation in the Americas," *American Historical Review* 112, no. 5 (December 2007), 1329–358, followed by three responses and a rejoinder in *American Historical Review* 115, no. 1 (February 2010), 123–71.
30. Laurent Dubois, "An enslaved Enlightenment: rethinking the intellectual history of the French Atlantic," *Social History* 31, no. 1 (2006): 1–14, 7, 12.
31. Ira Berlin, "From Creole to African: Atlantic Creoles and the Origins of African-American Society in Mainland North America," *William and Mary Quarterly* 53, no. 2 (April 1996): 251–88.

32. Linda M. Heywood and John K. Thornton, *Central Africans, Atlantic Creoles, and the Foundation of the Americas, 1585–1660* (Cambridge: Cambridge University Press, 2007), 311, 327.
33. For encounters in Europe, see: David Northrup, *Africa's Discovery of Europe, 1450–1850* (Oxford: Oxford University Press, 2002); T. F. Earle and K. J. P. Lowe, eds., *Black Africans in Renaissance Europe* (Cambridge: Cambridge University Press, 2005); and Alden T. Vaughan, *Transatlantic Encounters: American Indians in Britain, 1500–1776* (Cambridge: Cambridge University Press, 2006).
34. Anne Pérotin-Dumon, "French, English and Dutch in the Lesser Antilles; From Privateering to Planting, c. 1550–c. 1650," in P. C. Emmer and Germán Carrera Damas, eds., *General History of the Caribbean, Volume II: New Societies: The Caribbean in the Long Sixteenth Century* (London: UNESCO Publishing, Macmillan Education, 1999), 114–58, 146.
35. Brian Sandberg, "Beyond Encounters: Religion, Ethnicity, and Violence in the Early Modern Atlantic World, 1492–1700," *Journal of World History* 17, no. 1 (March 2006): 1–25, 7. Peter Linebaugh and Marcus Rediker, *The Many-Headed Hydra: Sailors, Slaves, Commoners, and the Hidden History of the Revolutionary Atlantic* (Boston: Beacon Press, 2000). Emma Christopher, *Slave Ship Sailors and Their Captive Cargoes, 1730–1807* (Cambridge: Cambridge University Press, 2006). Richard White, *The Middle Ground: Indians, Empires, and Republics in the Great Lakes Region, 1650–1815* (Cambridge: Cambridge University Press, 1991).
36. Joseph C. Miller, "History and Africa/Africa and History," *American Historical Review* 104, no. 1 (February 1999): 1–32, 30.
37. T. H. Breen, "Creative Adaptations: Peoples and Cultures in Early America," in Jack P. Greene and J. R. Pole, eds., *Colonial British America: Essays in the New History of the Early Modern Era* (Baltimore, MD: Johns Hopkins University Press, 1984), 195–232, 198.
38. Faren R. Siminoff, *Crossing the Sound: The Rise of Atlantic American Communities in Seventeenth-Century Eastern Long Island* (New York: New York University Press, 2004), 153.
39. Juliana Barr, "A Diplomacy of Gender: Rituals of First Contact in the 'Land of the Tejas,'" *William and Mary Quarterly* 61, no. 3 (July 2004): 393–434, 397–98, 412.
40. James Lockhart, "Sightings: Initial Nahua reactions to Spanish culture," in Stuart B. Schwartz, ed., *Implicit Understandings: Observing, Reporting, and Reflecting on the Encounters between Europeans and Other Peoples in the Early Modern Era* (Cambridge: Cambridge University Press, 1994), 218–48, 219.
41. Allan Greer, "Towards a Comparative Study of Jesuit Missions and Indigenous Peoples in Seventeenth-Century Canada and Paraguay," in Aparecida Vilaça and Robin M. Wright, eds., *Native Christians: Modes and Effects of Christianity among Indigenous People of the Americas* (Surrey, England and Burlington, VT: Ashgate, 2009), 21–32, 30.
42. Kenneth M. Morrison, *The Solidarity of Kin: Ethnohistory, Religious Studies, and the Algonkian-French Religious Encounter* (Albany: State University of New York, 2002).

43. Kenneth Mills, *Idolatry and Its Enemies: Colonial Andean Religion and Extirpation, 1640–1750* (Princeton, NJ: Princeton University Press, 1997), 248, 251.
44. Mary-Elizabeth Reeve, "Regional Interaction in the Western Amazon: The Early Colonial Encounter and the Jesuit Years, 1538–1767," *Ethnohistory* 41, no. 1 (Winter 1993): 106–38.
45. Claudio Saunt, ""Our Indians": European Empires and the History of the Native American South," in Jorge Cañizares-Esguerra and Erik R. Seeman, eds., *The Atlantic in Global History, 1500–2000* (Upper Saddle River, NJ: Pearson Prentice Hall, 2007), 61–75, 61. Taylor, *Divided Ground.*
46. Jeremy Adelman and Stephen Aron, "From Borderlands to Borders: Empires, Nation-States, and the Peoples in Between in North American History," *American Historical Review* 104, no. 3 (June 1999): 814–41, 816.
47. Robin Law, "'Here is No Resisting the Law': The Realities of Power in Afro-European Relations on the West African 'Slave Coast,'" *Itinerario: European Journal of Overseas History* 18, no. 2 (1994): 50–64, 52, 58.
48. Melanie Perreault, "'To Fear and to Love Us': Intercultural Violence in the English Atlantic," *Journal of World History* 17, no. 1 (March 2006): 71–93, 72.
49. Guillaume Aubert, "'The Blood of France': Race and Purity of Blood in the French Atlantic World," *William and Mary Quarterly* 61, no. 3 (July 2004): 439–78, 441–42.
50. Filipa Ribeiro da Silva, *Dutch and Portuguese in Western Africa: Empires, Merchants and the Atlantic System, 1580–1674* (Leiden: Brill, 2011).
51. David Eltis, "The English Plantation Americas in Comparative Perspective," in *The Rise of African Slavery in the Americas* (Cambridge: Cambridge University Press, 2000), 193–223. Cf. Russell R. Menard, "Law, credit, supply of labour, and the organization of sugar production in the colonial Greater Caribbean: a comparison of Brazil and Barbados in the seventeenth century," in John J. McCusker and Kenneth Morgan, eds., *The Early Modern Atlantic Economy* (Cambridge: Cambridge University Press, 2000), 154–62.
52. Camilla Townsend, *Tales of Two Cities: Race and Economic Culture in Early Republican North and South America* (Austin: University of Texas Press, 2000), 235. Mariana L. R. Dantas, *Black Townsmen: Urban Slavery and Freedom in the Eighteenth-Century Americas* (New York: Palgrave MacMillan, 2008).
53. J. H. Elliott, *Empires of the Atlantic World: Britain and Spain in America, 1492–1830* (New Haven, CT: Yale University Press, 2006). Patricia Seed, *Ceremonies of Possession in Europe's Conquest of the New World, 1492–1640* (Cambridge: Cambridge University Press, 1995).
54. J. R. McNeill, *Mosquito Empires: Ecology and War in the Greater Caribbean, 1620–1914* (New York: Cambridge University Press, 2010).
55. For the concept for entanglement, see Eliga H. Gould, "Entangled Histories, Entangled Worlds: The English-Speaking Atlantic as a Spanish Periphery," *American Historical Review* 112, no. 3 (June 2007): 764–86.
56. Claudia Schnurmann, *Atlantische Welten: Engländer und Niederländer im amerikanisch-atlantischen Raum 1648–1713* (Köln, Germany: Böhlau Verlag, 1998). April Lee Hatfield, *Atlantic Virginia: Intercolonial Relations in the Seventeenth Century* (Philadelphia: University of Pennsylvania Press, 2004). Christian J. Koot, *Empire at the Periphery: British Colonists, Anglo-Dutch Trade, and the*

Development of the British Atlantic, 1621–1713 (New York: New York University Press, 2011).
57. A.J.R. Russell-Wood, "Ports of colonial Brazil," in Alan L. Karras and J. R. McNeill, eds., *Atlantic American Societies. From Columbus through Abolition 1492–1888* (London and New York: Routledge, 1992), 174–211, 189.
58. Horden and Purcell, *The Corrupting Sea,* 133.
59. Kristin Mann, "Shifting Paradigms in the Study of the African Diaspora and of Atlantic History and Culture," *Slavery & Abolition* 22, no. 1 (2001): 3–21, 10.
60. David Barry Gaspar and David Patrick Geggus, eds., *A Turbulent Time: The French Revolution and the Greater Caribbean* (Bloomington: Indiana University Press, 1997). David P. Geggus, ed., *The Impact of the Haitian Revolution in the Atlantic World* (Columbia: University of South Carolina Press, 2001).
61. David Hancock, "'A Revolution in the Trade': Wine Distribution and the Development of the Infrastructure of the Atlantic Market Economy, 1703–1807," in John J. McCusker and Kenneth Morgan, eds., *The Early Modern Atlantic Economy* (Cambridge: Cambridge University Press, 2000), 105–53.
62. Daviken Studnicki-Gizbert, *A Nation upon the Ocean Sea: Portugal's Atlantic Diaspora and the Crisis of the Spanish Empire, 1492–1640* (Oxford: Oxford University Press, 2007), 94, 101. For three cases studies of Portuguese merchants residing in seventeenth-century Amsterdam, see Jessica Vance Roitman, *The Same but Different? Inter-cultural Trade and the Sephardim, 1595–1640* (Leiden: Brill, 2011).
63. Rosalind J. Beiler, "Dissenting Religious Communication Networks and European Migration, 1660–1710," in Bailyn and Denault, *Soundings in Atlantic History,* 210–36.
64. Banks, *Chasing Empire across the Sea,* 173. Ian K. Steele, *The English Atlantic, 1675–1740: An Exploration of Communication and Community* (New York and Oxford: Oxford University Press, 1986), 229–50.
65. Gordon K. Lewis, *Main Currents in Caribbean Thought: The Historical Evolution of Caribbean Society in Its Ideological Aspects, 1492–1900* (Baltimore, MD: Johns Hopkins University Press, 1983), 17–18.
66. John Garrigus, *Before Haiti: Race and Citizenship in French Saint-Domingue* (New York: Palgrave Macmillan, 2006), 37. Carl Bridenbaugh and Roberta Bridenbaugh, *No Peace beyond the Line: The English in the Caribbean 1624–1690* (New York: Oxford University Press, 1972), 313–314. Wim Klooster, "Subordinate but Proud: Curaçao's Free Blacks and Mulattoes in the Eighteenth Century," *New West Indian Guide* 68, nos. 3–4 (1994): 283–300, 291–93.
67. Gregorio Robles, *América a fines del siglo XVII: noticia de los lugares de contrabando* (Valladolid, Spain: Casa-Museo de Colón y Seminario Americanista de la Universidad, 1980), 35, 81.
68. Wim Klooster, "Inter-Imperial Smuggling in the Americas, 1600–1800," in Bailyn and Denault, *Soundings in Atlantic History,* 141–80. Wim Klooster, *Illicit Riches: Dutch Trade in the Caribbean, 1648–1795* (Leiden: KITLV Press, 1998).
69. Wim Klooster, "Le décret d'émancipation imaginaire. Monarchisme et

esclavage en Amérique du Nord et dans la Caraïbe au temps des Révolutions," *Annales historiques de la Révolution française,* no. 363 (2011): 109–29.

70. Julius Sherrard Scott III, "The Common Wind: Currents of Afro-American Communication in the Era of the Haitian Revolution" (PhD diss., Duke University, 1986), 117–18.
71. David Patrick Geggus, "Slavery, War, and Revolution in the Greater Caribbean, 1789–1815," in Gaspar and Geggus, *A Turbulent Time,* 1–50, 10. Matt D. Childs, *The 1812 Aponte Rebellion in Cuba and the Struggle against Atlantic Slavery* (Chapel Hill: University of North Carolina Press, 2006), 161.
72. Michael Craton, "Slave Culture, Resistance and the Achievement of Emancipation in the British West Indies, 1783–1838," in James Walvin, ed., *Slavery and British Society, 1776–1846* (Baton Rouge: Louisiana State University Press, 1982), 100–122: 103, 105, 120.
73. Emilia Viotti da Costa, *Crowns of Glory, Tears of Blood: The Demerara Slave Rebellion of 1823* (New York and Oxford: Oxford University Press, 1994), 177.
74. Gloria García Rodríguez, *La esclavitud desde la esclavitud: La visión de los siervos* (México City: Centro de Investigación Científica "Ing. Jorge L. Tamayo," A.C., 1996), 69–89. Laurent Dubois, *A Colony of Citizens: Revolution and Slave Emancipation in the French Caribbean, 1787–1804* (Chapel Hill: University of North Carolina Press, 2004), 106–7.
75. Gabriel Debien, *Études antillaises (XVIIIe Siècle)* (Paris: Librairie Armand Colin, 1956), 119. Robin Blackburn, *The Overthrow of Colonial Slavery 1776–1848* (London and New York: Verso, 1988), 172. Clarence J. Munford and Michael Zeuske, "Black Slavery, Class Struggle, Fear and Revolution in St. Domingue and Cuba, 1785–1795," *Journal of Negro History* 73 (Winter–Autumn, 1988): 12–32, 21. Carolyn E. Fick, *The Making of Haiti: The Saint Domingue Revolution from Below* (Knoxville: University of Tennessee Press, 1990), 91. Léo Elisabeth, *La société martiniquaise aux XVIIe et XVIIIe siècles, 1664–1789* (Paris: Éditions Karthala, Fort-de-France: SHM, 2003), 447, 450–53. Dubois, *Avengers of the New World,* 79–80, 106–7.
76. Neville A.T. Hall, *Slave Society in the Danish West Indies: St. Thomas, St. John, and St. Croix,* ed. B. W. Higman (Baltimore: Johns Hopkins University Press, 1992), 208–9.
77. Scott, "The Common Wind," 68.
78. See, for example, Leïla Maziane, *Salé et ses corsaires (1666–1717): Un port de course marocain au XVIIe siècle* (Mont-Saint-Aignan, France: Publications des Universités de Rouen et du Havre, Caen: Presses Universitaires de Caen, 2007).
79. Robert C. Davis, "Counting European Slaves on the Barbary Coast," *Past & Present* 172 (August 2001): 87–124, 121–22.

THREE

Tide, Beach, and Backwash: The Place of Maritime Histories

Nicholas Purcell

> "haec medium terris circumdat linea pontum
> atque his undarum tractum constringit harenis."[1]

The aim of this chapter is precisely that of the gathering at which its oral precursor was read—to investigate the nature and potential of "the new thalassography" as a scholarly initiative, while it is indeed still relatively new. Thalassography has hitherto been a maritime mirroring of geography, in that word's more literal meaning of the description of the land—a more local subdivision of oceanography. That more technical usage has become a little more familiar over the last years as a keyword in a geographically deterministic Grand Theory.[2] Rather, just as geography long ago escaped from disciplinary boundaries and became notable for being a hard-to-classify crossover zone of methods from many parts of the sciences and the humanities, so thalassography too has recently come to seem a suitable vehicle for the fertile intermingling of scholarly traditions.[3] It is a stimulatingly versatile idea, which it would be perverse to attempt to claim for any one tradition, but this chapter relates primarily to its historical manifestation. It offers a number of particular enticements to the historian who is concerned with how more local histories engage with the history of everything—universal history, *histoire à très large échelle*—not least because of its emphasis on the integration of history with its neighbouring disciplines.[4] It is important, though, to recognize a paradox at the outset. Although it is informed by maritime history in these more usual senses of the term, and has, indeed, grown from it, thalassography is not primarily concerned with the sea.

The impetus for these remarks derives from helping to write a work on Mediterranean history in which a constant concern was precisely not to revive or condone geographical determinism.[5] That book examined briefly how deterministic and exceptionalist theories of Mediterranean history had arisen, and oriented itself around the recognition that the subject-matter of the historian is the human behaviours which manipulate the conditions of life and invent the settings in which they do so. That was the reason for using as a title, *The Corrupting Sea,* a label derived from the hostility to mobility and connectivity which conservative ancient Greek thought linked closely with all that the sea stood for.[6] The aim of the book was to find elements which would make large-scale historical comparison both easier and more rigorous, starting within a large but reasonably well-defined space, looking for ways of overcoming the tyranny of period. The investigation entailed extended reflection on historical space, and on the nature of the microregions of which Mediterranean landscapes are composed, as well as the macroregions with which global history might be conducted, and of which the Mediterranean as a whole might be an instance.[7] The logical next step is to consider how the findings of that project relate to what was happening in adjacent macroregions, and to proceed to the *désenclavement* of Mediterranean history. The plan is to move on to see how sets of characteristics of the kind which Mediterranean history exhibits might be used comparatively in the task of explaining the interactions of even larger pieces in the jigsaw of global history, also across profound time-depths but, in addition, overcoming the barriers of space and scale. Now, this trajectory of Mediterranean research encourages the general observation that seas—including both maritime ensembles much larger than the Mediterranean and more locally focussed ones—can be very helpful for the investigation of global historical (and, by extension, other cultural) themes just because they are so good at eliding the boundaries of space and time, by promoting an emphasis on social, economic, and cognitive patterns which have different extensions and distributions. It is trivial to say that the emancipated historiography which results is "denationalized." What it offers, more constructively, is the possibility of new historical objects. The present discussion uses the phenomenon that I shall call "mobilization" to help display this.

Let us begin with a vignette from the years 49–46 BCE. In the difficult days of the Roman Civil War of that time, a eunuch (that is, a self-castrated) priest of the Great Mother named Soterides, in a harbor-town called Kyzikos on the south shore of the Sea of Marmara, set up a votive inscription to the deity whom he served.[8] He is anxious for his "life-partner" (the word is more usually employed of a spouse), Marcus

Stlaccius, son of Marcus, who had been a sacred flautist, presumably also in the entourage of the Great Mother. He has gone to war in the quadrireme *Saviour Goddess,* one of twelve ships from the region which joined in Julius Caesar's expedition to Africa in the war of Zama. The ship had been captured and its crew enslaved, and the Goddess revealed to Soterides in a dream that Marcus was one of those who had been taken prisoner. Alas, we shall never know what happened next . . . There are numerous histories into which we could insert Soterides and Stlaccius, and some of them are pretty clearly "sea-histories" in a large sense. The naval vocation of this port on the Sea of Marmara, and the navigational continuum which turns out, from the episode, to link it to distant Tunisia, suggest a pan-Mediterranean cosmos.[9] Still more, the capacity of a political power in Italy to engage in the mobilization of crews from northwest Anatolia for aggressions in the Maghreb conjures up a quite startling picture of maritime integration, and such an effect is clearly a significant base ingredient in thalassography. But the anecdote opens an important wider vista of "mobilization," since Stlaccius, mobilized as crew by Caesar's agents, was "remobilized"—and this is assuming that Soterides's dream was true, but that scarcely matters to this argument—in the slave markets of North Africa. Had his ship had better fortune, he might have brought home to Kyzikos considerable wealth from the spoils of a successful war; as it was, if he was ever heard of again, it would likely have required considerable outlay by his friends and family, and a considerable transfer of resources from the narrow seas where Turkey joins Europe to the shores of Tunisia to free him from slavery and bring him home. Violence, whether in peacetime or wartime, is capable of being as important a motor in the networks of movement of people and goods which are at the heart of maritime histories, as are the more or less legal operations of commerce.[10] It may seem coy to label such atrocious events so blandly as to call them mobilization in this larger sense, but there are advantages too in the relatively colorless terminology.

It is, therefore, the social relations and processes which render people and things mobile, as we see them, in two rather different ways, in this anecdote, rather than the mere fact of the connectivity, that should be emphasized. This more extended application of the term "mobilization," a deploying power over materials, people, and even ideas, is advanced here as a *fil conducteur* for a rewarding application of thalassographic history.

But the interest of this inscription doesn't stop there. Soterides's name tells us little except that his family had been long devoted to the religion that he too served, but Stlaccius's is another matter. It is an unusual and distinctive Italian or Roman family name. The presence of people bearing it in the port of Kyzikos is certainly related to the conquest of the Mediterranean by Rome and its expression through an extensive diaspora of agents of Roman economic and fiscal purposes across all the routeways—maritime and terrestrial—of the greater Mediterranean basin.[11] It is worth emphasizing that there is no way of telling, from Stlaccius's name, whether he was a descendant of a freeborn line of emigrants from Italy, or an ex-slave of such emigrants, since freed Roman slaves always took the family name of their former owners.[12]

Thanks to a relatively abundant harvest of inscriptions, reflecting a habit of self-publicising dedication related instructively to upward social mobility and high but precarious self-esteem, the panoply of Italian and Roman names visible across the economic continuum of Roman influence in the Mediterranean world and beyond it—from around 100 BCE to 100 CE—is extraordinarily rich, and the numbers involved in these structures of mobility were certainly large: an enemy of Rome had, a generation before the opening episode, massacred what could be believed to be 150,000 of them on the eastern seaboard of the Aegean Sea alone.[13] Not wholly without reason, as this diaspora, made possible by Roman power, served, wherever they went, and on highly favourable terms, to mobilize and garner resources for the ultimate benefit of Roman rulers. Flying in the face of the prevalent scorn for all things commercial in ancient societies, these men referred to themselves, with a certain pride, in Latin and Greek as "businessmen," but within their diaspora, voluntary movements of the self-determining free and coerced mobilities of their current slaves or obligatory deployments of their recent ex-slaves are impossible to distinguish.[14] The considerable legal duties owed by ex-slaves to their ex-owners formed the main legal basis of Roman business agency, which was patterned by long sequences of such dependents, thereby tying up all the organisms of financial and commercial practice with the institutional matrix of slavery—maintaining which, through the slave trade, was one of the principal objectives of that practice.[15] Nor is it easy to distinguish the private commercial or financial ambitions of any of the players from the role they played in the service of the Roman state, or the opportunities which should be located essentially in the sphere of developing private enterprise from ones arising directly from the consequences of recent subjugation to Rome. Once again, a larger "mobilization," of the kind identified above, lies at the heart of a history which gives a social-history expression to a unit in the geography of seas and coasts with greater clarity and eloquence than is usually the case before the early modern period.

A unit in maritime geography, indeed: but what? There is no doubt that the tale of Soterides and Stlaccius is a Mediterranean microhistory. As it happens, the strangely hybrid Italian-servile diaspora does indeed articulate a world of connectivity which maps onto what we call the Mediterranean world. The deploying power of the Roman imperial state itself has traditionally and plausibly been viewed by historians as being somehow intrinsically Mediterranean. Through the networks of these mobilizations, we can eventually hope, as we shall shortly see, to map the all-important sutures and capillaries where a world centred on Mediterranean connectivity is spliced with its neighbouring worlds, and with their connective systems. And yet before we adopt this kind of Mediterranean thalassography wholeheartedly, certain doubts need to be raised which may prove helpful in the analysis of the potential of this way of thinking.

Two Problems with Thalassography

As the appeal of the subject to the Bard Graduate Center attests, maritime histories speak directly to scholars of literature and the visual arts as well as to historians, and addressing this is, as has been suggested, a prominent aim of thalassography. As well as to historians? Well, perhaps one might venture, "even more than to" historians. Despite the respect in which some sea-oriented histories are generally held, there is also a certain tendency in parts of the profession to hold them somewhat at arm's length, or at least to undervalue their specifically maritime elements as somehow sentimental or populistic.

There is still something of a parting of the ways among those who aspire to large-scale treatments of the past, separating the cognitive and the perceptual on the one side from the ecological, the practical, and the economic on the other. There are some historians, of course, of, on the one side, a materialist or, on the other, a primarily doxological bent who are not much interested in bridging the gap between the *Realien* and the cognitive and cultural questions. But it would be productive if the division could at least sometimes be resolved, and maritime contexts offer one rather eloquent and persuasive instance of how some kinds of resolution can be achieved. The pursuit of this disciplinary rapprochement marks thalassography out from mere maritime history. But there are some outstanding issues.

Problem 1: Regionalism

While some maritime horizons facilitate the breaking of regional stereotypes, many maritime history projects retain strong regionalist associations. The seas of maritime history are frequently quite well-known objects of reflection.[16] Indeed, their place in much wider than academic discourses is ultimately what makes thalassography possible. But such popularly intuitable regionalities attract suspicion just because they are a commonplace, or because they come with cumbersome ideological baggage, because they smack of physical determinism, or because they seem a dull, old-fashioned, or arbitrary way of framing contemporary historical investigations.[17]

Now, directly against this, it has been said that units of maritime space don't have to be like this. They have been promoted precisely as liberating the regional historian from the tyranny exerted by most terrestrial regional units, and revealing just those entanglements and hybridities which are such fertile objects of modern reflection. Here again, where some have indeed limited themselves to how we can see "societies transformed by their contacts" across a sea, more ambitious scholars have sought, with Kären Wigen "worlds unified more by contacts among contrasting places than by any purported similarities across their shores."[18]

No one will quarrel with this, though perhaps there are fewer such unexpected maritime regional associations still to be conjured into existence than might be expected. The links and patterns of seaborne communications in fact often transcend regional conceptualization. Take for instance, accidents of very long-distance—very early trade, such as the pattern which brought Indonesian cinnamon to the Roman empire via the Kenyan coast, so that ancient authors believed it originated in the Horn of Africa.[19] The mobility of seafarers is such that the patterns traced by their habitual movements from time to time are apt to be too chaotic to construct regions with: in eighteenth-century Dunquerque, on the Narrow Seas of northwestern Europe, a quarter of the numerous foreign-born sailors were from the ports of distant northern Italy.[20] Maritime history does not need to be thinly disguised regional history with a wacky coloring. The "place of maritime histories" addressed in the present contribution is on the whole a smaller kind of unit, though there is every reason to examine large clusters of such small units where possible, and this is a point which the three terms of the subtitle—beach, tide, and backwash—are meant to evoke. Sea-based regional perceptions are mainly of interest in the context of the societies which invented and used them. Thus the study of the contested classification or self-attribution of geographical horizons is of prime interest to the thalassographer, and a nice instance of our new discipline's embracing both practical and cognitive agendas: taking the geographies too seriously ourselves can distort the analysis and impede this integrative goal.

Next, therefore, a word about comparative thalassographies. Here is an instance from a recent arresting plea by David Abulafia for the general multiplication of Mediterraneans as a tool of analysis. He argues that the way in which Buddhism reached Japan across what he posits as a 'Mediterranean of the North China Sea' deserves comparison with the reception of East Mediterranean culture in al-Andalus. Here, the comparison is enabled simply by the common role played in the two cases by the intervening void, a Mediterranean of the more general kind which he explores in his interesting piece.[21] The comparison is undoubtedly worthwhile, but there is a missing link in the argument. Abulafia is keen that the history of his numerous Mediterraneans should be one of 'human encounters', 'the way the inhabitants of the opposing shores of the sea interacted across the sea'.[22] He wishes to restore 'human beings' to a history which is apt to become abstract (and with the abstraction, essentialised, we might add). But people are in fact still largely absent from his account. A general attention to 'commercial, cultural and religious interaction' as he proposes, does not of itself restore a human dimension to a history which might otherwise concentrate on generalisations about social, cultural or economic process.[23] It is indeed precisely the nature of the behaviour of individuals—and above all the behaviour which is related, directly or obliquely to mobility, which must provide the foundation for the characterisations of the systems of contact between places. Only then can we do history with the physical regions which have patterned those contacts. One of the more important ingredients which the papers gathered in this collection might advance for future work in thalassography is the centrality to this approach of the detailed histories of the doings and thoughts of those whose movements give unity to the maritime. It takes more than a void to make a Mediterranean . . .[24]

This larger problematic of contact-zones and the specifics of mobility naturally brings in to the discussion, alongside real seas, other spaces of distinctive connectivity. The favourite has been deserts, just because of their extensive emptiness and uninhabitability.[25] An interesting question is thereby raised for the would-be thalassographer—is there actually anything distinctive to insist on about boat-borne transportation which in turn makes seas distinctive as connective spaces, or does the difference between caravan and cargo-ship, pack-mule and ferry-boat, not matter much?[26] We need to make up our minds about this. There can be a new historiography of mobility, in which we include the terrestrial on equal terms, but it will be quaint to go on labelling it thalassography. If we think, however, that there is something distinctive about the sea, thalassographic history becomes to a much greater extent than we do, or perhaps would like to, think, the distinctive domain of the history of the boat (which raises questions in turn about fluvial histories). While it remains true that an important reason for doing thalassography is the possibility of making thalassographies a building-block in larger problematics, and that is shared with other approaches which conduce to the big picture, the true thalassographer, informed by the ineluctable specificities of coping with the maritime, should constantly be in a position to instruct the student of terrestrial mobilities about the complexities of both thought and behaviour which relate to the media of mobility themselves.

The history in question thus becomes a history of the customs-post, of the shipbuilding yard, of the harbour tavern. It is one in which rhythm and scale are crucial: the arrival of a single shipload of luxury goods, for instance against the background of strict attempts to regulate seafaring, can be a thalassographical datum just as much as the customs-records of 2000 ships passing in a year—both instances come from medieval Japan.[27] The social relations of seafaring form a happier object of enquiry than a new thalassological regionalism: and it does not mean abandoning the grand goal of *histoires connectées*. The paradox is that to do the history of the whole globe you don't have to work through the history of regional building-blocks. There are alternatives—above all, bigger generalisations derived from longer series of smaller-scale observations, crossovers from the iterated smallscale structures to the larger ones which are of service to history on the biggest scale, and it is in facilitating these that the sea has a unique role to play.[28]

Problem 2: Cognitive and Sentimental Thalassographies

The second major reason for historians' reluctance to engage with thalassography may perhaps be found in the unavoidable prominence in discussions of the sea of the cognitive and the sentimental. People think with the sea; and they have the strongest of opinions about the sea. But the subjective and the ideological are not distractions from the main object of enquiry: indeed, thalassography helps show how they form an inextricable part of everyday experiences of the maritime. Further, the emotional charge itself should be a prominent object of enquiry.

The potential richness of research of this kind can be illustrated by geographical inventiveness. It is often remarked that learned geography has been the handmaid of imperial projects and the instrument of colonialism, whether created by the conquerors directly in their own interests, or appropriated in part or whole from the subjected. This relationship between knowledge and power is, however, only one specialised form of a much larger set of historical relationships between knowledge and social contexts which can be seen in action in past societies, favouring comparison and systematization of information. The simple fact of there being several, let alone many, similar players in parallel locations in networks of exchange of people and ideas encourages the drawing of conceptual parallels between the places in question and the theoretical convergence of their experiences, through the telling of similar tales about their pasts and their purposes, or the systematic description and classification of their spatial characteristics through a proto-geography. The natural, and perhaps almost spontaneous, drawing of simultaneous parallels in this manner makes it possible to take the step of recognizing the classification or the labelling as an object of reflection in its own right. Now, though once again there is no place for essentialising claims for the maritime in this context, it is here that the littoral makes a very important contribution.

The parallelism of the situations of coastal settlements mediating land to sea and sea to land, and sharing obviously in a lateral continuum born of this paralleled vocation, is one of the most widespread and easy to intuit (and compare) of such environmental configurations.[29] So here is a first crossover of scale, between local folk-thought and elite rationalization. In popular geographies of the maritime, we have a key to how the lived history of seafaring can be engaged with a much larger intellectual and cultural history, at precisely the same time as it is engaging with the more apparent social and economic ones. Parallels include the cross-fertilisation between ancient learned geography and the practical tradition of the coasting voyage or *periplous,* with its geographical lingua franca, and the oceanographical tradition which led from the portolan through the early chart.[30] Both conceptual geographies are necessary. The systematization, the perception of parallels, simultaneity, and lived classifications, conduce to social and political integration, and can serve ideological ends. They also have a vital part to play in economic life, by forming part of the load-bearing structures of economic information on which trust is founded, a theme of high contemporary interest among economic historians of all periods.[31] Networks of mobility are also networks of knowledge.

These consequences of reflecting on the processes of maritime mobility introduce the more general question of concepts of the sea itself. Coastlands and chains of harbors are patterned as iterative series as much by the repeated commemorations of marine disaster and the remedies for it, such as dedications to divine or saintly protectors of the seafarer, as they are by the topographies and labels of the mediation of land to sea. People conceptualise every kind of landscape, but the monstrosity and alienation of the sea put it in a special category. In the strange sixth century BCE epic poem, "The Arimaspea," an enigmatic traveler reaches a happy—and imaginary—continental people in the far distant north who have never dreamed of anything like the sea, and who are traumatised by his description of seafaring and the people of the Mediterranean "inhabiting the water, theirs eyes turned only to the stars, life and soul in the sea, praying without cease and perpetually seasick." For another ancient Greek writer, sailors are hard to classify between living and dead.[32] Once again, it would be an error to regard literary constructions such as these, and the popular thought to which they relate, as alien to the domain of the historian whose ultimate objective is a very large-scale understanding of world history. Such ideas are vitally and, in ancient thought explicitly, connected with the perception of the sea as a medium of appallingly dangerous differentiation—the "boundless sea of unlikeness" of Plato—and of alienation and deracination.[33] Alongside the simple fear of death at sea we find the more complex sense that the sea is a place of chaos where connexions founder—a fear that belies the actual function of the sea as a medium of connectivity.

There is thus a whole cultural history of the association of the sea with damage to good order—damage wrought through separating people from their homeland, through mixing them up, and through introducing uncontrollable numbers of newcomers. The ancient Mediterranean, it should be recalled, was a region in which a writer could assert that an observer will find that in every community—and not just the large and well-networked, but the small, poor, and remote—outsiders outnumber the people who really belong there.[34] It is hardly surprising that the sea could be shunned as a source of corruption (above, n. 6).

Now, we can agree that this type of question is central to thalassography, and show against the more materialist historical schools, how informative it actually is.[35] But unfortunately, that makes much harder some of the comparisons and methods which are most widespread among those interested in doing new kinds of history with seas. It is clear that the more irreducibly marine your analysis of a maritime context becomes, and the investigation of the ways in which people have thought through and about the sea can hardly tend in any other direction, the less easy it is to use maritime spaces and their social relations as general, rather loosely metaphorical elements in what can actually become a postmodern regional historiography. Ultimately, there must be certain characteristics of the really maritime that we should not be prepared to squander through a facility for metaphor which enables too ready comparisons. Proper thalassography must really take seriously the nature of human engagement with seas, and accept no substitutes. And I am very happy to insist on the plural—seas —as the essentialising of a global maritime element must be left to the people and societies we are studying, and not embraced by the historian: a universal and unitary idea of The Sea is one of the ways in which the people we study have constructed the maritime, but by no means a given of our own work. There is a whole new grand romanticism of the sea—a trio of historical geographers have recently enthused about the "more than human scope" of its historical study, for instance—a claim that should set off alarm bells.[36] There is a whole new romanticism of the scholarship of the sea, and when the object of our study is so potent a sentimental subject that it even manages to romanticize scholarship, we really need to pay attention.[37] Investigating the actualities of maritime social relations, however, in the fullness of their human complexity, of thought as well as deed, will in the long run make possible more focussed, rigorous, and productive historical comparisons than the looser forms of generalisation.

Here, the argument about how this can be done is developed under the rubrics of the three impressionistic words of the subtitle—beach, tide, and backwash. The beach, to begin with, stands for a vital theme which might be called maritime onset, or some such, but for which I prefer to use the term that Greek and Roman historians used to describe the sudden reorientation of terrestrial societies toward seafaring—"becoming maritime."

The Beach: Maritime Onset—Becoming Maritime

The genesis of popular histories of the maritime is a second essential instance of crossover between the actualities of lived histories which are strongly patterned by the oceanographical givens, and the multiple layers of narrative and explanation which make up the historiographical continuum which includes our own efforts. *Becoming maritime* is the transformation through which a community which has to various extents not previously been engaged with the sea undergoes some kind of structural revolution and emerges, typically after a strikingly short time, as one which excels at some or all of the maritime vocations, such as naval war, piracy, seaborne trade, shipbuilding, even systematic fisheries. This is a fascinatingly transhistorical notion. If its earliest occurrence is the bizarre claim by Herodotus of Halicarnassus that the Athenians had gone so far—remember those Greek images of marine desolation—as to become people of the sea (it is a very strong formulation, "became thalassic") in order to defeat the fleets of the Persian invader, and if it was famously resumed by Polybius with slack-jawed wonder at the way in which the Romans suddenly took to all the ways of the sea to beat the Carthaginians, there are many other familiar occurrences—brigands becoming pirates, barbarian invaders coming out of the steppes and taking to the waves, Mu'awiya and the Arabs, the Sicilian Normans, and the Turks; Kublai Khan in 1281, or Henry the Navigator.[38]

Now, here is a tangle of material which really justifies the existence of the thalassographer. Some of these cases are part of a tradition, some are spontaneously generated; most echo or arise out of essentially popular narratives; all are laden with ideology, and above all with the encoding of a double apologetic—we do not belong on the terrifying alienating and corrupting seas, and we only became involved because we were absolutely compelled to, and what we do at sea is to be distinguished from all the more suspect things that the kind of sea-peoples that we are not get up to. The set of commonplaces is related to the parallel tradition of the good thalassocrat, the land-dwellers who have a healthy organic relationship with a sea to which they bring order and civilisation. That an analysis of the genesis of these narratives, and their cross-fertilisation, needs to be conducted in close dialogue with accurate assessment of the realities of maritime engagement at the time goes without saying. But the thalassographer is also the expert who is needed to remedy the problems which beset the professional historiography of the various peoples of whom "becoming maritime" is predicated. For it is the case, unsurprisingly, that representations of this kind are almost never true in any literalist sense—the time frame of the revolution, to identify only one implausibility, can be interestingly protracted, as with the author who remarks that Mongol warriors, "used only to traveling on horseback, and having no nautical knowledge or experience," came to challenge China for maritime mastery "within the space of a few decades"—but that has not prevented them from being woven deeply into the heart of naval and national histories, even by historians who should know better.[39]

The thalassographer, on the model advanced in this chapter, will bring two parallel frames of reference to the elucidation of this type of value system: first, a wider, deeper, and more sophisticated experience of the ways in which the representation of the maritime relates to the genesis, acceptance, and propagation of ideologies, and, second, a sensitivity to the practicalities of what is involved in each society's engagement with the sea and, alongside it, the ability to assess precisely how difficult it is to attain each of the individual steps needed to design and build ships, to find and train crews, and to manage their deployment at sea. Such processes are part of a wider world of mobilization—of wood and wealth, iron and expertise, canvas and pitch, and (above all), of the labor of very many people, with a whole array of intellectual and artisanal skills, which forms the subject of our second metaphorical subheading—"tide."

The Tide: Maritime Mobilization

Underlying these clustered narratives of maritime-terrestrial reorientation lie questions of that "mobilization" which we met earlier in the anecdote of Soterides and Stlaccius. A common denominator between many thalassographies can be found in the varying and changing ability of a disposing power, which can be a larger or smaller state, a community, or a locally or more widely powerful individual, to marshall and concentrate the worlds of labor, of expertise, and of suitable materials, which are necessary for the intensification of maritime activity. That intensification of effort, and the abatement which counterpoints it where a relaxation or diminution of the inputs of energy or time into social and economic processes occurs, are important variables in the history of primary production.[40] Transforming systems in a rising and falling rhythm, these patterns of variation recur in history on the longer timescale with tide-like effect.

Such questions of the concentration of demographic resources underlie a theme which is dear to the Mediterranean historian, and of likely relevance to many other parts of the world too—the supposedly determinative shortages connected with seafaring which are still frequently invoked as major historical causes. In a strangely—and hardly appropriately—skittish passage in Braudel's La *Mediterranée,* for instance, the master advanced a theory of the intrinsic shortage of *gens de mer* in Mediterranean history, which he compared to the lack of fisheries in an oligotrophic sea—but likening his missing sailors to the rarity of what he called, in so many words, *frutti di mare*.[41] It is hard, when one reflects on the matter, to see how there can be an intrinsic shortage of something which training can produce—the analogy fatally confounds nature and nurture, in a way to which pre-thalassographic maritime histories are all too subject. "L'homme ne devient pas marin tout à coup," observes a memoir of the French government of 1758, recommending the naval recruitment of fishermen. Maybe—but we should not exaggerate the difficulty of the transition. It takes even less than "only a few decades," in the case of most who have been born into preexisting maritime vocations and milieux.[42]

There is no need to acquiesce in the essentialising assumptions on which both the sudden onset of maritime skills and the perception of a shortage of potential seafarers are both based. Instead, it is the play of the structures of mobilization, culture by culture, which invites careful analysis. Such organizing institutions of mobilization are adapted to many different purposes, not all maritime. But the common denominators of the maritime cases are indeed common to many different times and societies, and include precisely those distinctively maritime elements which are the thalassographer's business: the imbrication of mobilizations with other seaborne activities, and the overarching background, in all except the most sheltered lagunar, fluvial, or lacustrine waters, of the repugnant dangers of living at sea.[43]

One might go so far as to say that the apparent scarcity of recruits to early modern navies actually derives from the perception that what was wanted was a specialized figure who already belonged to the world of the sea. Create an excessively sharp dichotomy, and you risk forfeiting the room for maneuver which a more relaxed classification might give you. The older state of affairs, at least in the Mediterranean, appears to have been much more fluid: if far more people were available to participate in maritime activities, it was because no one expected such activities to be only the domain of the proper *gens de mer.*

The opening vignette introduced at least two separate types of deployment of people, and in this more extended consideration of mobilization, it is now appropriate to return to the other kind of possibility offered by engagement with maritime networks. If the sea calls for a distinctive type of engagement from people who will be coerced to face its dangers in order to realise—for whatever purpose—the potential of connectivity, it also makes possible, through that connectivity, the equally forced redeployment of people for all the other social, economic, and political purposes that leaders of communities can devise. High connectivity, to repeat what the Greeks feared about the mobility of seaborne communications, is a good deracinator, and though other spaces of easy communications also facilitate mass resettlement, none does it quite as effectively as the sea, both moving people and obliterating ties with ruthless efficiency. It is not just Atlantic thalassographers who must quickly come to understand that the spaces in which they have become expert are spaces of slavery.[44]

Corrupting Sea limited itself to a premodern past, with a caution which some of its critics have censured. By 2000 it was already becoming clear that a modern thalassography of the Mediterranean as a space of dislocation would shortly become a necessity, and it is now clear that, as a two-shored sea of transition—the object of the hopes of the many thousands of people who attempt to cross it clandestinely from south to north each year—it has many points of comparison with the Mediterranean of the early modern period, with its predatory slaving between the Christian and Muslim littorals, or the ancient Mediterranean with its circulatory system of the unfree.[45] Such a maritime history, let it be said once more, cannot be written exclusively at a high level of abstraction. It will not be possible until it can be informed by the microhistories of the backwater harbors in which the crumbling boats of today's new outlaw connectivity originate, or the remote islands and coasts onto which, if the exiles are lucky, they are cast away. For general thalassography the moral is clear: here is a phenomenon of surpassing historical interest and importance which will not be visible in the history of the high-profile modernities of the Mediterranean littoral. Thalassographies in all periods, in all parts of the world, will be incomplete unless they take care to be histories from below.

The Backwash: Spaces of Recursivity

"Becoming maritime," then, is a way of talking about a set of processes which are patterned by the need to engage with the sea, but not in any sense determined by its characteristics, though those have profoundly affected the ways of talking themselves. This penultimate section observes—rather briefly—how the play of mobilization, which is the structure underlying the phenomenon of maritime onset, need not only be directed from the land toward the sea. Coastal histories can be patterned either by offshore or by inland points of deploying power, and in many cases exhibit frequent reversals between the two.

Brent Shaw has used the term "recursivity" to describe these reciprocal alternatives, and recursive alternations of polarity between maritime and terrestrial deploying powers are a vital part of Mediterranean maritime history, to be observed as much in medieval Provence, between the Muslim beachhead of Garde-Freinet and the crusading base of Louis IX's Aigues-Mortes, or the early modern history of Bougie in the Maghreb as in any Greek or Roman context.[46] But similar effects have been used in modelling littoral histories in other parts of the world too. Thus, the anthropologist Jean-Claude Penrad has poetically applied the term *ressac* ("backwash"), to the history of a coast which has been profoundly marked by such ambiguous changes in polarity between land and sea, the Indian Ocean facade of Africa, so providing the epigraph for this section. The insular configurations of communities poised between the competing polarities of the land and the sea are a very rich thalassographic subject.[47] As a term, "recursivity" has its advantages, however. Perhaps it is unfair to criticise "nomenclature," which is notably colorful and effectively memorable, but *ressac* suffers from some of the sentimental associations which have made maritime history seem less serious. It attributes to the purely physical and the wholly oceanographic phenomena which are all too rooted in the grim ordinary business of human history—phenomena such as colonization, in all too many instances, for example.[48] The resources of the coastal zone are a normal target of predatory engagements: being strung out between the connective and deracinating milieu of the sea and the terrestrial interiors makes the communities of the littoral uniquely vulnerable to mobilization of every intrusive kind from both directions.

Once again in Indian Ocean spaces, the term "littoral history" has also been employed, by Michael Pearson, to describe the engagements of settlements on coasts of this sort, and there is a great deal to be said for drawing together the experiences of communities which in parallel mediate the land to the sea.[49] Drawing it together, that is, as an exercise in comparison, but also studying the history of parallel gateways, the traffic of each of which is constantly in dialogue with the traffic of the others—sometimes directly in competition with it, more often perhaps sharing in other ways a common experience of the transitory. What does it mean to look, as it might be, at the three cities of the Tripolitanian littoral, outlets of the fertile interior, precisely as a trio, and not one by one? The collective history of series of settlements is different from that of more isolated communities just because of the fact of mutual participation in the series. Littoral and maritime locations, promoting serial conceptualizations, thus lend themselves to the genesis of geographical generalisations and new ways of perceiving regularities in space (and so in political and social institutions), but also to new ways of recounting the shared—or contested—past. This parallelism is therefore a factor in the development of one of the historically most important aspects of recursive zones, and that is state-formation—a subject too large for the present discussion, though it is highly relevant to the understanding of the zone of mediation between West Asia and the Mediterranean world with which it concludes below.

The serial configuration of coastal societies offers the wherewithal for a further comparative exercise. For it is by no means clear that Janus-faced zones between two milieux of connectivity, intermittently made into an advance base for mobilization of resources in relation to the one by the inhabitants of the other, are limited to the tracts of shoreline where land is distinguished from sea. Even if we demand a more specialised collectivity, in which a space of distinctive high or low connectivity or mobility is surrounded by a "littoral" of settlements which have the greater mutual self-consciousness and higher level of shared social and cultural experience that comes from being a full or part circle rather than just a line, we can satisfy this criterion quite well in places where important uplands and lowlands abut. Indeed, there is a sense in which coasts which are sandwiched between a littoral mountain range or upland edge and the sea might be said to experience a fasciated coastal history, in which there are multiple parallel zones of shared positioning between deep interior and open sea. The gateways to the Mediterranean coastlands through the cordilleras which hem them in from the interior have played arguably as great a part in a distinctively Mediterranean history as do the ports which mediate between the sea and the land—though where the coastlands are especially narrow, of course, the same settlement may have both functions.

The parallel which concludes this survey, though, is to be found further east. The plains and basins of West Asia, centred on the valleys of the Euphrates and Tigris, but extending far beyond them, form a macroregion surrounded on three sides by significant mountain chains, and centering, except in the valleys of the Two Rivers and certain oases, on an extensive arid desert. Now, this desert has often been compared to a sea, and indeed, in the manner criticized above, to the Mediterranean. Its oases are islands, and the gateways—Aleppo or Damascus, or Nineveh—from which caravans set out, are its ports.

The analogy is certainly suggestive. This huge continental space, which Garth and Elizabeth Fowden have called "the mountain arena," has, it is true, a sort of littoral, in that those communities in its periphery share a vocation and draw important mutual resemblances from their sharing in it.[50] And that vocation has been primarily concerned with the mobilization of what can be moved across the routeways of the arid plains, and still more especially with the deployment of the human resources of the surrounding upland areas, through both an immemorial practice of recruitment of military manpower, but also through an equally ancient slave-trade. The periphery has, at a number of different periods, also suffered from recursive changes in the direction from which its resources were exploited, now being dominated by empires of the lowlands, now by states formed out of the advantages enjoyed by the parts of the periphery best suited to controlling the mediation of the highland to the low. In all of this the history of the old Babylonian state and its first-millennium successors offers an extraordinary prequel for many of the vicissitudes of the successors to the Abbasid dominion in the same zone.[51]

And yet comparison is more interesting when it diverges. It is worth insisting on the importance of understanding the particularities rather than being content with the lofty and impressionistic level of description. We might well hesitate to call this great lowland a Mediterranean of the continent. The West Asian macroregion, circumscribed though it is, susceptible of definition in the terminology of connectivity, mobility, and mobilization, is not a sea. That is, first of all, visible in the essentially more habitable nature of much of the arid and semi-arid steppe land. The sea, in its intrinsic hostility to life, in being the home of no one, has no nomads either. But West Asia is the place for which Michael Rowton elaborated the theory of the social dimorphism of the semi-arid pastoralist who is at once part of the adjacent world of the sown, and the complex set of intensifying and extensifying changes to the balance between pastoral and arable strategies, according to economic and political conditions, which provide a spectrum of peripheral variability as interesting as the recursivity of littoral histories, and analogous to it, but at the same time specifically different.[52] One might make a parallel point about the nature of the caravan. Convoys of ships engaged in ancient or medieval trade are sometimes impressionistically compared to the caravans that moved long-distance goods across West Asian or Saharan space. But the differences between the economic and social dynamics of shipborne trade and those of packhorse or camel caravan trade are enormous. The advantage of clinging to such distinctions becomes apparent when we come to the issue with which this rather breakneck survey concludes.

A Place of Thalassography: The West Asian Hinge

The work on which this chapter is based began in search of the *désenclavement* of the Mediterranean. To put it another way, it derived from the attempt to find ways of studying areas which drew their character from maritime spaces which would encourage a breaking out from tightly bounded definitions. In the case of the Mediterranean this did not prove difficult, and the experience may encourage us to assert that thalassographic enquiry properly so called will always militate against regionalisms which exclude. The seas, with which its numerous varieties engage, help formulate global histories that are resistant to enclosure and boundedness.

In the end, then, the only way of attaining such a regionalism is to base the study on what weakens regionalism, what elides the boundaries, and what renders the apparent frontiers permeable. That has ultimately to mean modeling movement—the individual micro-level movements of people, and the things and ideas that they carry with them. If larger structures seem to cross the boundaries and to render them obsolete, it is because at a less visible level, actual people have moved to convey the raw material out of which the bigger structures are built. A model example of what can be achieved is the recent work that has been done on what they call *circulation,* principally in South Asia, the basic patterns and conditions of the movements of people connected with economic life through which the larger formations of cultural history were established and maintained.[53] Their *circulation,* as they have so far explored it, has, in the circumstances, relatively little to do with maritime histories or thalassographies, but it illustrates exactly what we as thalassographers are called to investigate. That is why the mobilization of people to serve as sailors of different kinds, which underlies the transhistorical notion of becoming maritime, is so promising as a basis for a properly sea-based comparative history of some of the conditions of a rich and important mobility.

If sea histories need to be compared with, but differentiated from, terrestrial ones, and the comparison is at its most interesting when it is apparently closest, then a challenge with rich potential rewards, is to be sought in the historical modeling of how land-spaces, which are congeners of maritime ones, actually abut with the worlds of the sea. In such places, the social—and, one might say, political and cultural—consequences of the meetings of the different kinds of *circulation* will take potentially fascinating forms. A signal instance is the strange hinge-like region of the Levant, where the non-sea of the West Asian lowland meets the maritime facades of the Mediterranean and the Red Sea. But the area to the north of that region, where both abut with the ecologically different universe of the steppes, will also be a key margin with which to assert the importance of the maritime history. And the Levantine hinge has a vital interface with the seas to south and east as well, making it reasonable to speak of with the promising term "Central Eurasia."[54]

The themes outlined in this chapter, then—maritime engagement, mobilization, and recursivity—illustrate what thalassography can, and does, do. It brings its distinctive expertise to bear on a range of concepts which can be extended beyond the purely maritime. Even more important, it offers a set of methodological tools which must be imitated or adapted if history on the very large scale is to be grounded in a realistic sense of human behavior, incorporating the cognitive and the cultural alongside the social and economic. The world of the unfortunate Romans of the opening vignette was a wider than maritime one, though the sea was freedom and slavery, life and death, to Soterides and Stlaccius. In that age, the Roman diaspora reached far from the Mediterranean into the passes of the mountain armature of the Mediterranean basin, over the Caucasus, across the West Asian plain, even to the Tamil littorals of southern India and Sri Lanka. Thalassography and its developments can help us model phenomena of that kind, which are too big for the regional and period specialisms of the historical order, and which have defied older approaches to global history such as world-systems theory. The new discipline also aspires to make use of each more or less closely framed piece of research as an ingredient in a growing understanding of interactions on a much larger, and potentially universal, scale. In the case of the West Asian "hinge," the overlap zone of seaborne networks reaching their filaments inland with the systems of redistribution which grew up in and around the West Asian arena, was marked by a long series of parallel cities and entrepôts from Antioch, Ugarit, Carchemish, and Aleppo, to Jerusalem, Petra, Gaza, and the cities of the Hejaz. Often independent, each always interactive with the rest of the threshold series, as well as with both differently complex zones to east and west, this region produced the most complex and influential political, religious, and cultural sets of *conjonctures* in world history. What higher praise for the "new thalassography" could be devised than that it has serious potential to illuminate the long-term history of this all-important zone of cultural entanglement, and how it came to be braided with the histories of other places near and far.

NOTES

I am very grateful to Peter Miller for the invitation to what proved a very stimulating discussion in the hospitable setting of the Bard Graduate Center.

1. Manilius, *Astronomica* IV, 628, "this line, of coastwise journeying, circumscribes the sea in the heart of the lands, shutting it in with its beaches."
2. Cosandey 2007. Ferretti 2010 reflects thoughtfully on the place of this theory in the history of geography. Neither this nor the other papers in this collection concern this use of the label.
3. Peters 2003 used the term "thalassology" of Horden and Purcell 2000, providing a cue for the title of Horden and Purcell 2006. The need to differentiate the new frame from Cosandey's work (n. 2) might recommend this label over "thalassography," but the precise name of itself matters little.
4. Chartier 2001, 122. "Connected histories": Subrahmanyam 2004; *histoires croisées:* Werner and Zimmermann 2003.
5. Horden and Purcell 2000. The work frequently engages with this problem, which is fully indexed. Purcell 2003 also concerns the problem of Mediterranean exceptionalism.
6. Horden and Purcell 2000, 278, 300; Purcell 2003.
7. For the contemporary spatial turn in history, Torre 2008. On macroregions, long a familiar, if controversial, concept in Chinese history, Bin Wong 2001.
8. Cagnat 1901, vol. IV, 1375.
9. Hasluck 1910 remains an essential introduction to the site.
10. Horden and Purcell 2000, 383–91.
11. Hatzfeld 1919; Wilson 1966.
12. Rizakis 1996; Müller and Hasenohr 2002.
13. Plutarch, *Life of Sulla* 24, 7.
14. An example from the province of Africa: *ILS* 9495 "to the God Augustus, this is dedicated, under the Curatorship of L. Fabricius, by the Roman Citizens who do business at Thinissut" (*Augusto deo cives Romani qui Thinissut negotiantur curatore L. Fabricio*).
15. Aubert 1994; Fabre 1981.
16. Wong 2001, 22 on the "lisibilité physiographique" of the Mediterranean, contrasted with terrestrial macroregional spaces as postulated for Chinese economic history.
17. These are among the more frequent criticisms of Horden and Purcell 2000: see Harris 2005; Fentress and Fentress 2001.
18. Abulafia 2005, 85; Wigen 2006, 720. M. Greene gave a paper "Off-center in the Mediterranean: on writing the history of marginal places."

19. Thus Miller 1969; see, however, De Romanis 1996.
20. Cabantous 1980, 81.
21. Abulafia 2005, 66.
22. Abulafia 2005, 67.
23. Abulafia 2005, 93.
24. On the question of the emptiness of the sea, see Purcell 2003.
25. Thus again Abulafia 65, and 75–6 specifically on the Sahara.
26. Wigen 2006 clearly takes it for granted that there is such a distinction.
27. For the ancient world, something of a *locus classicus* of this kind is provided by the cargo of a single ship from India assessed for customs duty in a Red Sea port in the second century BCE, which a papyrus shows was worth the property qualification of seven Roman senators: Sidebotham 2011, 217.
28. The discussion thus relates closely to the debate on microhistory, its methods and advantages: see Revel 1996, and the review Fabbiani 1998. For the necessity of interaction of scales, see also Fusaro 2003, 607; Trivellato 2003, 584 n. 7.
29. Desert-edge or piedmont valley-opening settlements could be advanced as further instances of readily intuitable concept-generating chains of settlements, or better, chains of parallel locations for mobility-related social formations.
30. The term *lingua franca* is apt: Kahane, Kahane, and Tietze 1958.
31. See Trivellato 2003 and 2009 for an example.
32. Arimaspeia from 7 Bolton (= Longinus 10.4), see Romm (1992), 70–75. Parallels from "Anacharsis," at Diogenes Laertius 1 104, or in Letters 36–51.
33. Purcell 2003.
34. Seneca, *Consolatio ad Helviam,* 6, 2–4.
35. Among comparisons for the Mediterranean, we may note the case of the Caribbean, described by postmodernists as a "meta-archipelago"—Benítez Rojo 1992.
36. Lambert, Martins and Ogborne 2006.
37. Matvejević 1999.
38. Herodotus 7, 144; Polybius I, 20, 8.
39. Filesi 1972, 26.
40. Horden and Purcell 2000, 263–70.
41. Braudel 1972, 138.
42. Cabantous 1980, 54.
43. See, among many possible examples, Columella, *On agriculture,* preface, 8, on the "lottery of sea and business": "Breaking the covenant of Nature, at the mercy of the wrath of winds and sea, shall he trust himself to the currents, and evermore wander over an unknown world, stranger on a distant shore, like the birds?"
44. See James Warren's essay in this volume.
45. Godart 2003, and more general reflections in Ilbert 2000.
46. Shaw 2001 for the term recursivity. Bougie, Valérian 2006, a spectacular instance of successive turning toward or away from the sea.
47. Penrad 1994; see Margariti in this collection. The interaction of littoral and maritime spaces is an important theme also in Bentley, Bridenthal, and Wigen 2008.

48. For this theme, see Purcell 2005.
49. Pearson 2006.
50. For the term, Fowden 1993.
51. Briant 1982 is especially sensitive to these long-term themes.
52. Rowton 1973 and 1974.
53. For example, Markovits, Pouchepadass, and Subrahmanyam 2003. Trivellato 2003, 583 for the need to model trade alongside cultural exchange. Carras 2004 provides a modern Mediterranean example.
54. For a longer account of the place of this hinge-region in global history conceived in this manner, Purcell 2013, forthcoming. For the mirroring of Indian Ocean and Mediterranean in the *geniza* and in Shlomo Goitein's oeuvre, Miller, in this collection.

BIBLIOGRAPHY

Abulafia, D. "Mediterraneans." In *Rethinking the Mediterranean,* edited by W. V. Harris, 64–93. Oxford: Oxford University Press, 2005.

Aubert, J.-J. *Business Managers in Ancient Rome.* Leiden: Brill, 1994.

Balzaretti, R., M. Pearce, and C. Watkins. *Ligurian Landscapes: Studies in Archaeology, Geography and History in Honour of Edoardo Grend.* London: Accordia Research Institute, 2004.

Benítez Rojo, A. *The Repeating Island: The Caribbean and the Postmodern Perspective.* Durham, NC: Duke University Press, 1992.

Bentley, J. H., R. Bridenthal, and K. Wigen. *Seascapes, Littoral Cultures, and Transoceanic Exchanges.* Honolulu: University of Hawaii Press, 2008.

Braudel, F. *The Mediterranean and the Mediterranean World in the Age of Philip II.* London: Collins, 1972.

Briant, P. *État et pasteurs au Moyen-Orient ancien.* Paris: Maison des Sciences de l'homme, 1982.

Cabantous, A. *La mer et les hommes: pêcheurs et matelots dunkerquois de Louis XIV à la Révolution.* Dunkerque:Westhoek-Editions, 1980.

Cagnat, R. et al., eds. *Inscriptiones Graecae ad res Romanas pertinentes.* Paris: E. Leroux, 1901.

Carras, I. C. "Topos" and Utopia in Evgenios Voulgaris' Life and Work (1716–1806)." *Mediterranean Historical Review* 1 (2004): 127–56.

Chartier, R. "La conscience de la globalité." *AnnHSS* 56, no. 1 (2001): 119–23.

Cosandey, D. *Le secret de l'Occident: vers une théorie générale du progrès scientifique.* Paris: Flammarion, 2007.

De Romanis, F. *Cassia, cinnamomo, ossidiana: uomini e merci tra Oceano Indiano e Mediterraneo.* Rome: "L'Erma" di Bretschneider, 1996.

Fabiani, J.-L. Review of Revel 1996. *Annales ESC* 53, no. 2 (1998): 444–47.

Fabre, G. Libertus: recherches sur les rapports patron-affranchi à la fin de la République romaine. Paris: de Boccard, 1981.

Fentress, J., and E. Fentress. "The Hole in the Doughnut." *P&P* 173 (November 2001): 203–19.

Ferretti, F. "Articolazione costiera ed egemonia europea nella geografia del XIX secolo." *Storicamente* 6 (2010), http://www.storicamente.org/05_studi_ricerche/summer-school/ferretti_articolazione_costiera.htm.

Filesi, T. *Relazioni della Cina con l'Africa nel Medio-Evo.* Translated by David L. Morison. *China and Africa in the Middle Ages.* London: F. Cass, 1972.

Fowden, G. *Empire to Commonwealth: Consequences of Monotheism in Late Antiquity.* Princeton, NJ: Princeton University Press, 1993.

Fusaro, M. "Les Anglais et les Grecs. Un réseau de coopération commerciale en Méditerranée vénitienne." *Annales HSS* 58 (2003): 605–25.

Fusaro, M., C. Heywood, and M.-S. Omri. *Trade and Cultural Exchange in the Early Modern Mediterranean.* London: Tauris Academic Studies, 2010

Godart, L. "I flussi migratori nel Mediterraneo tra passato e presente." In *Il fenomeno coloniale dall'antichità ad* oggi, 9–15. Rome: Academia nazionale dei Lincei, 2003.

Harris, W. V. "The Mediterranean and Ancient History." In *Rethinking the Mediterranean,* 1–42. Oxford: Oxford University Press, 2005.

Hasluck, F. W. *Cyzicus.* London: Cambridge University Press, 1910.

Hatzfeld, J. *Les trafiquants italiens dans l'Orient Hellénique.* Paris: de Boccard, 1919.

Horden, P., and N. Purcell. *The Corrupting Sea: A Study of Mediterranean History.* Oxford: Wiley-Blackwell, 2000.

Horden, P., and N. Purcell. "The Mediterranean and 'the New Thalassology.'" *American Historical Review* 111, no. 3 (June 2006): 722–40.

Ilbert, R. "De Nankin à Marseille: la Méditerranée et l'épreuve du monde." In *Mégapoles méditerranéennes, géographie urbaine rétrospective,* edited by C. Nicolet, R. Ilbert, and J.-C. Depaule, J.-C, 117–30. Paris: Maisonneuve et Larose, 2000.

Kahane, H., R. Kahane, and A. Tietze. *The Lingua Franca in the Levant: Turkish Nautical Terms of Italian and Greek Origin.* Urbana: University of Illinois Press, 1958.

Lambert, D., L. Martins, and M. Ogborn. "Currents, Visions and Voyages: Historical Geographies of the Sea." *Journal of Historical Geography* 32, no. 2 (2006): 479–93.

Markovits, C., S. Subrahmanyam, and J. Pouchepadass. "Introduction." In *Society and Circulation: Mobile People and Itinerant Cultures in South Asia, 1750–1950,* 1–22. New Delhi: Permanent Black, 2003.

Matvejević, P. *Mediterranean: A Cultural Landscape.* Berkeley and Los Angeles: University of California Press, 1999.

Miller, J. I. *The Spice Trade of the Roman Empire, 29 B.C. to A.D. 641.* Oxford: Clarendon, 1969.

Müller, C., and C. Hasenohr, eds. *Les Italiens dans le monde grec. IIe siécle av. J.-C.—Ier siécle ap.J.-C., circulation, activités, intégration,* Athens: Ecole française d'Athènes, 2002.

Pearson, M. N. "Littoral Society: The Concept and the Problems." *Journal of World History* 17, no. 4 (2006): 353–73.

Penrad, J.-C. "Societies of the Ressac: The Mainland Meets the Ocean." In *Continuity and Autonomy in Swahili Communities: inland influences and strategies of self-determination,* edited by D. Parkin, 41–48. London: School of Oriental and African Studies, 1994.

Peters, E. "*Quid nobis cum pelago?*: The New Thalassology and the Economic History of Europe." *Journal of Interdisciplinary History* 34 (2003): 49–61.

Purcell, N. "On the Significance of East and West in Today's 'Hellenistic' History: Reflections on Symmetrical Worlds, Reflecting through World Symmetries." In *The Hellenistic West,* edited by J. Prag and J. Quinn. Cambridge: Cambridge University Press, 2013.

Purcell, N. "Colonization and Mediterranean History." In *Ancient Colonizations: Analogies, Similarity and Differences,* edited by H. Hurst and S. Owen, 115–39. London: Duckworth Publishing, 2005.

Purcell, N. "The Boundless Sea of Unlikeness? On Defining the Mediterranean." *MHR* 18, no. 2 (December 2003): 9–29, reprinted in I. Malkin, ed., *Mediterranean Paradigms and Classical Antiquity.* London: Routledge, 2005.

Revel, J. *Jeux d'échelles: la micro-analyse à l'expérience.* Paris: Gallimard, Sevil, 1996.

Rizakis, A. D. *Roman Onomastics in the Greek East: Social and Political Aspects, Meletemata* 21, Athens: de Boccard, 1996.

Romm, J. *The Edges of the Earth in Ancient Thought.* Princeton, NJ: Princeton University Press, 1992.

Rowton, M. B. "Enclosed Nomadism." *JESHO* 17 (1974): 1–30.

Rowton, M. B. "Autonomy and Nomadism in Western Asia." *Orientalia* 42 (1973): 247–58.

Sallares, R. "Ancient Greece: Some General Considerations." In *Production and Public Powers in Classical Antiquity,* edited by E. Lo Cascio and D. Rathbone, 5–13 . Cambridge: Cambridge Philological Society, 2000.

Shaw, B. D. "Challenging Braudel: A New Vision of the Mediterranean." *JRA* 14 (2001): 419–53.

Sidebotham, S. E. *Berenike and the Ancient Maritime Spice Route.* Berkeley and Los Angeles: University of California Press, 2011.

Subrahmanyam, S. *Explorations in Connected History: From the Tagus to the Ganges.* Delhi: Oxford University Press, 2004.

Torre, A. "Un 'tournant spatiale' en histoire: paysages, regards, ressources." *AnnHSS* 63 (2008): 1127–144.

Trivellato, F. "Juifs de Livourne, Italiens de Lisbonne et Hindus de Goa: réseaux marchands et échanges culturels à l'époque moderne." *Annales HSS* 58 (2003): 581–603.

Trivellato, F. *The Familiarity of Strangers: The Sephardic Diaspora, Livorno, and Cross-Cultural Trade in the Early Modern Period.* New Haven, CT: Yale University Press, 2009.

Valérian, D. *Bougie, port maghrébin, 1067–1510.* Rome: Ecole Français de Rome, 2006.

Werner M., and B. Zimmermann. "Penser l'histoire croisée: entre empirie et réflexité." *Annales HSS* 58 (2003): 7–36. "Beyond Comparison: *Histoire Croisée* and the Challenge of Reflexivity." *History and Theory* 45 (2006): 30–50.

Wigen, K. "Introduction: Oceans of History." *American Historical Review* 111, no. 3 (2006): 717–21.

Will, E. *Les Palmyréniens : la Venise des sables (Ier siècle avant-IIIème siècle après J.-C.).* Paris: A. Colin, 1992.

Wilson, A. J. N. *Emigration from Italy in the Republican Age of Rome.* Manchester: Manchester University Press, 1966.

Wong, R. Bin "Entre monde et nation: les régions braudéliennes en Asie." *AnnHSS* 6, no. 1 (2001): 5–41.

FOUR

The East Asian "Mediterranean": A Medium of Flourishing Exchange Relations and Interaction in the East Asian World

Angela Schottenhammer

The present contribution uses the macro-geographical space of an "East Asian Mediterranean" in order to characterize multilateral exchange relations in the East Asian world that have traditionally been investigated as relations of various independent nation-states or countries (territorial area studies), or have been analyzed with concepts such as center and periphery (for example China versus the Ryūkyū Islands), and that, in addition, have largely neglected human-environment interaction. The concept of the East Asian Seas[1] as a medium of exchange is borrowed from Fernand Braudel's work on the European Mediterranean, *La Méditerranée* (Ch. *dizhonghai* 地中海).[2] Although by far not all conceptual elements of Braudel's model can and should be transferred to the Far Eastern context, the concept can be applied in order to underline different forms of political, commercial, and cultural exchange between China, Taiwan, the Ryūkyū Islands, Japan, and Korea in order to put the emphasis of analysis on the history of exchange relations rather than that of more or less isolated nation-states or countries.[3] Braudel's concept can at least be used to enlarge our view on a macro-region that was only superficially characterized by a strict antagonism between a "center" (China) and "peripheries" in East Asia. After introducing the methodological frame and approach, I provide the reader with a few examples of how this approach can be used to view the East Asian Seas and which conclusions can be drawn.

Methodology and Approach

Fernand Braudel has argued that conventional frameworks for historical analysis—notably territorial entities such as nation-states, empires, and continents—were inadequate because they largely ignored human-environment interaction. Thus, while Paris exercised political dominance over southern France, the existence of people who lived there had much more in common with that of other inhabitants of the Mediterranean littoral, including northern Africa, being largely shaped by the Mediterranean Sea and its climate, than with those in Paris or northern France. In the same context, a Fujianese inhabitant of the Chinese littoral often had more in common with other coastal inhabitants settling further south or north or even with inhabitants of littorals abroad, beyond the seas, than with those of the political center in for example Kaifeng (Northern Song), or Beijing (Ming), or with Chinese settling along the northern border regions.

Braudel's theories inspired the development of the *Annales* School of historians in Europe,[4] and more recently a new school of historians of Asia, such as K. N. Chaudhuri,[5] Anthony Reid,[6] and Abu-Lughod,[7] who stress the monsoon system as a facilitator of trans-oceanic, intra-Asian, sail and trade. The monsoons comprise a system of regularly alternating winds and currents unique to the Indian Ocean and South and East China Sea. From April to September, as the Asian land mass heats up, hot air rises producing a vacuum, which sucks in the air from the ocean, creating the southwest monsoon. During the other six 'winter' months of the year, the opposite reaction occurs, creating the northeast monsoon. The conventional view is that these winds enabled vessels to engage in purposeful two-way trans-oceanic trade that by about the tenth to thirteenth centuries connected the major productive areas of Asia, China and India, and the Persian Gulf (present-day Iran and Iraq) in a system of trade that endured to the mid-eighteenth and possibly the nineteenth century.

However, these studies suffer from major limitations. First, they largely ignore human-environmental interaction beyond accommodation of a simplistic monsoon model. Second, conventional histories are imbued with interpretative preconceptions that obscure major developments and sectors in the history of exchange relations. These include the nation-state and regional studies approaches, which analyze the East Asian world in terms of component modern states and geographical zones, such as the Far East, Southeast Asia, or South Asia. Yet these derive from Eurocentric[8] and colonial-era classifications and serve to imprison Braudelian concepts within conventional frameworks of nation-states or 'territorial' area studies. In addition, these histories tend to regard exchange merely in the terms of official relations and diplomacy, thus neglecting reality and developments that took place behind the surface. They do little justice to the complex trans-frontier, trans-oceanic exchange of commodities, monies, technologies, ideas, and people that characterized the early modern era. Furthermore, whereas previous research has basically documented the history of a specific region or the trade relations between two areas, the concept of an East Asian Mediterranean enables one—on the basis of sources of all integrated regions and countries,[9] above all China, Taiwan, Japan, Korea, and the Ryūkyūs—to analyze and expound the supra-regional and international socioeconomic, scientific, and cultural exchange within the region of the East Asian world. Research should, consequently, focus on "inter-East-Asian" exchange relations.[10] A third pervasive preconception is that economic modernisation and state formation are closely correlated—economic development, as a rule, being most advanced in centralized hierarchical states like China and least advanced in decentralized stateless societies. This approach has resulted in the assumption that countries like China and India constituted the 'core' or the 'centers' of the Asian maritime world, while regions such as the Indonesian Archipelago (outside Sumatra and Java) or the islands in the East Asian seas like the Ryūkyūs, Taiwan, or Tsushima were of marginal significance and constituted little more than "peripheries"; that following European intrusion into the region from circa 1500, European agents largely controlled the most valuable aspects of production and trade; and that with the imposition of European colonial rule, indigenous systems of long-distance exchange collapsed. In reality, however, a vivid intra-East-Asian exchange already existed long before the advent of the Europeans. And a historic approach too much orientated along centers, such as China, obscures vivid exchange relations in and along the so-called peripheries.

Fernand Braudel's concept of *La Méditerranée* should, thus, be understood as one characterizing the dynamics of exchange in a geographical and geopolitical area whose neighboring countries were linked and related with each other by sea routes. This dynamics of exchange did not just start with the appearance of Western powers in the East Asian waters, although it was the latter, namely the British, who eventually successfully weakened but not destroyed the original national mode of production in a country like China[11]—in contrast to the Dutch, who dominated trade with Japan and who did not exert a destructive influence on Japan's national economy. Historical changes in East Asia cannot be reduced simply to such impulses from "the outside," or to the developments of exchanges with the European nations. Neither were they a result of supraordinate, teleological developments allegedly inherent in history (such as a trend toward modernity). The East Asian countries had their own particular dynamics of change and their own overseas and land-based characteristics of exchange relations. Even during times of seemingly fixed power relations an inner dynamics prevailed; for example, within the regulated tribute relations between China and other East Asian countries in the first half of the Ming dynasty (1368–1550), or the official seclusion policies that at first sight give the impression that no exchange was going on between "secluded" countries.[12] Many scholars in East and West have contributed in the last decades to redirect our attention to China's intensity of exchange relations throughout East Asia and it is impossible to provide an overview here.[13]

This chapter, based on Braudel's concept, discusses the three above-mentioned premises viewing the East Asian world with China as the alleged center or core region.

Human-Environmental Interactions

Environmental changes are of significant importance for human activities. They may themselves be results of human activities or of natural phenomena. For maritime trade of course events such as taifuns 大風, cyclones, or storms in general, earthquakes with consequent tsunamis 津波, and inundations, or the silting of ports and river estuaries arc of particular importance. The history and development of the East Asian "Mediterranean" World, for example, or in the larger perspective the Indian Ocean World (IOW) comprising eastern Africa, the Middle East, Southeast Asia, China, and India—an arena of primary geopolitical importance whose economic, social, and political foundations date back more than three millennia—can only be understood within the context of the development of human-environment interaction: primarily of human interaction with the monsoons, a system of regularly alternating winds and currents unique to the Indian Ocean and South and East China Sea that exerted a huge influence over agricultural production in the lands bordering those seas, and created the possibility of direct transoceanic sail. However, many other natural factors—at times positive, at times negative—influenced the material life of peoples in the Indian Ocean and Asian world, including the El Niño-Southern Oscillation and stochastic events such as cyclones and volcanic eruptions. In East Asian Studies we have only just begun to examine these relations in more depth and detail, Mark Elvin's *The Retreat of the Elephants* being a welcome pioneer study.[14] That this aspect has become more important and gradually more and more attention is being paid to human-environment action may be seen from recent publications.[15]

Development of the Ports of Liuhe 劉河 and Shanghai 上海

Here, I address an example that may show how human-environment interaction changed the particular development of a region and eventually made Shanghai the major port on the southeast Chinese coast taking over many functions that originally were served by other ports. We briefly introduce the development of the Liuhe port, located near the Yangzi estuary in Southeast China, that combined overseas and domestic trade—with the city of Suzhou 蘇州 as a major commercial center—and compare it with Shanghai. Suzhou functioned as the commercial center and axis between domestic and overseas trade during the Ming and Qing dynasties. It lay strategically favorable only some sixty *li* away from the Yangzi River, close to its estuary. Liuhe originally was the hub between overseas and domestic distribution of products, a door toward China's inland market and cities such as Suzhou. But it was eventually given up due to what we may call human-environment interaction. One reason was the silting of the river. Thus, on the one hand, environmental changes forced the people to give up this port. On the other hand, human action also contributed to the fact that Liuhe disappeared as a center of distribution and Shanghai took its position.

The Liuhe port had both easy access to the sea and was strategically located for a further distribution of ocean-shipped goods within China. The *Liuhe zhenji lue* 劉河鎮記略, compiled during the Daoguang reign (1821–1850) by Jin Ruibiao 金瑞表, provides us with interesting information on this port.[16] As it shows, originally, in Ming times, sea routes were divided into northern and southern routes, with a clear distinction of ports responsible for either the northern or the southern routes. The Liuhe port was not an unproblematic location. It is considered a place lying secluded (*pi* 僻) in the corner of the sea that experienced great changes over time—the *Liuhe zhenji lue* even describes it with the words of the old Chinese saying of "*canghai sangtian* 滄海桑田," which literally means "turning the sea into mulberry fields," but has been widely used to indicate great transformations. The abridged term "*cangsang* 滄桑" is now often used as a synonym for hardships experienced as a result of unexpected and undesirable changes.[17] This is exactly what happened with the Liuhe port. The Liuhe River, with its tides and currents, is even designated the *Weilü* 尾閭 (commonly translated as "the ultimate drain" or "cosmic cloaca")[18] of the three rivers—an ocean current pushing up from the Equator and flowing past the Philippines, Taiwan, Ryūkyū, and Japan and diverging into the Arctic and North Pacific currents. It was considered a "point of no return"; a current which flowed to a drain in which everything irrevocably disappears. These difficult circumstances apparently also made it an ideal spot for pirates. Because of the problems with piracy, from the late Ming until 1653 (Shunzhi 10), the Liuhe was not taken care of for more than thirty years. Only in 1657 it was dredged again.[19] Why, however, we may ask, was the Liuhe port at the Liuhe River maintained when it was such a delicate or difficult location?

In this context, the *Liuhe zhenji lue* informs us about different sea routes, ships, and products. It discusses the reason for the mentioned division of sea routes between southern and northern or eastern ones respectively. The following passage clearly shows the importance of environmental factors and how they were taken into consideration.

> In 1685 (*kangxi* 24), the provincial governor (*xunfu* 巡撫) of Jiangsu "personally went to the seaports to inspect the local circumstances and suggested to fix (the following) regulations: trading certificates are granted for sailing overseas; merchant ships from Min province (Fujian) are called "bird ships" (*wuchuan* 烏船); they are proficient in (sailing) in the open sea between Zhe(jiang) and Tai(wan), but do not enter the Northern Waters (*beiyang* 北洋); when they come to Jiang (province), they are all duly received at the port of Shanghai. The merchant ships from Jiang (province) are called "sand ships" (*shachuan* 紗船); they are proficient in sailing in the Eastern Sea, but do not enter the Southern Seas; when they come to Jiang (province), they are all duly received at the port of Liuhe 劉河. For both of them a customs station has been established to levy taxes.[20]

Consequently, according to geographical and environmental circumstances, different ship types were used and two distinct ports of taxation established. Ships that came to Liuhe were so-called sand ships, with a shallow bottom that could easily use the inland waterways, including the Grand Canal. Commodities they transported were primarily precious and valuable goods from the northeast (Dongbei 東北). During the Qing dynasty in particular, soy beans (*dadou* 大豆) and soy-cake fertilizer, required for the local cotton fields, were transported from Dongbei via Shandong to Jiangnan. They were shipped via Liuhe and the Liuhe River further to Suzhou—the major commercial distribution centers for domestic trade. Liuhe was, thus, a flourishing port, which was consequently also attractive to pirates and smugglers. But environmental changes and human reaction gradually led to the insignificance of Liuhe.

Shanghai (Huangpu), on the other hand, originally a small, peaceful harbor for so-called bird ships with sharp bottoms that mostly transported coarse and heavy goods would have capsized in the shallow waters off the northeast coast and along the Liuhe River. Consequently, there was a clear distinction of routes, products including their origin, and taxation.

During Qing times, however, Shanghai gradually emerged as the major commercial and trading center important not only for imports and exports from abroad and to foreign destinations, but also for the further distribution of goods within China; for example, to Suzhou. Historically, as Linda Cooke Johnson has noted, Shanghai was only one of a succession of ports located in the vicinity. "The alluvial soil and shifting waterways of the delta region formed an unstable terrain where streams and rivers frequently changed course, creating new land where water or marsh had formerly been, and cutting new waterways where none had existed before."[21] In this process, Shanghai had already developed into an important port for overseas trade in the late Song dynasty, but environmental changes sealed its further destiny—as those of the Liuhe port.

Inland access from Shanghai was, for example, provided by the Wusong River, which lead to Suzhou only some seventy miles west of Shanghai. The Wusong River, today no more than a creek, "was a major waterway in the Tang dynasty, then called the 'Songjiang.'"[22] Originally, the silting of the Wusong River had favored the development of Shanghai. In the Ming dynasty the channel of the Wusong had shifted to the north of Shanghai city where it joined the Yangzi at Wusongkou 吴淞口 instead of emptying into the sea east of Shanghai. Between the mid-fifteenth century and the Kangxi reign, Shanghai, consequently, lost most of its former port functions, becoming simply a county seat.[23] By Qing times, the Wusong River was still navigable but obviously only with great caution and under strict attendance of lieutenant-colonels.[24] Ships that originally used to anchor at Ming and Qing period Shanghai could hardly navigate northward or enter the Yangzi estuary with its tides and currents, not to speak of the Liuhe River. Liuhe and Wusong had become "peaceful" waterways.[25] But after the shift of Wusong's river channel to the north, in addition due to the continuous widening and deepening of the Huangpu 黄浦 River and the development of Huangpu into a big port,[26] the ships that used to sail only in the southern ocean and to anchor at Shanghai could then not only more easily navigate in the Yangzi estuary but generally had a better access to a domestic commercial center like Suzhou.[27] In addition, after the opening of maritime trade in 1684, commodity transportation was increasingly shifted to overseas routes and consequently the larger overseas vessels started to sail north, too, and took over part of the business that was formerly served by Liuhe junk masters with their sand ships. The importance of Liuhe and the junk masters who were involved in this northern trade with their sand ships consequently decreased.

Closely related with environmental changes is also the establishment of customs stations (*haiguan* 海關) in Jiangsu province. The traditional transport route to the north went via the Grand Canal and inland waterways. Customs regulations during the Ming period had consequently divided the collection of taxes according to these two routes: Commodities and products from the north, transported via the Grand Canal or coastal routes, were taxed at Liuhe, while Shanghai was responsible for trade from the southern regions. After the lifting of the maritime trade proscriptions in 1684, however, increasing volumes of merchandise were shipped on the sea route, a fact that weakened the importance of Liuhe as the hub between overseas and domestic trade even further. Bird ships and vessels with a larger tonnage in general could now transport the products that were formerly shipped by sand ships and, using the sea route, they could then be taxed at Shanghai.

That means that until the seventeenth and early eighteenth centuries ships called at *both* Shanghai *and* Liuhe, although already great efforts had to be undertaken to guarantee the functioning of Liuhe and Shanghai had already overtaken Liuhe's role as the hub between maritime and domestic trade. The regulations dividing tax collection according to the two original southern and northern routes, however, continued to exist in Qing times, although they had gradually become obsolete.[28] Between the Kangxi and the early Qianlong (1736–1795) reigns, the Liuhe port experienced its last heydays; gradually, it silted. In particular, during the

Qianlong era, increasing problems arose. How did local officials and merchants react on this situation, or did they simply accept it?

With the increasing signs of the gradual sanding up of its river estuary, in 1727 (Yongzheng 5), the official Fan Shiyi 范時繹 (d. 1741), governor-general of Liangjiang (*Liangjiang zongdu* 两江總督) between 1726 and 1730, sent a memorial to the Yongzheng 雍正 Emperor (r. 1723–1736), drawing the emperor's attention to the problem and the potential consequences of a further silting of the river bed for the local population—many of whom would lose their economic basis, as they depended on this northern trade and the business between Liuhe and Suzhou. Still, during the Qianlong reign, for example, 290 shops of Shandong merchants existed in Suzhou.[29] Not few were engaged in the soybean trade with Shandong. The problem with the shipment of soybeans and soy-cake fertilizer consisted in that they were transported on sand ships via the Grand Canal or coastal routes to Liuhe and then to Suzhou. As a consequence of the silting and the increasing competition from Shanghai junk masters, many of the smaller junk masters and merchants in Liuhe had to give up their business. But, of course, only those among the disenchanted merchants who could afford it *peu à peu* shifted their business to Shanghai—that consequently developed not only as a transshipment port for commodities from the south but also from the north, shipped on the northern ocean routes.

Therefore, Fan Shiyi requested official funds (*gongjin* 公金) for a dredging of the Liuhe River mouth in order to maintain the functionality of the Liuhe River leading to the city of Suzhou. In 1734 (Yongzheng 12), the subsequent governor-general of Liangjiang, Zhao Hong'en 趙宏恩, reported to the Yongzheng emperor about the problems related with the shipping of soybeans and grains along maritime routes, suggesting that the former routes (shipped by sand junks) be restored. He reported that the dredging works in the Liuhe had been completed that year with the help of the official funds.[30] This measure would prevent many merchants and households involved in the northern trade from bankruptcy.

Until the Qianlong era boats could still commute between Suzhou and Liuhe, but it became more and more difficult. By the early Jiaqing reign (r. 1796–1820), port and riverbed had silted again. Consequently, the process of a shift of commercial activities from Liuhe to Shanghai that had already set in on a larger scale in the early seventeenth century (recalling that until 1653 no dredging works had been carried out for more than thirty years), was accelerated.[31]

The fact that with the shift of the Songjiang (Wusong) River's channel to the north, Shanghai bird ships could also easily make northern voyages[32] emerged yet as an additional competition factor for the Liuhe junk masters, who anyhow had big problems due to the silting of the port. Increasing economic problems in 1730 (Yongzheng 8) had for example prompted the junk master Zhang Qiu 張球 and four others from Liuhe forcibly to transport merchants' goods and urge the merchants to pay for additional fees, as we know from a stone inscription preserved in the Tianhougong 天候宮 at the Liuhe port.[33]

In Shanghai, on the other hand, Shanghai junk masters and shipping merchants competing with those from Shandong in the cotton and fertilizer transport had formed the "Sand-junk-merchants *huiguan* 會館" as early as 1685. "The early success of these merchants and their continuing activities testify to a lively trade [at Shanghai, A.S.] in spite of the [then still existing, A.S.] nominal restrictions and competition from Liuhekou."[34] With the silting up of the Liuhe port, the center of commerce shifted even faster to Shanghai. By 1804 (Jiaqing 9), suddenly 3,500 to 3,600 sand junks assembled at Shanghai.[35] As a result, Shanghai, originally a port designed for trade along southern sea routes, emerged as the major transshipment center for maritime and domestic trade along both southern and northern routes—a shift that subsequently also had consequences for other Chinese ports.

The Shift from Civil to Military Supervision

The commercial upswing of Shanghai affected not only the location of tax and administrative authorities but also a shift from civil to military supervision. The main customs office of the *Jianghai* 江海 Customs Station was in the beginning located at Chongque 漴闕 in Huating 華亭 in the vicinity of Shanghai, but later they moved to the Baodai 寶帶-Gate in Shanghai.[36] By the final years of the Kangxi period, the administration and management of the customs offices underwent an important reformation. As a consequence of tax deficits, Emperor Kangxi ordered that all tax-collection stations should be directed by the provincial governors in order to increase official supervision. In 1720, the responsibility of the Jianghai maritime customs was, for example, transferred to the provincial governor of Jiangsu in order to guarantee tighter control.[37]

The Yongzheng emperor continued these reforms. Going along with the increasing importance of Shanghai, the tax authority of the provincial governors, and supervision, too, was gradually moved from Suzhou to Shanghai. In 1725 (Yongzheng 3), Zhang Kai 張楷,[38] approved to dispatch Zhu Yifeng 朱一鳳 (*jinshi* 1709; c. 1685–1770; Han official)[39] of the SuSong 蘇松 (i.e., Suzhou and Songjiang) General Surveillance Circuit to manage the Jianghai *haiguan*. From the fifth month of 1725, the *haiguan* in Jianghai was, consequently, managed by a branch office of the SuSong General Surveillance Circuit, sign of the intent to submit it to an even tighter control. In 1730, following a petition of the Provincial Governor, Injišan 尹繼善 (1696–1771),[40] the General Surveillance Circuit of SuSong was moved from Suzhou to Shanghai, and it was transformed into a Military Defense Circuit (*bingbei dao* 兵備道)—a measure that reflects both the increasing commercial importance of Shanghai and the necessity for controlling commerce. The first official in charge of the Jianghai *haiguan* based in Shanghai, and concurrently of the Shanghai Military Defense Circuit, was Wang Chenghui 王澄惠 (Han official).[41] By 1730 (Yongzheng 8), the main customs office had, thus, completely shifted to Shanghai by a decree of the Yongzheng emperor. Both this decision and the fact that control was transferred to military officials attest to the necessity of a tighter control of maritime trade and the collection of taxes, of so-called unlawful elements,[42] as originally proposed by the Jiangsu Provincial Governor Zhang Kai. The rise in the volume of trade is, thus, also reflected in "illegal" smuggling activities and the necessity to bring them under control. In 1800, Li Tingjing 李廷敬 (*jinshi* 1775) of the Military Defense Circuit (*bingbei dao* 兵備道) transferred the port arsenal from Liuhe to Shanghai and stopped the dredging works that would have been necessary to regain access to the port[43]—certainly a consideration that took into account that Shanghai had anyhow already overtaken Liuhe's former functions.

These political changes resulting in the relocation of the circuit and customs offices as a consequence of socioeconomic and environmental factors, thus, sealed the decline of the Liuhe port and the further future of Shanghai as the major overseas trading port and an exclusive customs office for external commerce.[44] This, conversely, also had consequences for other ports like Ningbo or Quanzhou that were still open to foreign trade, although the decline of Quanzhou as an international transshipment port has at least partly also to be traced back to the silting of the harbor. We can thus clearly see the importance and consideration of environmental factors when decisions were made concerning the establishment of customs stations or reactions to environmental changes, both in terms of socio-commercial and natural developments. Local people such as junk masters, merchants, shopkeepers, etc., who had profited from the northern trade going through Liuhe reacted on natural and socioeconomic changes, as did officials. But eventually in this case they had to accept environmental and commercial shifts that had made Shanghai the undisputed commercial center in the region. The fate of the ports of Liuhe and Shanghai can, thus, be satisfactorily explained only by taking into account both natural and socioeconomic human-environment interaction.

East Asian Nation-States or "Territorial" Areas versus the East Asian World

Here, we examine some examples that attest to the complex trans-frontier and trans-oceanic exchange of commodities, monies, technologies, ideas, and people that characterized the early modern era. We see that describing inter-Asian exchange relations simply in terms of an exchange between nation-states (or countries) characterized by their officially pronounced policies fails to see the real extent and complexity of exchanges.

Let us take a look at Ming China. Part of the Ming period tribute and foreign trade system, for example, was the maritime trade proscription, which was initiated in 1372 and officially prohibited all private maritime trade activities.[45] It is, however, important to note that the Ming rulers permitted a certain quantity of "business," or private trade for profit, to be conducted by private merchants during the phase of strict overseas trade prohibition; provided that foreigners strictly followed the rules and rites set up by the Chinese authorities, a certain extent of private trade was also possible. Above all, though, it is necessary to see that in spite of this official proscription very vivid trade and exchange relations were actually maintained in the East Asian waters; for example, through clandestine channels. Superficially speaking, thus, China was secluded from the world and only official tribute relations were permitted. In reality, however, a vivid private trade was going on along with and beyond official diplomatic and tribute relations. Therefore, only by considering China as a part of the East Asian networks, at the same time paying attention to political, socioeconomic, and cultural changes both within China and in the East Asian world in general, do we get a more precise picture of the story.

Pirates and Smuggling

The Ming maritime trade proscription lasted until 1567. Of course, it decisively influenced trade and commerce across the East Asian "Mediterranean" and beyond, but it did not bring trade to a standstill, as it might appear at first sight. The ban enabled merchants from other countries, primarily the Ryūkyūans, to take over intermediary roles or replace Chinese merchants. Southeast Asian ceramics suddenly appeared on the scene to fill the gap of Chinese ceramics, for example. Many Chinese merchants were driven into smuggling, either becoming pirates, emigrating, or both, and many trading networks were only maintained with the help of Chinese residing overseas.[46]

> "We have great numbers of people who make their living on the sea, drifting wherever they do fishing and gathering. The prohibition policy unsettled them. As long as they have a chance to work for the merchants, they will be engaged in trade and settle themselves down to a career, and they will have no other intention but to follow the pirates and smugglers. We should be reminded by our experience (that is, that the maritime prohibition engenders piracy, A.S.): The pirates from the r*enzi* 壬子 year (1552) were frustrated maritime merchants, the "*guichou* 癸丑 pirates" (1553) were people from other businesses aiming at making more income, but the "*jiashen* 甲申 pirates" (1554) were common people and the "*yimao* 乙卯 pirates" (1555) were people from everywhere. The tendency indicates that we have to stop the origins of the piracy from its very beginning (that means, providing the coastal people with a way to make a living and allow maritime trade, A. S.)."[47]

The reason for the trade proscription was, as Li Kangying has shown,[48] not piracy, as it has sometimes been claimed—which vice versa mainly emerged as a result of the trade ban—but a new political-ideological *raison d'état* initiated by the first Ming emperor. Ideally, thus, under Ming maritime policy, tribute trade should represent all Sino-foreign economic interflow. This was but an official ideal. This system did not consider people's economic interests—not to speak of the fact that former merchants were from one day to the other decreed illegal traders by government decision. If tribute trade failed to satisfy economic demands of the people, other ways were opened as substitutes. Market demand was subsequently satisfied via an informal economy created by smugglers. It is little wonder, as Chang Pin-tsun has emphasized, that from the mid-fifteenth century onward, maritime smuggling outweighed tribute trade to become the major way of Sino-foreign economic exchanges.[49] More and more high officials, such as Tang Shu 唐樞 (1497–1574), the director of the Ministry of Justice during the Jiajing reign, argued for lifting the trade prohibition.

> "We are now in a situation where our government has had no policy to deal with piracy for the last four to five years. The war on the coast has exhausted our central and local financial resources . . . but trade is (yet) unstoppable, as the foreigners are bound to come to harass; (this is) because both our peoples are of the same human nature and need the exchange of local products. The maritime prohibition is unpractical . . . (Furthermore) smuggling by our own illicit merchants has also created an enormously powerful movement that the coastguards dare not to resist . . . Prohibiting trade will turn merchants into pirates, as the policy drives them (into a situation) where they have no way to maintain their subsistence . . . The stricter the maritime prohibition is preserved, the more intensified piracy will become . . . It is much easier to go with the desire of the people by permitting trade, while it is more distressful to go against the trend by the prohibition policy."[50]

Eventually, in 1567, the government decided to lift the trade prohibition. In 1567, a new emperor, the Longqing 隆慶 emperor (r. 1567–1573), was enthroned. Guided by a group of practical officials headed by Xu Jie 徐階 (1503–1583) as senior grand secretary, the new reign began with measures of reform.

One may also direct the reader's attention to the offshore islands along the Chinese littoral—for the political center mostly the seeming periphery. During the time period from the late Ming through the Ming-Qing transition until the early Qing, smuggling was omnipresent in these Chinese waters. In 1653, Shang Kexi 尚可喜 (1604–1676), the governor of Canton—who himself had been able to amass a great fortune through maritime trade—gave the responsible person in the salt supervisorate, Bai Wanju 白萬舉 (also Bai Yuheng 白有珩) and the salt merchant Shen Shangda 沈上達 ample monthly provisions and salaries, established a supervising office at its former place where the Ming period office had been located, and had these two persons manage foreign (tribute) trade according to the old Ming regulations.[51] Shen Shangda also permitted some Chinese merchants to carry on illegal private trade.[52] The system initiated by Shang Kexi was later continued under his son, Shang Zhixin 尚之信 (1636–1680).[53] Smuggling had obviously reached such an extent that the Kangxi Emperor in 1684 had to state: "Although we have a strict maritime prohibition, as for privately conducted trade, how can this be interrupted?"[54] As Li Shizhen 李士楨 (1619–1695) informs us, after 1662 no ships arrived officially, but the clique around Shen Shangda privately conducted trade; at one time they could make a profit of 40,000 to 50,000 silver *liang,* within a year a thousand ships were coming and going and their profits amounted to 400,000 to 500,000 *liang.*[55]

Smuggling also increased at Macao in relation with the Kangxi's decision to evacuate the local coastal population to the hinterlands (*qianjie* 遣界),[56] although Macao still possessed a special status. "By the end of 1668, the evacuation laws had been revoked, and commissioners were touring the Kwangtung coastal areas to supervise their re-population. The gate was opened every five days with few exceptions. Macao was not exempted from the continuing prohibition of maritime trade, but its enforcement seems to have been considerably relaxed. Chinese merchants came to Macao, often at night, and Portuguese ships arrived from overseas and departed again."[57] Between 1680 and 1690, Canton merchants continued to trade and smuggle with the Portuguese at Macao as well as with the Dutch among the offshore islands.[58]

Sino-Japanese Relations

That we have to consider and analyze exchange networks beyond the strict division of states becomes obvious when we go a bit further back to the origins of an intensive and active exchange networks in the East Asian waters—a process that can be traced back to the Song period at the latest. Solely to examine official Sino-Japanese relations; that is, the relations between China and Japan, would for example only lead to relatively general ideas and to a picture of two isolated states in East Asia. Thus, it is important to look behind the curtain and examine these relations in their complexity. We have to explore who sent the ships and what kind of societies produced and shipped the products exchanged. During the Song period, regional society developed well, stimulated by active local economy and other activities. Many Japanese Buddhist monks, for example, traveled to China. They were sent not by the official Japanese government, but by local feudal lords and powerful temples; that is, private institutions. During this time, significant changes also occurred in Japanese society. Thus, we need to know what was happening in Song China in order to learn how the two societies were synchronizing with each other's changes. Regional societies were becoming more mature and active in both nations, their influence expanding to local economies on small village units or to local religious activities. Studying and analyzing such changes leads us to understand the real conditions of maritime cross-cultural exchange. Economic development in villages both in

China and Japan made it possible for them to support the construction of temples. Villagers were no longer blind believers and began to take initiatives in their religious pursuit. Such changes and developments in both nations collectively contributed to promote and support the cross-cultural exchange. Although we must also take into consideration that many Chinese people crossed the ocean out of poverty, it was essentially the changes in Song society that promoted cross-cultural exchange and not official diplomatic relations between China and Japan.[59] An analysis of the social fabrics of both countries—and not solely their official foreign policy—is, thus, indispensible to understand the complexity of relations between the two countries.

Also during Ming and Qing times mutual relations were far more complex than it might appear at first sight.[60] In our *East Asian Maritime History* project we have drawn attention to the fact that, far beyond official diplomatic relations, the kidnapping or smuggling of Chinese physicians to Japan has been a kind of popular sport in seventeenth-century East Asia.[61] The trend of Chinese physicians going or being taken to Japan continued throughout the early and mid-eighteenth century. This has to be seen in direct relation with Tokugawa Yoshimune's 徳川吉宗 (1684–1751) policy, proclaimed in 1718 (Kyōhō 3), ordering ship captains to bring good Chinese physicians to Japan. In this context, we can discern an interesting exchange in the fields of medicine, medical knowledge, doctors, veterinarians, horses, archers, and books[62]—that was not part of the two countries' official relations. The historian of East Asia will once and again have to realize how important it is to analyze a broad variety of sources from if possible all countries involved in order to learn more about the complexity of exchange relations in the East Asian world.[63]

Illegal trade or smuggling is attested to not only in Chinese and Japanese sources. As Chinese merchants very often had to leave the port of Nagasaki with many goods unsold, time and again they tried to sell these things to Japanese smugglers in the nearby waters, as Engelbert Kaempfer observed.[64] A particular role within the smuggling network in Japanese waters at that time was played by Satsuma, as many examples in the *Tsūkō ichiran* 通航一覧 (1853) may show.[65] Smuggling (*nukeni* 抜荷) was, thus, nothing extraordinary.[66] Dutch sources also attest to this phenomenon. A local interpreter, after the Dutch had repeatedly complained about the rampant smuggling of the Chinese, noted: "The Chinese have to smuggle in order to render their business profitable."[67]

All these illegal and clandestine activities beyond official interstate relations have to be considered to understand mutual relations. Officially, for example, until 1871, no diplomatic relations existed between Qing China and Japan.

Sino-Ryūkyūan Relations

Ming and Qing Sino-Ryūkyūan relations were officially monopolized and part of China's alleged tribute system. Still, as a closer investigation of the exchange relations shows, private elements and private exchange were omnipresent and quite influential during both time periods. Private trade could be pursued either legally, as a part of the tribute missions that always also constituted an opportunity for trade, or illegally, as smuggle. Following an entry in the *Chouhai tubian* 籌海圖編, the Ming court permitted the Ryūkyūs to carry local products (*fangwu* 方物) and officially established brokers to trade with the people: "*qi lai ye, xu dai fangwu, guan she yahang, yu min maoyi, wei zhi hushi* 其來也, 許帶方物, 官設牙行, 與民貿易, 謂之互市."[68] As for the Ryūkyūan side, the *Lidai bao'an* speaks of thirty-six surnames from Min who entered the country (*Minren sanshiliu xing ruguo* 閩人三十六姓入國).[69] Descendents of these thirty-six families later served as foreign interpreters (*yi tongshi* 夷通事) in bilateral relations. It must be emphasized that the role and functions of these interpreters went far beyond the task of interpreting and that they were also directly involved in the management of Sino-Ryūkyūan diplomatic and trade relations.[70] In reality, a broad variety of both private and official—what is not always easy to distinguish—persons including merchants, officials and others were involved in this trade beyond the official state-state relationship.[71]

We know also that the Ryūkyūans always brought official or "royal silver" (*ōgin* 王銀) with them to China, which was called *to-Tō gin* 渡唐銀 (lit. silver for sailing to China) in Japanese. It amounted to 151 *kanme* or 15.100 taels (*liang* 兩) of silver for a tribute ship—that means 302 *kanme* for two tribute ships. It consisted of three parts: the silver used for trade, for ship repairs, and for gratuities (*qianyin* 遣銀) to Chinese officials in Fujian and Beijing who had something to do with Ryūkyū affairs.[72] It was also to be used in emergency situations; for example, when Ryūkyūans needed to bribe local officials for favors or to reach their goal with more speed. The complexity of these bilateral relations, the blurring between private and official—otherwise simply designated as tribute trade—may only become evident when we take a look at the networks working behind the surface, behind official interstate relationships.

In 1634 (Chongzhen 7), the tribute envoys Cai Jin 蔡錦, Mao Shaoxian 毛绍賢, and Liang Tingqi 梁廷器[73] had a sum of 1,000 *kanme*—that is, 100,000 *liang*, of *to-Tō gin* on board; subsequently they provided thirty-one Chinese merchants with silver in order to purchase Hu 湖 silk—that is, best-quality silk from Huzhou 湖州 in Zhejiang, for them. In the course of events, a quantity of 4,998 *liang* of silver for the purchase of 4,594 *jin* of Hu silk was misappropriated from the original *to-Tō gin*. It is unclear if these thirty-one merchants were private or official brokers or simply private merchants. Only two years later, in 1636 (Chongzhen 9), about 40,000 *liang* out of 120,000 *liang*; that is, one-third, of the *to-Tō gin* were misappropriated, once again for the purchase of Hu silk.[74] This becomes evident from a comparison of entries in the *Chūsan seifu* 中山世譜 (1636),[75] and in decrees of King Shō Hō 尚豐 (1621–1640) compared with the quantity of silks that had been confiscated by the Chinese government, as the exportation of silk was prohibited at that time.[76] Clearly there must have existed close networks and a kind of cooperation between the private and official sector to make these activities possible that were, of course, not always detected. And although Sino-Ryūkyūan trade was officially controlled and monopolized by the government both in Ming and Qing times, although it functioned as part of China's alleged tribute system, private merchants and characteristics of private trade were not only the starting point of its management in the early Ming period but remained present and important throughout the centuries into the late Qing dynasty—and considering this trade merely as an exchange between two countries or nations, China and the Ryūkyūs would have obscured, or at least blurred, all these ongoing developments.[77]

Sino-Korean Relations

The illegal trading activities within the triangle of Ming China, Korea, and the newly founded dynasty of the rising Manchus in northeast Asia may serve as yet another example. So-called secret merchants (*qianshang* 潛商), who were both Chinese and Korean[78] and often worked closely together with officials, carried out a vivid trade in this region.

> As for secret merchants, they are not only in Korea, but Ming China also has secret merchants, even until the Qing dynasty there still were activities of secret merchants. They only look for places where they can make profit, but secret merchants often do not distinguish between enemies and ourselves (*chang bufen di wo* 常不分敵我). The meaning of "secret merchants" and of "smugglers" is very close. At a

> time, when Yuan Chonghuan 袁崇煥 (1584–1630), a high Ming official, practiced his policy of a commercial blockade against the Later Jin Dynasty, there were activities of secret merchants between the Later Jin (i.e. the Manchus) and Ming China. The designation of "secret merchant" is also closely related with the "Pidao merchants" (merchants from Kado Island). Their activities can be designated in general terms, such as officials and private merchants working together (*guansi heying* 官私合營), half official and half private (*ban guan ban si* 半官半私), both official and private (*yi guan yi si* 亦官亦私), officials and merchants forming a league (*guanshang goujie* 官商勾結), and officials supervising and merchants carrying out (*guan du shang ban* 官督商辦). Among them are also commercial activities of merchants with a sincere and honest (*chun* 純) disposition, but their commercial activities and transactions are illegal (*bu hefa de* 不合法的). After Korea was oppressed and permanently cut off from Ming China in 1637 (Chongde 2), it was nevertheless permitted that secret merchants commuted (*yue* 躍) between Ming China and Korea under the condition that (they traded) commodities produced in Ming China.[79]

Officially, all this kind of trade was interdicted, but examples like the above mentioned show, once again, the complexity of interaction in the East Asian waters that go far beyond official diplomacy and official relations between countries. In particular, personal networks between merchants and merchant groups active in East Asian waters as well as their relations to official and religious institutions have only started to be investigated in more detail. Recently published archive material from the Qing period (huge quantities of documents from the First Historical Archive in Beijing, the Diyi lishi dang'an guan 第一歷史檔案館), for example, provides us with much more valuable source material that will hopefully enable us to obtain a better view of these networks. And it may show that at first sight seemingly unrelated persons, institutions, and organizations actually had much closer relations with each other than it might appear.

Center and Periphery

For centuries, China's political, economic, and cultural success and superiority had been reflected in the consciousness of the Chinese social and ruling elites in the traditional concept of the "Middle Kingdom" (*zhongguo* 中國).[80] Superficially speaking, the idea of being situated in the middle of the world also characterized and defined China's official relations to her neighboring countries: These were "barbarians" (*yi* 夷), which could only learn from China and otherwise were required to subject themselves to and to accept Chinese suzerainty as tribute states. This was but a political ideology as promoted by the ruling and at least great parts of the intellectual elites, which found its expression in the traditional tribute system and the idea of a suzerain-vassal relationship.[81] China was considered the center that was surrounded by peripheries—the small and weak kingdom of the Ryūkyū Islands, or the "tiny dwarf" Japan (*cui'erguo* 蕞爾國).

Actually, this concept represented rather an official self-reflective attitude of the Chinese elite and does not tell us much about the reality of exchange relations. Not only did reality of China as the center versus the peripheries change throughout history. As the example of the Qing dynasty may show, this self-reflective attitude is not even that easily applicable to the practice of China's everyday foreign policy—neither Japan nor the Ryūkyūs were, for example, simply considered and treated as distant maritime periphery, although official language suggests this.[82] And although we certainly have to admit that China, if it was not always the political center, it at least functioned as the economic motor of the premodern East Asian world, as an economic center—a methodological approach that divides its object of research into center and peripheries fails to see the broad variety of exchange relations that existed. We, therefore, have to look beyond these categories. A merely China-centered view of the history of China's relations with her neighbors concentrating on the official Chinese ideology of center and periphery rather tends to obscure the vivid and diverse history of her exchange relations and the important impact non-Chinese, partly "peripheral" societies, cultures, and states had on Chinese civilization.

Even adopting the concept to China herself—the political center versus the peripheries—one can clearly see that frequently a large divergence existed between central policies and actual local activities that attests to various differences, conflicts, and even contradictions between the political center and local actors. In terms of China's maritime trade this become perhaps most evident in the case of the Ming court's official maritime trade proscription (*haijin*) and merchants, people, and local authorities at the coasts. A classical distinction between political center and local periphery cannot do justice to the complex supra-regional networks that existed despite an official prohibition and to the divergence of interests. A huge body of written evidence from the later Ming dynasty attests to the fact that while the center still favored the trade proscription, more and more officials pronounced themselves in favor of liberalizing trade.[83] Neglecting these differences and conflicts would only obscure the historian's view to the actually very vivid, sometimes perhaps even liberal, trade and exchange relations that actually existed during a period of official prohibition of maritime activities—which frequently involved those who officially should have interdicted them, like local officials, tax supervisors, etc.

The Ryūkyū Islands

In terms of international relations, the example of the Ryūkyūs—the "obvious" periphery—may serve to show the methodological boundedness of dividing regions into center and periphery. Since early Ming times, the Ryūkyūs had been firmly integrated into China's tribute system.[84] The island country continued to function as probably China's most faithful tribute country[85] throughout the Qing dynasty until Meiji Japan in 1875 prohibited the Ryūkyūans to continue with tribute payments.

The Ryūkyūs are generally described in sources as a relatively poor country. In this context, her relations to China were not solely those of vassal to master. Throughout the Ming and Qing periods, China "transferred" commodities, wealth, knowledge, and people to the islands and gave Ryūkyūans opportunities to obtain technical, medical, philosophical, and other knowledge in China. But it is perhaps less known that due to China's maritime prohibition policy the Ryūkyūs were temporarily able to take over the role of an intermediary in East Asian waters.[86] They provided Japanese merchants, for example, with tropical goods from Southeast Asia, and the latter reexported a part of these to Korea. This intermediary role enabled the islands to enjoy considerable independence. Consequently, their overseas trading network reached from China, Korea, and Japan to Siam, Annam, Palembang, Java, Malacca, Aceh, and Sunda in Southeast Asia. The fifteenth and early sixteenth centuries have even been described as the "golden age" of the Ryūkyūs.[87]

After the invasion of Satsuma by Toyotomi Hideyoshi 豊臣秀吉 (1536–1598) and the subduing of the Ryūkyūs by the *daimyō* of Satsuma, Shimazu Iehisa 島津家久 (1576–1638), the archipelago was subsequently divided into a northern part placed directly under Satsuma administration and a more indirectly controlled southern part; although the kingdom remained *de jure* independent, it was forced to send tribute to Satsuma. Satsuma and Japan, consequently, maintained vivid exchange relations with China via the Ryūkyūs. Every trip made by Ryūkyūans to China was carefully supervised by Satsuma officials. When Ryūkyūan merchants went to Fuzhou in China, they were advised not to mention their Satsuma connection.[88] Commodities brought to Naha by Chinese merchants would mostly be shipped to Satsuma for resale on Japanese markets.[89] Both the Satsuma clan government and the merchants from Kagoshima invested in the China trade, entrusting the trade to middlemen from Naha "in order to preserve the elaborate pretence required to satisfy Chinese sensibilities."[90] Consequently, a complex network of exchange was maintained even between China and Japan via Satsuma and the Ryūkyūs, although officially no diplomatic contacts existed.

Japan

Qing China's official picture of Japan provides us with another good example of how relying on a distinction between center and periphery can blur our view against real developments. Officially, Qing China considered herself the center against which Japan was the periphery—the "tiny dwarf" Japan (*Riben cui'er* 日本蕞爾) that "is completely dependent on the commodities sold by merchants of our Heavenly Dynasty for its provisions."[91] In official Chinese eyes, Japan did not even dispose of an able ruler but simply of a "weak Japanese king who has no authority," and thus transfers all his business to the *shōguns* who then manage the affairs with Chinese merchants (*Riben guowang rounuo wu neng, quan jie gui yu guanshi zhi jiangjun zhangwo* 日本國王柔懦無能權皆歸於管事之将軍掌握).[92] Also, the fact that until 1871 no official diplomatic relations between China and Japan existed may nourish the idea of a China as the undisputed strong center with the Japanese periphery. Reality—as close as we can come to it—however, speaks a different language. Not only was Japan very active in the East Asian waters and not as weak as claimed, but it was also China that, on the other hand, was dependent upon a Japanese commodity—namely copper. Only a methodological approach that drops the traditional view of center and periphery can open the historian's view to the real extent and quality of mutual exchange relations.

With the beginning of the Tokugawa rule, Japan's relations with China received a fundamental change and starting with the first Tokugawa ruler, Tokugawa Ieyasu 徳川家康 (1542–1616; r. 1603–1616), Japan tried to strengthen itself and become more independent from China—in both political and economic terms. This led to a series of measures for stricter control of the trade with China. As we have described these mutual relations in detail in an earlier contribution, only a few most important developments shall be introduced.[93]

Chinese merchants were, for example, gradually controlled more strictly. In 1635 (*kan'ei* 12), a so-called *nengyō ji* 年行司 (a position for the administration of the Chinese settlement filled by representatives and officials of the town, *Tō nengyō ji* 唐年行司, *Tō* meaning "Chinese") was established.[94] In 1688, the Japanese government also started to restrict the total number of ships being allowed to enter the harbor of Nagasaki, and around 1698 the "Nagasaki Accounting House" (*Nagasaki kaisho* 長崎会所, or *Geldkamer* in Dutch) was established.[95] In 1715, the *Shōtoku* export restrictions (*Shōtoku shinrei* 正德新例 or *kaihaku gōshi shinrei* 海舶互市新例)[96] were promulgated, to reorganize the Nagasaki trade. Two commissioners of foreign trade were appointed to Edo and Nagasaki and thirty Chinese and two Dutch vessels permitted entry each year.[97] At the same time, the export of copper was restricted—a product that China urgently needed for its domestic economy, above all for the casting of copper cash.[98] In addition, trade permissions (*shinpai* 信牌, in Chinese sources also referred to as *Wozhao* 倭照) for foreign ships were issued,[99] which, according to the *Huangchao wenxian tongkao* 皇朝文獻通考, were sold for 8,000 to 9,000 silver *liang* each.[100] All these measures did, therefore, not mean that mutual exchange was not flourishing anymore. Only its character had changed qualitatively. It shows that we certainly cannot speak of a weak Japanese periphery that is absolutely dependent on the Chinese center.[101]

The Japanese measures conversely raised concern within the Chinese ruling elite. The Kangxi emperor was the first to send an secret agent, Morsen 莫爾森, into Japan, who, however, only confirmed the traditional picture of the "weak neighbor."[102] The following Yongzheng Emperor was more concerned and at least twice sent secret agents to Japan to inquire about the situation there. In 1728 (Yongzheng 11), the governor-general of Zhejiang, Li Wei 李衛 (1687?–1738),[103] in a memorial reported the emperor about all kinds of illegal activities on the East Asian waters. At the end of his memorial he comes to a clear conclusion.

> Investigating Japan, although it is a tiny dwarf island barbarian, it relies on its big copper casted cannons, which can attack from a faraway distance and its extraordinarily sharp swords; already in the former Ming [dynasty] [Japan] belonged to the evils of the seas (*haihuan*

> 海患), and in the Eastern Seas (*Dongyang* 東洋) it was considered a violent invader (*qiangkou* 強寇). Our dynasty, with her majestic spirit fearfully suppressed and avoided [this topic] for many years and devoted herself entirely [to the idea] that [Japan] would never intentionally offend China (*zhonghua* 中華).[104]

This was a clear warning to pay more attention to developments in Japan and the Yongzheng emperor reacted accordingly. When reading Morsen's report about Japan, he openly stated that it contained "many fabrications and empty words," like claiming that Japan was "weak and obedient". "Subsequently," he continued, "no more heed was paid to that, and the opening up of the ocean [by the Japanese] actually started from that time" (*cihou sui bu jieyi, kaiyang zhi ju, yi you ci qi* 此後隨不介意開洋之舉亦由此起).[105] Yongzheng was also concerned about the close and obviously friendly relations between Japan and Korea (*qie fu wen Riben yu Chaoxian wanglai jiao hao zongji shen mi* 且復聞日本輿朝鮮往來交好蹤跡甚 密).[106]

The idea that Japan used its copper for casting cannons, most probably has to be traced back to the previously mentioned *Shōtoku* export restrictions. Stagnating copper outputs forced the Japanese government to restrict copper exports, as a consequence of which Chinese imports from Japan fell drastically and merchants responsible for the provision of copper were continuously clocking up deficits. The Qing government subsequently not only changed its complex "copper managing system," but undertook increasing efforts to promote domestic copper mining. Only under the Qianlong emperor did the Chinese court again become more and more passive in face of Japan.[107] These developments, thus, show how complex bilateral relations actually were and that it would be completely misleading to treat them as those between a center and the periphery.

Taiwan

Taiwan consitutes another example. The immediate cause for Taiwan's emergence as an international trading rendezvous has to be seen in the opening of the port of Haicheng 海澄 near Xiamen in 1567.[108] Chinese trading junks now could sail overseas without risking in smuggling. Similar as in the case of the emergence of the Ryūkyūs, Taiwan's rise has to be seen in direct relation with Ming China's maritime prohibition policy. As China remained closed to foreigners, business transaction had to be carried out elsewhere overseas. Taiwan came to be the closest country for Chinese junks sailing from Haicheng to visit. With the ban on trade to Japan remaining in effect, they stopped at Taiwan to meet their Japanese partners. Jilong 鷄籠 and Danshui 淡水, which had long been familiar to sailors travelling between Fujian and Japan, won out as their favorite ports to service their business. Chinese authorities in Fujian may have suspected their maritime merchants were trading with Japanese in Taiwan, but they simply cast a blind eye to what happened on that foreign land and sanctioned the trade there in 1575. With Chinese junks visiting Taiwan, Japanese, and later on, also Europeans were attracted to exchange merchandise on this island. Thus, the seeming periphery developed as a center of exchange.[109]

Officially, Taiwan continued to be considered a remote periphery in Qing times. The island is described as "lonely hanging beyond the seas" (*gu xuan haiwai* 孤懸海外),[110] an expression (*haiwai*) that is also used to designate a foreign country, as a "lonely island hanging beyond the seas with the evil of cruel foreign people" (*gu xuan haiwai you xiong fan zhi huan* 孤懸海外有兇番之患) and robbers,[111] as a "far away place beyond the deep ocean" (*yuan zai zhongyang zhi wai* 遠在重洋之外).[112] The local inhabitants are repeatedly referred to as cruel, ruthless, and stubborn. In early Qing Chinese poetry the island was considered "a remote outer island that never has been a part of Chinese territory (*zi gu bu ru bantu* 自古不入版圖), but has strategic importance to shield Min and Yue 粵 (modern Fujian and Guangdong) from outside"[113]—a *frontier* in other words—despite its growing importance as an international trading center and regardless of the question how much trade and exchange actually took place between Taiwan and the mainland.

In this context, many of China's so-called peripheries—at least in the official understanding and terminology—were actually vivid centers of commerce and exchange. We should, therefore, free ourselves from this traditional concept of center and periphery, if we wish to understand East Asian exchange relations in their complexity.

Conclusion

The concept of the East Asian seas as an "East Asian 'Mediterranean'"—despite all its differences with the European Mediterranean—enables us to see East Asia as a macro-region that was connected by the sea as a medium of flourishing exchange relations, no longer treating the countries surrounding the East Asian seas as more or less isolated entities (or nation-states) with only official and otherwise hardly any interchange. Taking this concept as a starting point for research, we should look beneath the surface—freeing ourselves from traditional views of nation-states or countries as well as centers and peripheries, paying attention to all parts of the macro-region, including the seemingly peripheries. In particular, during certain time periods, the latter may have played a major role in the area. Finally, we have to pay due attention to human-environment interactions, be they man-made or natural, in order to better understand changes and developments.

Equipped with relevant source materials, these tools or methodological concepts may help us to obtain a better and more detailed picture of what was going on in a macro-region like the East Asian "Mediterranean" or the Indian Ocean World, and why and how qualitative changes in East Asia's maritime history occurred.

NOTES

This study is part of the MCRI (Major Collaborative Research Initiative) project sponsored by the Social Sciences and Humanities Research Council of Canada, carried out at the History Department of McGill University since 2010.

1. "East Asian seas," or the "East Asian Mediterranean", is used here to include the East China Sea, the Yellow Sea (*Huanghai* 黃海), the southern section of the Japanese Sea, and parts of the South China Sea (now usually *Nanhai* 南海).
2. Fernand Braudel, *The Mediterranean and the Mediterranean World in the Age of Philip II* (Berkeley and Los Angeles: University of California Press, 1996).
3. Even on a micro level it can also be used for a number of regional "scenarios." Braudel himself, when referring to the inner Japanese sea—called "Seto no uchi 瀬戸内海" (Seto Inland Sea), spoke of a flourishing Japanese "Mediterranean"—mostly an account of its important role in local trade and commerce. Fernand Braudel, *Sozialgeschichte des 15.–18. Jahrhunderts. Der Handel* (München, Germany: Kindler Verlag, 1986), 646. Several scholars have seen these similarities and adopted this expression to the Far Eastern context. Liu Shiuh-feng 劉序楓 from the Academia Sinica, however, applied the concept of the "Mediterranean" to an even greater regional "entity" that includes Southeast Asia and parts of the Indian Ocean. See his "Qingdai huan Zhongguo haiyu de hainan shijian yanjiu—yi QingRi liangguo jian dui waiguo nanmin de jiuzhu ji qianfan zhidu wei zhongxin 清代環中國海域的海難事件研究—以清日兩國間對外國難民的救助及遣返制度為中心 (1644–1861)," in Zhu Delan 朱德蘭, ed., *Zhongguo haiyang fazhanshi lunwenji* 中國海洋發展史論文集. (Taibei: Zhongyang yanjiuyuan, 2002), vol. 8,

173–238. For some additional references to works which discuss Braudelian ideas within the context of East and Southeast Asia, see the contribution by Christine Moll-Murata in Angela Schottenhammer and Roderich Ptak, *The Perception of Maritime Space in Traditional Chinese Sources* (Wiesbaden: Otto Harrassowitz, 2006), 109–23.

4. *Annales. Histoire, Sciences sociales* (Paris: L'Éditions de l'École des Hautes Études en Sciences Sociales), edited by Braudel from 1956–1968.
5. K. N. Chaudhuri, *Trade and Civilisation in the Indian Ocean. An Economic History from the Rise of Islam to 1750* (Cambridge: Cambridge University Press, 1985); Chaudhuri, *Asia Before Europe. Economy and Civilisation of the Indian Ocean from the Rise of Islam to 1750* (Cambridge: Cambridge University Press, 1992).
6. Anthony Reid, *Southeast Asia in the Age of Commerce, 1450–1680. Vol. I: The Lands below the Winds* (New Haven, CT: Yale University Press, 1998); Reid, *Southeast Asia in the Age of Commerce, 1450 1680. Vol. II: Expansion and Crisis* (New Haven, CT: Yale University Press, 1993).
7. Janet Abu-Lughod, "The World System in the Thirteenth Century: Dead-End or Precursor," in Michael Adas, ed., *Islamic & European Expansion. The Forging of a Global Order.* (Philadelphia: Temple University Press, 1993); Abu-Lughod, *Before European Hegemony. The World System A.D. 1250–1350* (New York: Oxford University Press, 1989).
8. Several recent articles and studies discuss problems of Eurocentrism or new approaches to history, such as "entangled history," "shared history," or "histoire croisée." See especially Sanjay Subrahmanyam, "Connected Histories: Notes towards a Reconfiguration of Early Modern Eurasia", *Modern Asian Studies* 31 (1997): 735–62; Sebastian Conrad, Shalini Randeria (Hrsg.), *Jenseits des Eurozentrismus. Postkoloniale Perspektiven in den Geschichts- und Kulturwissenschaften.* (Frankfurt am Main: Campus Verlag, 2002).
9. One will of course have to take into account that for some regions we possess no or only few written sources. In this context archaeological evidence has become more and more important.
10. See for example Angela Schottenhammer, ed., *The East Asian Mediterranean—Maritime Crossroads of Culture, Commerce, and Human Migration* (Wiesbaden: Otto Harrassowitz, 2008); Schottenhammer, ed., *The East Asian Maritime World 1400–1800: Its Fabrics of Power and Dynamics of Exchanges* (Wiesbaden: Otto Harrassowitz, 2007); Angela Schottenhammer and Roderich Ptak, *The Perception of Maritime Space in Traditional Chinese Sources* (Wiesbaden: Otto Harrassowitz, 2006); Schottenhammer, ed., *Trade and Transfer across the East Asian Mediterranean* (Wiesbaden: Otto Harrassowitz, 2005).
11. The British also succeeded to weaken the country politically but not to destroy its political power.
12. For what this meant, for example, for Japan: Tashiro Kazui, trans. Susan Downing Videen, "Foreign Relations during the Edo Period: Sakoku Reexamined," *Journal of Japanese Studies* 8 (1982): 282–306.
13. Recent scholarship on maritime history, especially in China and Japan, is simply too comprehensive to be listed here in detail. Among English-language publications reference should be made especially to the works of

(in alphabetical order) William Atwell, Leonard Blussé, Chang Pin-tsun, Ng Chin-keong, Roderich Ptak, Geoffrey Wade, Wang Gungwu, John E. Wills, but there are many other names that could be mentioned here. If we subsequently put emphasis on studies resulting from a research project on "The East Asian 'Mediterranean'" carried out at the East Asia Department of Munich University between 2002 and 2009 under my supervision, it is not only to draw the reader's attention to most recent research directly linked to the topic, as it was our aim to continue and intensify the idea of East Asia and its seas as a macro-region of intense interchange; we would also like to introduce studies of scholars that are still little known and have contributed aspects that have so far been little studied. By no means does this intend to disqualify prior research as less important. Conversely, without all the important investigations by other scholars who have been working in the field for decades, we would have been unable to carry on our investigations to the point we did.

14. Mark Elvin, *The Retreat of the Elephants: An Environmental History of China* (New Haven, CT: Yale University Press, 2004). I would like to raise the reader's attention to a new project entitled "The Indian Ocean World: The Making of the First Global Economy in the Context of Human-Environment Interaction" that is being carried out at the Indian Ocean World Centre, History Department, McGill University, Montreal, under the supervision of Prof. Dr. Gwyn Campbell since 2010. In this project particular emphasis is paid to human-environment interactions and collect relevant data.
15. In this context, I would like to emphasize the new series published by Harvard University Press, entitled "The History of Imperial China." See, for example, Timothy Brook, *The Troubled Empire. China in the Yuan and Ming Dynasties* (Cambridge, MA: Harvard University Press, 2010), or Dieter Kuhn, *The Age of Confucian Rule. The Song Transformation of China* (Cambridge, MA: Harvard University Press 2010), who repeatedly turn the reader's attention to environmental questions.
16. There is also an investigation of Liuhe's port history by Makoto Ueda 上田信, "Ryūkakō monogatari Shindai Kōnan no ichi kōekikō ni ikita hitobito 劉河港物語 - 清代江南の一交易港に生きた人々," *Chūgoku kindaishi kenkyū* 中國近代史研究 6 (1988): 1–19. I am extremely grateful to Shiba Yoshinobu 斯波義信, who recently turned my attention to this article.
17. *Liuhe zhenji lue* 劉河鎮記略 by Jin Ruibiao 金瑞表, *j.* 3, 1a–b (337), 14 *juan,* in *Zhongguo difangzhi jicheng* 中國地方志集成, vol. 9, 286–482; also in *Shanghai tushuguan cang xijian fangzhi congkan* 上海圖書館藏稀見方志叢刊 (Beijing: Guojia tushuguan chubanshe, 2011), vol. 71.
18. Joseph Needham, *Science and Civilisation in China,* vol. 4: *Physics and Physical Technology,* part III: *Civil Engineering and Nautics.* (Cambridge: Cambridge University Press, 1971), 549, notes that the term even had a usage in Taoist mystical anatomy as it referred to the rectum.
19. *Liuhe zhenji lue, j.* 3, 2a (337).
20. *Liuhe zhenji lue, j.* 3, 16a–17a (344–45).
21. Linda Cooke Johnson, "Shanghai: An Emerging Jiangnan Port, 1683–1840," in Linda Cooke Johnson, ed., *Cities of Jiangnan in Late Imperial China* (New

York: State University of New York Press, 1993), 151–81, 154. She provides an excellent survey on the historical development of the emergence of Shanghai as the major port in Southeast China.

22. See Linda Cooke Johnson, "Shanghai: An Emerging Jiangnan Port, 1683–1840," 153–54.
23. Linda Cooke Johnson, "Shanghai: An Emerging Jiangnan Port, 1683–1840," 157.
24. *Liuhe zhenji lue, j.* 3, 16b (344).
25. *Liuhe zhenji lue, j.* 3, 16b (344).
26. *Liuhe zhenji lue, j.* 3, 16b (344).
27. See also the description of the historical development in Linda Cooke Johnson, "Shanghai: An Emerging Jiangnan Port, 1683–1840," 153–60.
28. Linda Cooke Johnson, "Shanghai: An Emerging Jiangnan Port, 1683–1840," 160.
29. Zhang Haipeng 張海鵬, Zhang Haiying 張海瀛 (Hrsg.), *Zhongguo shida shangbang* 中国十大商帮 (Wuhu: Huangshan chubanshe, 1993), 185.
30. Makoto Ueda, "Ryūkakō monogatari Shindai Kōnan no ichi kōekikō ni ikita hitobito," 15. See also "Yongzheng shi'er nian bayue chu bari Shandong xunfu chen Yue Rui (?–1753) qinzou wei beichou jicha haikou douchuan shiyi zouqing 雍正十二年八月初八日山東巡撫臣岳濬謹奏為備籌稽察海口豆船事宜奏請," in *Shizong Xianhuangdi zhupi yuzhi* 世宗憲皇帝硃批諭旨 by Yongzheng 雍正 (1677–1735) and Qianlong 乾隆 (1711–1799), j, 201, xia, in *Siku quanshu,* fasc. 423.
31. Makoto Ueda, "Ryūkakō monogatari Shindai Kōnan no ichi kōekikō ni ikita hitobito," 15.
32. Linda Cooke Johnson, "Shanghai: An Emerging Jiangnan Port, 1683–1840," 160.
33. Makoto Ueda, "Ryūkakō monogatari Shindai Kōnan no ichi kōekikō ni ikita hitobito," 1–2.
34. See Linda Cooke Johnson, "Shanghai: An Emerging Jiangnan Port, 1683–1840," 160.
35. Makoto Ueda, "Ryūkakō monogatari Shindai Kōnan no ichi kōekikō ni ikita hitobito," 16.
36. *Songjiang fuzhi, j.* 28, 46a–b (639).
37. *Qinding Da Qing huidian shili* (*jiaqing*) 欽定大清會典事例 by Li Hongzhang 李鸿章 et al. (rev.). (Lithography of the *guangxu* period; Shanghai: Shanghai yinshuguan, copy of the *guangxu*-ed.), *j.* 189, 6b–7a, 7b.
38. Linda Cooke Johnson, "Shanghai: An Emerging Jiangnan Port, 1683–1840," 160–61.
39. *Da Qing jifu xianzhe zhuan* 大清畿輔先哲傳, *j.* 31, 21a (*QDZJCK* 200 - 559–60). It is reported that he deceased as a poor man at the age of eighty-five.
40. He was a member of the Manchu Bordered Yellow Banner and came from the Janggiya 章佳 clan, which had settled in Liaodong long before the rise of Nuharci. In 1727, Injišan was first appointed sub-expositor and subsequently acted as director on a department of the Ministry of Revenue. Later, still in 1727, he was sent to Canton to conduct the trial of two corrupt officials and was then appointed as provincial judge of Guangdong. In 1728,

employed as acting governor of Jiangsu, he was raised to a position of a full provincial governor in 1729. *Qing shigao* 清史稿 by Zhao Erxun 趙爾巽 (1844–1927). (Beijing: Zhonghua shuju, 1976), j. 313, 1a. *Man mingchen zhuan* 滿名臣傳 edited by the Guoshi guan 國史館, Republic of China. (Taibei: Tailian guofeng chubanshe, 1970). 6 vols., *j.* 46, 33b; *Qing shi liezhuan,* 清史列傳 edited by the Guoshi guan 國史館, Republic of China. 王鍾翰. (Shanghai: Zhonghua shuju, 1928; reprint Beijing: Zhonghua shuju, 1987; Taibei: Taiwan Zhonghua shuju, 1964), *j.* 18, 26b–34b; *Guochao mingchen yanxing lu,* 國朝名臣言行錄 by Wang Bngxie 王炳燮. (Taibei: Wenhai chubanshe, 1966), *j.* 14, 12a.

41. Huang Guosheng 黄國盛, *Yapian zhanzheng zhiqian Dongnan sisheng haiguan* 鴨片戰爭之前東南四省 海關. (Fuzhou: Fujian renmin chubanshe, 2000), 48, quotes the *Shanghai yanjiu ziliao xuji* 上海研究資料續集. (Shanghai: Shanghai shudian, 1984), 62–63.
42. Linda Cooke Johnson, "Shanghai: An Emerging Jiangnan Port, 1683–1840," 161.
43. Makoto Ueda, "Ryūkakō monogatari Shindai Kōnan no ichi kōekikō ni ikita hitobito," 15.
44. See also Linda Cooke Johnson, "Shanghai: An Emerging Jiangnan Port, 1683–1840," 160–61.
45. *Ming Taizu shilu* 明太祖實錄, *j.* 70, 3b. (Taibei: Zhongyang yanjiuyuan lishi yuyen yanjiusuo, 1967), 133 vols., vol. 3, 1300: "*Reng jin binhai min bu de si chu hai* 仍禁濱海民不得私出海"; see also Fan Zhongyi 範中義, Tong Xigang 仝晰綱, *Mingdai wokou shilüe* 明代倭寇史略. (Beijing: Zhonghua shuju, 2004), 71–72.
46. See for example, Ng Chin-keong, *Trade and Society. The Amoy Network on the China Coast 1683–1735* (Singapore: Singapore University Press, 1983).
47. "Fu Hu Meilin lunchu Wang Zhi 復胡梅林論處王直," in *Yu Wo zazhu* 禦倭雜著 by Tang Shu 唐樞, *j.* 3, 5a, in *Ming jingshi wenbian* 明經世文編 by Chen Zilong 陳子龍 et al. (1608–1647) (comp.). (Beijing: Zhonghua shuju, 1997), fasc. 270.
48. See the excellent analysis by Li Kangying, *The Ming Maritime Trade Policy in Transition, 1368–1567* (Wiesbaden: Otto Harrassowitz, 2010).
49. Chang Pin-tsun, "The Emergence of Taiwan as an International Trading Rendezvous in the Sixteenth Century," in Angela Schottenhammer, ed., *Taiwan—A Bridge between East and South China Seas* (Wiesbaden: Otto Harrassowitz, 2011), 9–24, with reference to Kōbata Atsushi 小葉田淳, *Chūsei nantō tsūkō bōekishi no kenkyū* 中世南島通交貿易史の研究 (Tōkyō: Tōkō shoin, 1968), 350–60. Li Jinming 李金明, *Mingdai haiwai maoyishi* 明代海外貿易史 (Beijing: Zhongguo shehui kexue chubanshe, 1990), 80–108.
50. "Fu Hu Meilin lunchu Wang Zhi", in *Yu Wo zazhu, j.* 3, 3a–4a.
51. *Guangzhou fuzhi* 廣州府志 by Shi Cheng 史澄 (1814–?), *j.* 162, 3b (杂識); Peng Zeyi 彭澤益, "Qingdai Guangdong yanghang zhidu de qiyuan 清代廣東洋行制度的起源", *Lishi yanjiu* 1 (1957), 1–24, 5.
52. Peng Zeyi, "Qingdai Guangdong yanghang zhidu de qiyuan," 8.
53. Shang Zhixin joined the rebellion of Wu Sangui 吳三桂 (1612–1678) in 1676. He later regretted having joined the rebellion and started negotiations with the government forces in Jiangxi His allegiance was but an act, for

he subsequently refused to engage in any further operations against rebels, ignoring all orders sent to him by the government. In 1680, he was permitted to commit suicide. Arthur W. Hummel, *Eminent Chinese of the Ch'ing Period*, 2 vols. (Taibei: SMC Publishing Inc., 1991), vol. 2, 635.

54. *Da Qing lichao Shengzu Renhuang shilu* 大清歷朝聖祖仁皇實錄 (Kangxi). (Taibei: Huawen shuju, 1964), *j.* 116, 4a (1548).
55. Li Shizhen 李士楨 (1619–1695), *Fu Yue zhenglue* 撫粵政略, *j.* 7, 16a–b, in *Jindai Zhongguo shike congkan sanbian, di 39 ji* 近代中國史科叢刊三编第39輯, fasc. 382–84. (Taibei: Wenhai chubanshe, 1988).
56. Between 1661 and 1662, the complete relocation from the coastal region was ordered. This measure included the suspension of all trade and industries along the coast.
57. John E. Wills Jr., *Embassies and Illusions. Dutch and Portuguese Envoys to K'ang-hsi, 1666–1687.* (Cambridge, MA: Harvard University Press, 1984), 101.
58. As for Shang Zhixin's relations with the Dutch, see for example John E. Wills Jr., *Pepper, Guns and Parleys: The Dutch East India Company and China 1622–1681* (Cambridge: Cambridge University Press, 1974), 153, 154, 157–59, 194–98. In company trade and diplomacy between 1676 and 1679, "Shang's chief client-merchants [were] . . . Tsonqua or Tsjongqua and Lichoncong [Lin Qifeng 林奇逢] in the Dutch sources" p. 158.
59. In a recent article, Sue Takashi 須江隆 shows how important an involvement in shrines and temples—in other words, in aspects of socio-religious life, was also for mundane affairs. In this context, the article emphasizes what I would call the indirect importance of religion for social life and commercial activities as a basic precondition for success in such non-religious affairs. Sue also shows that the temples and shrines, originally places of worship, began to acquire economic value themselves and gradually more or less physically turned into centers of commerce. See Sue Takashi 須江隆, "The Structure of Regional Society as Seen from Local Historical Documents: Focusing on the Southeastern Shoreline Areas, Song to Ming," in Angela Schottenhammer, ed., *Trading Networks in Early Modern East Asia* (Wiesbaden: Otto Harrassowitz, 2005), 27–50. Developments in Song regional societies are also discussed in Ihara Hiroshi 伊原弘, "Numerical Indices that Can Reveal the Life of Song Commoners," in Angela Schottenhammer, ed., *Trading Networks in Early Modern East Asia*, 5–26.
60. For recent relevant studies concerning Sino-Japanese relations during the Ming period, see Oláh Csaba, *Räuberische Chinesen und tückische Japaner. Die diplomatischen Beziehungen zwischen China und Japan im 15. und 16. Jahrhundert* (PhD diss., Wiesbaden: Otto Harrassowitz 2008). *East Asian Maritime History,* 7; Oláh Csaba, "Troubles during trade activities between Japanese and Chinese in the Ming period," in Angela Schottenhammer, ed., *The East Asian Mediterranean—Maritime Crossroads of Culture, Commerce, and Human Migration,* 317–30.
61. Wang Su, "Sino-Japanische Beziehungen im Bereich der Medizin: Der Fall des Xu Zhilin 徐之遴 (c. 1599–1678)," in Angela Schottenhammer, ed., *Trade and Transfer across the East Asian "Mediterranean"* (Wiesbaden: Otto Harrassowitz, 2005), 185–234.
62. See Angela Schottenhammer, "Japan—the tiny Dwarf? Sino-Japanese Rela-

tions from the *Kangxi* to the Early *Qianlong* Reigns," in Angela Schottenhammer, ed., *The East Asian Mediterranean—Maritime Crossroads of Culture, Commerce, and Human Migration*, 331–88.

63. In a recent study I have just been compelled by the fact that much more information for example concerning Qing period Sino-Japanese copper trade and its organization comes actually from Japanese and not Chinese sources; see my "Brokers and 'Guild' Organizations (*huiguan*) in China's Maritime Trade with Japan in High Qing," in Billy So Kee Long and Harriet Zurndorfer, eds., *Ming Qing shiqi Jiangnan shichang jingji de kongjian, zhidu yu wangluo guoji taolunhui* 明清時期江南市場經濟的空間、制度與網絡國際研討會 *International Symposium on the Market Economy of the Lower Yangzi Delta in Late Imperial China: Space, Institutions and Networks.* (London: Routledge, 2011), chapter 11, 278–99.

64. Engelbert Kaempfer, *Geschichte und Beschreibung von Japan.* Aus den Originalhandschriften des Verfassers herausgegeben von Christian Wilhelm Dohm. Unveränderter Neudruck des 1777–1779 im Verlag der Meyerschen Buchhandlung in Lemgo erschienen Originalwerks. Mit einer Einführung von Hanno Beck. 2 Bde. (Stuttgart, Germany: Brockhaus Komm.-Gesch., 1964), vol. 2, 127. He reports on cases in 1690 and 1691.

65. See for example *Tsūkō ichiran* 通航一覽 (1853) Hayashi Fukusai 林復斎 (1800–1859), ed. et al. (Ōsaka: Seibundō shuppan, 1967), vol. 5, *j.* 198, 228.

66. For numerous examples see *Tsūkō ichiran,* vol. 4.

67. Paul van der Velde and Rudolf Bachofner, *The Deshima Diaries. Marginalia 1700–1740* (Tōkyō: The Japan-Netherlands Institute, 1992), 69. *Deshima Series,* ed. by J. L. Blussé and W. G. J. Remmelink, *Japan-Netherlands Institute Scientific Publications of the Japan-Netherlands Institute* no. 12. Opperhoofd Harmanus Menssingh observed in December 1705: "While the Chinese were shipping their copper, they were blatantly smuggling. The smuggling is condoned by the governor of Nagasaki, otherwise it could not take place."

68. *Chouhai tubian* 籌海圖編 by Hu Zongxian 胡宗憲 and Zheng Ruoceng 鄭若曾, annotated by Li Zhizhong 李致忠. (Beijing: Zhonghua shuju, 2007), *j.* 12, 852. The "*Fujian shibo tiju si zhi* 福建市舶提舉司志" of Gao Qi 高岐, who served as a superintendent of maritime trade in Fuzhou in 1554 (*jiajing* 33), notes that "as for brokers (*yahang* 牙行), originally twenty-four persons (*ming* 名) were engaged differently from year to year. (Subsequently) nineteen were dismissed, so that today only five have remained." *Fujian shibo tiju si zhi* 福建市舶提舉司志 by Gao Qi 高岐 (Ming). (Copy of the Harvard Yenching Library and of the library of the Shifan daxue, Fuzhou), 14a–15a.

69. See Wu Aihua 吳靄華, "Ming Qing shidai Liuren xingming suo shou Hanren xingming de yinxiang 明清時代琉人姓名所受漢人姓名的印象," in *Di'erhui ZhongLiu lishi guanxi guoji xueshu huiyi ZhongLiu lishi guanxi taolunhui* 第二回中琉歷史關係討論會 (Naha: Kabushiki kaisha, 1989), 369–402, 371; also Xie Bizhen 謝必震, "Guanyu Ming ci Liuqiu minren sanshiliu xing de ruogan wenti 關於明賜琉球閩人三十六姓的若干問題," in *Disanjie ZhongLiu lishi guanxi guoji xueshu huiyi lunwenji* 第三屆中琉歷史關係國際學術會議論文集. (Taibei: Sanmin shuju, 1991), 997–1013.

70. An excellent analysis of the role and function of interpreters in Sino-

Ryūkyūan relations is Kikō Nishizato 西里喜行, "ChūRyū kōshashi ni okeru to tsūshi to yakō (Kyūshō) 中琉交涉史における土通事と牙行(球商)," Ryūkyū daigaku kyōiku gakubu kiyō 琉球大學教育学部紀要 1/2:50 (1997), 53–92; for the role of interpreters in East Asian maritime trade in general see also Liao Dake 廖大珂, "Qingdai haiwai maoyi de tong-shi chutan 清代海外貿易的通事初探," in Angela Schottenhammer, ed., *The East Asian Maritime World 1400–1800*, 249–81.

71. See my argumentation in Angela Schottenhammer, "Brokers and 'Guild' (*huiguan* 會館) Organizations in China's Maritime Trade with her Eastern Neighbours during the Ming and Qing Dynasties," *Crossroads* 1/2 (2010): 99–150.
72. *Ryūkyū-kan monjo* 琉球館文書, vol. 1, 5–9; see Sakihara Mitsugu 崎原貢, "To-Tōgin to Sa-Ryū-Chū bōeki 渡唐銀と薩琉中貿易," *Nihon rekishi* 日本歷史4, no. 323 (1975): 27–47, 37.
73. The mission under the supervision of Cai Jin 蔡錦 went to China twice, on 11/9/1634 (*chongzhen* 7) and on 7/3/1642 (*chongzhen* 15).
74. Kikō Nishizato, "ChūRyū kōshashi ni okeru to tsūshi to yakō (Kyūshō)," 68.
75. The *Chūsan seifu* refers to a total of 120,000 *kan* that were used to purchase silks and other products; from these products one-third (三分の一) was confiscated; see *Chūsan seifu* 中山世譜, *fu* 附, *j.* 1, *Ryūkyū shiryō sōsho* 琉球史料叢書, according to Kikō Nishizato, "ChūRyū kōshashi ni okeru to tsūshi to yakō (Kyūshō)," 68–69, 90n53.
76. Kikō Nishizato, "ChūRyū kōshashi ni okeru to tsūshi to yakō (Kyūshō)," 69.
77. For more details see also Wang Qing, "Trade Relations between China and the Ryūkyūs During the Reign of Emperor Kangxi (1662–1722)," in Angela Schottenhammer, *Trading Networks in Early Modern East Asia*, 156–94.
78. Kenneth R. Robinson, "From Raiders to Traders: Border Security and Border Control in Early Choson, 1392–1450," *Korean Studies* 16 (1992): 94–115.
79. *Chosŏn wangjo shillok* 朝鮮王朝實錄 (Seoul: Kuksa p'yonch'an wiwŏnhoe, 1950–1954). 48 vols., here *Renzu shillok, j.* 37, 27 (Renzu 16th year, 7th month). This quotations has also been translated in Angela Schottenhammer, *The East Asian maritime world 1400–1800: Its fabrics of power and dynamics of exchanges—China and her Neighbours*, 61–62.
80. The term *Zhongguo* 中國 can be found in texts as early as Zhou times, but in the sense of *guozhong* 國中 it always referred to the political center of Chinese rule—the royal domains in contrast to the area where the feudal lords had their lands. It was actually only in the nineteenth century that the expression emerged as the name for the country. It first appeared in a formal document for the chief Manchu negotiator at the Treaty of Nerčinsk with Russia in 1689. See Endymion Wilkinson, *Chinese History. A Manual.* 中國歷史手冊 (Cambridge, MA: Harvard University Press, 2000), 132.
81. I would like to draw the reader's attention to an interesting article discussing the history of this suzerain-vassal relationship; that is, Zhang Shimin, "A Historical and Jurisprudential analysis of Suzerain-Vassal State Relationship in the Qing Dynasty," *Frontiers of History in China* 1, no. 1 (2006): 124–57. The article also discusses the different use of characters, for example the use of *fan*, written as 藩, 蕃 or 番. "The difference between the two ways [藩

or 番, A.S.] of writing 'fan' embodies the difference between respectful and disrespectful language," 136. The former was preferably used during Qing, the latter during Ming times.

82. See the argumentation in Angela Schottenhammer, "Empire and Periphery: The Qing Empires Relations with Japan and the Ryūkyūs (1644- c. 1800), a comparison," originally presented at the conference "Empires and Networks: Maritime Asian Experiences, 9th to 19th Centuries" (International conference convened by the Nalanda-Sriwijaya Centre, ISEAS, Singapore and Ritsumeikan University Japan, Singapore, 21–22–02.2011) forthcoming in *The Medieval History Journal* 15, no. 2 (October 2012).
83. See the analysis provided by Li Kangying, *The Ming Maritime Trade Policy in Transition, 1368–1567*; also Lin Renchuan, "Fukien's Private Sea Trade in the Sixteenth and Seventeenth Centuries," in Eduard B. Vermeer, ed., *Development and Decline of Fukien Province in the 17th and 18th Centuries* (Leiden: E. J. Brill, 1990), 167–215.
84. Xie Bizhen 謝必震, *MingQing ZhongLiu hanghai maoyi yanjiu* 明清中琉航海貿易研究 (Beijing: Haiyang chubanshe, 2004).
85. Officially, Ryūkyū was ranked next after Korea and before Burma (Xianluo 暹羅) and Annam. See for example Zheng Liangsheng 鄭樑生, "Liuqiu zai Qingdai cefeng tizhi zhong de dingwei shitan—yi Shunzhi Kangxi Yongzhen sanchao wei li 琉球在清代冊封體制中的定位試探—以順治康熙雍正三朝為例," in Disihui Liu Zhong lishi guanxi guoji xueshu huiyi 第四回琉中歷史關係國際學術會議, ed., Liu Zhong lishi guanxi lunwenji 琉中歷史關係論文集 (Naha: Nansei yinsatsu, 1993), 219–41, 229.
86. For a recent study on the role of Ryūkyūan merchants during this time, see Roderich Ptak, "The Ryukyu Network in the Fifteenth and Early Sixteenth Centuries," *Revista de Cultura/Review of Culture* 6 (2003): 7–23.
87. Ta-tuan Ch'en, "Investiture of Liu-ch'iu kings in the Ch'ing Period," 136.
88. Robert K. Sakai, "The Satsuma-Ryukyu Trade and the Tokugawa Seclusion Policy," *Journal of Asian Studies* 23, no. 3 (1964): 391–403.
89. Ta-tuan Ch'en, "Investiture of Liu-ch'iu kings in the Ch'ing Period," in John K. Fairbank, ed., *The Chinese World Order. Traditional China's Foreign Relations.* (Cambridge, MA: Harvard University Press, 1968), 135–64, 152: "The circumstances of these transactions are not known, nor is it possible to tell whether Liu-ch'iu was able to profit from the resale. . . ."
90. George H. Kerr, *Okinawa, the History of an Island People.* (Rutland, VT : C. E. Tuttle Co., 1958; paperback edition 1964), 182.
91. *Shizong Xianhuangdi zhupi yuzhi* 世宗憲皇帝硃批諭旨 by Yongzheng 雍正 (1677–1735) and Qianlong 乾隆 (1711–1799), *j.* 177, 30b., in *Siku quanshu,* fasc. 418–23.
92. *Shizong Xianhuangdi zhupi yuzhi, j.* 174, section 8, 43b.
93. Angela Schottenhammer, "Japan—the tiny Dwarf?"
94. *Nagasaki-ken shi* 長崎県史, *Taigai kōshō hen* 対外交渉編. (Tōkyō: Yoshikawa kōbunkan, 1986), 393. The *Tō nengyō ji* was a kind of self-administration of the Chinese settlement in Nagasaki managed by rich Chinese merchants who were appointed by the Bakufu to supervise the local Chinese community and take care that everybody respected Japanese law and was first estab-

lished in 1635. This authority, thus, became a second element in the local administration after the most important one—the interpreters' office (*Tō tsūji* 唐通事; the Dutch of course had their own interpreters). Even though they were interpreters in name, their actual functions were much wider. In all, there were about 150 Dutch and 200 Chinese interpreters. For details see for example, Li Xianzhang 李献璋, *Nagasaki Tōjin no kenkyū* 長崎唐人の研究 (Sasebo: Shinwa ginkō, 1991), 179–259, esp. 277.

95. *Nagasaki-ken shi,* 553. The office was responsible for the purchase of Chinese and Dutch imports and was managed by local officials who were appointed by the *Nagasaki bugyō.*

96. Named according to the *shōtoku* era (1711–1716). See *Tsūkō yichiran,* vol. 5, 311, for regulations concerning the Chinese.

97. George Samson, *A History of Japan, 1615–1867* (Stanford, CA: Stanford University Press, 1963), 145. The Confucian scholar and advisor of the Tokugawa *Shōgun* Ienobu 徳川家宣 (1662–1712, r. 1709–1712), Arai Hakuseki 新井白石 (1657–1725), played an important role in drafting these new regulations. He was also concerned about Qing China's expansionist policy and suspected that the Kangxi emperor aimed at weakening Japan. See Miyazaki Michio 宮崎道生, *Arai Haruseki no kenkyū* 新井白石の研究 (Tōkyō: Yoshikawa kōbunkan, 1969), 188–92, 380–81.

98. The increasing demand for copper in China has to be traced back to growing local markets going along with rising population in general that required an equivalent of value in a much smaller denomination than silver, a means of circulation for the everyday purchase on the markets.

99. *Tsūkō ichiran,* vol. 4, *j.* 164, 375–76. Mizuno Norihito, "China in Tokugawa Foreign Relations," 142–43, suggests that the use of the term "*Tō* 唐" on these "trade credentials" implies that the Japanese Bakufu regarded Chinese merchants as unrelated to the Chinese state and government. The trade credentials finally "took the form of being issued by Chinese lanuage interpreters (*Tō tsūji* 唐通詞)" and actually were "a contract made between the interpreters, not the Nagasaki Magistrate's Office, and the Chinese merchants, and was like a private pledge" (143, 144). See also *Tsūkō ichiran,* vol. 4., 430–31. There is also a Japanese study of Emperor Kangxi's response to the new regulations. See Matsuura Akira 松浦章, "Kōkitei to Shōtoku shinrei 康熙帝と正徳新例," in Yanai Kenji箭内健次, *Sakoku Nihon to kokusai kōryū* 鎖国日本と国際交流 (Tōkyō: Yoshikawa kōbunkan, 1988), vol. 2, 29–53.

100. *Huangchao wenxian tongkao* 皇朝文獻通考 by Qianlong 乾隆 (1711–1799), Zhang Tingyu 張廷玉 (1672–1755) et al. (comp.), *j.* 17, 30b: 昔年各商初往東洋時倭人設立倭照每張約費銀八九千兩不等.

101. Ōba Osamu 大庭脩, such as his *Edo jidai ni okeru Chūgoku bunka juyō no kenkyū* 江户时代における中国文化受容の研 (Kyōto: Dōbōsha, 1984); his *Edo jidai no Nitchū Chū hiwa kenkyū* 江戸時代の日中秘話研究 (Tōkyō: 1984); or his *Tokugawa Yoshimune to Kokitei: Sakokuka de no Nitchū kōryū* 徳川吉宗と康熙帝: 鎖国下での日中交流; Joshua Fogel ed, *Sagacious Monks and Bloodthirsty Warriors: Chinese Views of Japan in the Ming-Qing Period* (Norwalk: EastBridge, 2002), here especially Guo Yinjing's chapter on "Views of Japan and Policies Towards Japan in the Early Qing," 88–108; Hamashita Takeshi's

濱下武志, *Kindai Chūgoku no kokusaiteki keiki: choko boeki shisutemu to kindai Ajia* 近代中国の国际的契机：朝贸易システムと近代アジア (Tōkyō: Tōkyō daigakku shuppansha, 1990); Matsuura Akira's 松浦章, *Shindai kaigai bōkishi no kenkyū* 清代海外貿易史の研究 (Kyōto: Tomodachi shoten, 2002); or Matsuura Akira, *Edo jidai Tōsen ni yoru Nitchū bunka kōryū* 江戸時代唐船による日中文化交流 (Kyōto: Shibunkaku shuppan, 2007).

102. Disguised as a merchant, Morsen sailed from Shanghai in the summer of 1701 and returned to Ningbo in the autumn of the same year. See "Suzhou zhizao Li Xu zouzhe 蘇州織造李煦奏摺" (*kangxi* 40, 6th month) by Li Xu 李煦 (born c. 1650), in Gugong bowuyuan wenxianguan weiyuanhui 故宮博物院文獻委員會, ed., *Wenxian congbian* 文獻叢編 (Taibei: Tailian guofeng chubanshe, 1964), 2 vols., vol 2 (xia), 856–57.
103. He has a brief entry in Arthur W. Hummel, *Eminent Chinese of the Ch'ing Period.* 2 vols. (Taiwan: SMC Publishing, 1991), vol. 2, 720–721.
104. *Shizong Xianhuangdi zhupi yuzhi, j.* 174, section 8, 25 a–b (201).
105. *Shizong Xianhuangdi zhupi yuzhi* 世宗憲皇帝硃批諭旨 by Yongzheng 雍正 (1677–1735) and Qianlong 乾隆 (1711–1799), *j.* 174, section 8, 28a–b (203), in *Siku quanshu,* fasc. 423.
106. Ibid.
107. This is discussed in "Empire and Periphery: The Qing Empires Relations with Japan and the Ryūkyūs (1644- c. 1800), a comparison." For general details on the copper management see Helen Dunstan, "Safely Supping with the Devil: The Qing State and its Merchant Suppliers of Copper," *Late Imperial China* 13, no. 2 (1992): 42–81.
108. Chang Pin-tsun, "The Emergence of Taiwan as an International Trading Rendezvous in the Sixteenth Century."
109. Ibid.
110. *Shizong Xianhuangdi zhupi yuzhi, j.* 176, section 5, 27b.
111. *Shizong Xianhuangdi zhupi yuzhi, j.* 13, section xia, 61a.
112. *Shizong Xianhuangdi zhupi yuzhi, j.* 13, section xia, 22b; also *j.* 72, 1b (*yuan ge haiyang difang* 遠隔海洋地方).
113. See the contribution by Christian Soffel, "Taiwan in Early Qing Poetry," in Angela Schottenhammer, ed., *Taiwan—A Bridge between East and South China Seas* (Wiesbaden: Otto Harrassowitz, 2011), 117–32, 129–30.

FIVE

Metaphorical Perspectives of the Sea and the Sulu Zone, 1768–1898

James Francis Warren

My initial encounter with the Sulu Zone, located between the Asian mainland and the large islands of Mindanao, Borneo, and Sulawesi, began forty-five years ago. I first learned of the Sulu Archipelago and the maritime world of the Samal Bajau Laut when I received my Peace Corps posting in 1967, assigning me to Semporna, Sabah, on the east coast of Borneo, in the area where present-day Malaysia, Indonesia, and the Philippines meet. The search for historical antecedents to contemporary social change among the Samal Bajau Laut, whose maritime nomadic culture was being inexorably extinguished by political and economic development, initially involved me in a historical study of the interaction between a relatively weak quasi-colonial agency, the North Borneo Chartered Company, and the maritime nomadic people of the northeast coast of Borneo between 1878 and 1909. Much of the research was performed during a vacation period while I served in Sabah as a Peace Corps Volunteer.[1]

On Seeing the Zone: Framing, Commodity-Chain Analysis and the Metaphorical Perspective

Framing

The maritime populations that were dependent on the sea for their economic pursuits—trading, raiding, collecting, and fishing—had remained largely caricatures in the historiography of the region. I gradually realized, that the Samal Bajau Laut's marginalized view of Taosug, or Sulu history and interpretation of past events, was a historical perspective from the very edge of a wide-ranging trading sphere that centered on the Sulu Sultanate. To be released from the conceptual constraints of conventional historical geography, I called this web of economic influence and interpersonal relations a "zone."

Following in the footsteps of Fernand Braudel, I abandoned the insular geographic perspective of earlier historians of the Philippines, Indonesia, and Borneo, for a more dynamic definition of the Sulu Sultanate's boundaries based on larger scale systemic processes of socioeconomic change and a borderless history of a wide-ranging maritime trading network oriented toward China, Singapore Europe and the United States (fig. 1).[2]

I argued that in the late eighteenth and nineteenth centuries there existed a loosely integrated political system that embraced island and coastal populace, maritime nomadic fishers, and slash and burn agriculturalists of the coastal rim and interior foothills in the Sulu Archipelago, the northeast coast of Borneo, the foreland of southern Mindanao, and the west coast of Celebes. Coastal *datus* forged political links and trade pacts with riverine swidden cultivators on Borneo's east coast. Upland tribes, hunters and gatherers, and nomadic boat people formed subject groups who performed procurement roles for Sulu's export trade with China and the West. This network of interpersonal relations which was fluid and subject to disruption across time was integrated by the commercial marauding patterns which came to be concentrated in the Sulu Sultanate as the key redistributive center for the zone in the late eighteenth century.

By foregrounding the conceptual-theoretical importance of the "zone" (as Braudel did in his masterpiece on the Mediterranean), I was able to place the various maritime peoples—the Taosug, Balangingi Samal, Samal Bajau Laut, Iranun, and Magindanao—in their sociohistorical contexts and thus begin to understand the process of how maritime trading states evolved and expanded in Southeast Asia. By altering the spatial and temporal dimension of my research, a far more comprehensive history of how a maritime "frontier" region resisted and responded to the impact of the global economy and Western imperialism began to emerge. I had created a radically distinctive cultural-ecological "framework" for reinterpreting and re-presenting the interregional history of a Malayo-Muslim sultanate situated at the edges of the territorial boundaries of three colonial empires. My attempt to experiment with such a different form of framing and contextual presentation enabled

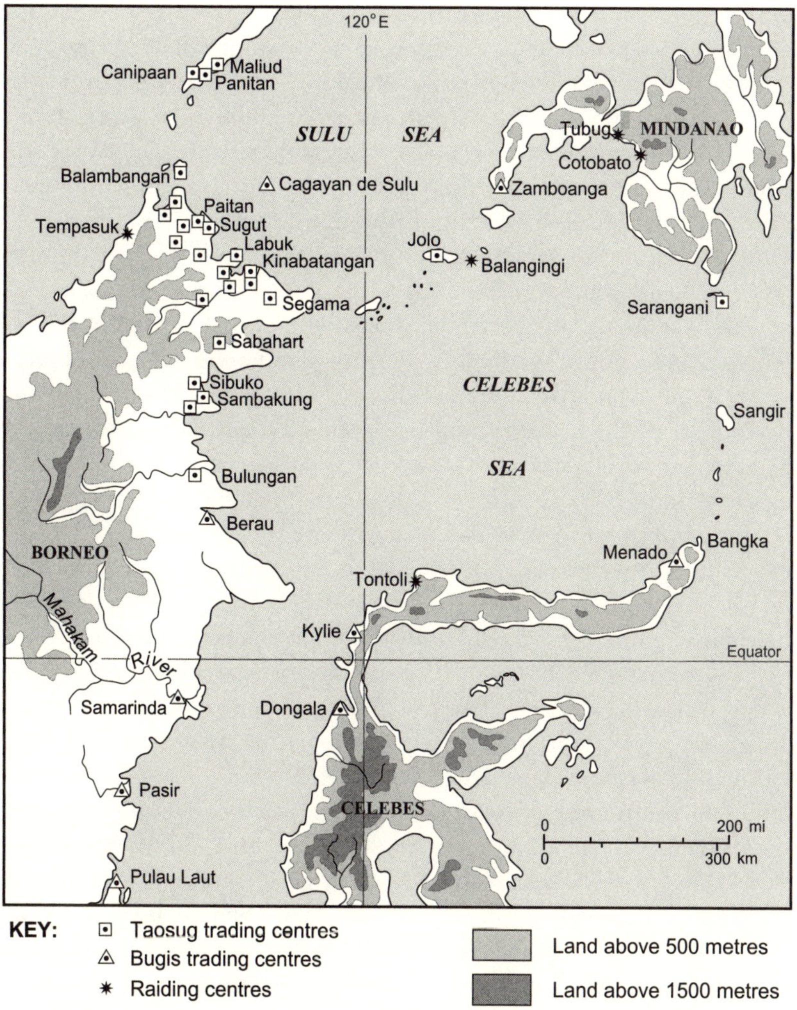

Fig. 1. Map of the Sulu Zone. The Sulu Zone constituted a Southeast Asian economic region with a multiethnic, precolonial Malayo-Muslim state and an ethnically heterogeneous set of societies of diverse political backgrounds and alignments. These diverse ethnic groups were set within a strategic hierarchy of kinship-oriented stateless societies, maritime nomadic fishers, and forest dwellers. In terms of world commerce and economic growth, the Sulu Zone was not an important region until the end of the eighteenth century. Map courtesy of the author.

me to concentrate on the historical complexities and connections at work then, and to construct a more natural, or autonomous angle of vision for interpreting maritime society and culture in the Mindanao-Sulu region in the late eighteenth and nineteenth centuries. The "zone" constituted the essential point of entry, as well as a metaphor, to help explain complex conditions of cultural difference and ethnic diversity, commodity chains, and the process of transition and transformation in the daily lives of maritime traders, marauders, and fisher folk, as well as the nature of a system of enslavement and the character of the state that supported it.[3]

A fundamental principle of this thalassographic framework is that researchers must not allow the boundaries of contemporary nation-states to shape the way they view, or study the past. Instead, they must map out the connections that people and the circulation and flow of commodities actually had with one another across oceans and seas. Thus, much of my work on the Sulu Zone is devoted to a painstaking reconstruction of the trade in *tripang* (*bêche-de-mer,* sea cucumber or sea slug), bird's nests, opium, guns, mother-of-pearl, and slaves centered on Jolo—the seat of the sultan of Sulu. The broad conceptual schema also enhanced understanding of global systemic links and interactions between geopolitical core areas, namely China and Europe, and a strategically positioned "zone"—a place with a loosely structured polity, strong trading bases, and a thin population, which encompassed a variety of economic subregions and extremely specialized territories.[4]

Equally significant in writing about the sea and maritime peoples from a methodological standpoint, is the need for the historian to engage in fieldwork akin to anthropologists. For example, the history of the Balangingi Samal and the Samal Bajau Laut had been shrouded in confused ethnic distinctions, and compounded by the dominant myth that the Samal slave raiders were forced into piracy because European trade had largely destroyed the Malay maritime coastal economy.[5] If I had lacked a firsthand working knowledge of the society and culture of these maritime strand communities, I would not have been in a position to expose the damaging myth. Having discovered the astronomical size of the demand for marine products in the Spanish archival sources (especially *tripang* and mother-of pearl-shell destined for Chinese markets), I immediately recognized just how much Samal Bajau Laut labor would have been required to procure and process it, and this figure far exceeded the existing population. I was not able to visit the Balangingi Samal in the 1830s, but I was able to "pass over" into a Samal Bajau Laut coastal community in the 1960s, and observe firsthand the daily life and subsistence-based economy of these maritime people over an extended period of time. With such detailed knowledge came an intuitive sense for asking the proper questions about traditional practices, earlier history, folklore, and ethnic stereotypes, and also a cultural recognition to what did not sound right in the historical record. I also created an exceedingly powerful image in my mind's eye of the Balangingi Samal, a people once terrifyingly referred to as the "Vikings of Asia."[6]

Commodity Chain Methodology

The commodity-chain methodology is related to approaches variously called commodity systems, commodity circuits, commodity networks, value chains, and production networks, to which the word *global* is often attached.[7] The Sulu Zone project was concerned with commodity chains because it was meticulously concerned with the social embeddedness of *tripang,* pearls, pearl shell, shark fin, and bird's nest procurement, production, distribution and consumption,[8] that clearly operated across national borders where the processes of the modern world system, or globalization were influential. Tea, gunpowder weapons, opium, and textiles were some items that closed this commodity chain.

The zone project's innovation resided in both its methodology and its conceptual underpinnings. It combined a unique spatial focus and commodity-chain methodology to study the dynamics of cross-cultural trade, slavery, and the accomplishment of ethnicity. It extended the international literature by analyzing Sulu's rapid rise to regional prominence by concentrating on the distinction between various ethnic groups, several economic roles, and their creation and maintenance by the incorporation of others. Existing studies of the history of the Sulu Sultanate had not utilized detailed interdisciplinary social science understandings of the historical, cultural, political, and economic contexts of specific commodity circuits.

Metaphorical perspective

In literature and history, one of the most common and effective forms of language syntax is the use of metaphor. A metaphor is a figure of speech which designates an object by implicit comparison or analogy.[9] In this chapter, I use the sea as a partially structured metaphor for the concepts of abundance, fear, opportunity, change, and history—all of which aid our perception and understanding of the rise and decline of the Sulu Zone from the late eighteenth to the late nineteenth centuries. These concepts are used to represent processes and domains that are not always easily accessed, understood, or quantifiable. Furthermore, my initial research on the Sulu Zone was an attempt to piece together the view from the other side of history and to understand how a maritime "frontier" region responded to Western expansion. It also sought to document the independent activities of several seafaring peoples in a large trading region—peoples who were stimulated by European expansion and commerce, but who developed their own patterns of trade and slaving, often in defiance of the interests of three European empires: the Spanish, the Dutch, and the British.

The Sea Is Abundance: Cross-Cultural Trade

During the eighteenth century, tea had replaced ale as the natural beverage in England and was especially popular among the poorer classes. China was almost the sole supplier of tea to England, and China was thereby opened to a wider trade involving both the Indian subcontinent and Southeast Asia. Asian, European, and American merchants based in Calcutta, and later Manila, Macao, and Salem, were unable to provide sufficient suitable commodities for the burgeoning Canton market. These merchants were quick to recognize the potential of participation in the Sino-Sulu trade as a means of redressing the flow of silver from India. The traffic in Indian opium to the Chinese provided a partial answer, but traders also exploited the huge demand in China for the marine and jungle products of Southeast Asia. Thus, on their way to China the country traders could stop at Jolo to exchange firearms, Indian textiles, and opium for *tripang,* pearls, mother-of-pearl shell, shark fin, bird's nests, and other local products that they could sell in China. Now, in this new globalised world, Jolo, Balangingi, Calcutta, Canton, and London were all intimately interconnected.[10]

The Focus of Exchange: Exotic Commodities and Marine Gardens

Pearls

The abundance of Sulu's pearl fisheries were renowned in Asia. Chinese records as early as 1349 mention the pearls found in the "marine gardens" of the Sulu Archipelago as important objects of trade.

> The Su-lu pearls, which are better than those produced at Sha-li-pa-chou, Ti-la san-kan (third port) and other places, are white and round. Their price is very high. The Chinese used them for head ornaments. Their colour never fades, and so they are considered the most precious rarities. There are some about an inch in diameter. Even in the place of production the large pearls cost over seven or eight hundred "ting." The medium ones cost two to three hundred "ting," the small ones ten or twenty "ting."[11]

Alexander Dalrymple's account of Taosug trade and society is the clearest exposition of the economy of Sulu's pearl fisheries. He was amazed by the richness and extent of Sulu's pearl banks—in places over twenty-five miles wide. Numerous pearling sites in the Laparan, Pangutaran, and Pilas groups were recorded by him in addition to the pearl beds near Balabac and Maratua. Dalrymple noted that the shoals and reefs in the environs of Tawi-Tawi "were extremely intricate and narrow," but invaluable to the commerce of the Sultanate—stating, "These guts (the intricate and narrow channels) are the most valuable pearl fishery in the world"[12] and they attracted a host of Samal-speaking people, their kindred and slaves.[13]

Sultan Bantilan (Muizz un-Din) furnished Dalrymple with a list of thirty-seven products of Sulu and its immediate dependencies that English traders could expect to purchase at Jolo. Among the more important items, for which there was a constant demand in China between 1760 and 1840, were "pearls, tortoise shell, *tripang,* mother-of-pearl, shark's fin, bird's nest, wax, camphor, cinnamon, pepper, cocoa, rattans and ebony."[14] For private traders these were to remain the principal items in their traffic from Sulu to Canton. This quest for marine products and other desirable commodities was to have a profound impact upon the various peoples of the zone and their way of life, as well as the rest of the eastern archipelago, constituting one of the most dramatic and fascinating episodes in the history of China's tea trade and the world capitalist economy.

Tripang, Sharks Fin, and Bird's Nest

In order to understand how these commodities altered the course of cross-cultural trade and historical events it is necessary to explore the banquet halls of the emperor of China, and the fisheries and forests of the Sulu Zone.

By the mid-eighteenth century, many types of exotic foods or commodities from the Nanhai or Southern Sea are mentioned as standard fare of Chinese imperial cuisine. Particular cooking styles and special foods such as *tripang* and bird's nest seem to have made a considerable impact on the tastes, variety of dishes and extravagance of Manchu court life.[15] T. C. Lai, in a historical survey of Chinese cuisine and culture, demonstrates from the accounts of literary persons, food connoisseurs and critics of the time how the burgeoning demand for a number of these commodities can be explained by the term *bu*, foods rich in high-quality protein, which were also said to possess the essence of the sheer vitality of life.[16]

The demand for these exotic commodities for the Chinese table led to a comparatively high incidence of violence and warfare, conducted by the same competing groups that organized and did the procurement. The evolution of the China tea trade also resulted in rapid commodity depletion and environmental degradation on a hitherto unprecedented scale. The reefs and forests of the Sulu Zone were deemed commons, or "open access" sites, where natural commodities such as *tripang* and bird's nest were systematically plundered. The Taosug *datus* and their rivals tended to over-collect, failing to conserve for the future in the heat of local-regional competition and an insatiable market demand. Faced with a range of changing global economic demands and Chinese food and eating habits, the Taosug destroyed parts of the natural world of the zone and abused its environment in exchange for other commodities—textiles, opium, and war stores.[17]

The Sea Is Fear

While the sea, in its abundance, provided singular commodities for the expanding global trade, at the end of the eighteenth century it was the labor power required to procure the produce that created fear among many coastal populations in the region. As the sultanate organized its economy around the collection and distribution of *tripang*, shark fin, pearls, and bird's nests for the trade in China tea, the issue of the nature of productive relations in the Sulu Sultanate—slavery—suddenly became of primary importance and affected the allocation and control of labor and the demand for fresh captives throughout the Sulu Zone.[18]

The Need for Labor and the Rise of the Iranun

The need for a reliable supply of labor-power was met by the Iranun and Balangingi slave raiders who annually captured several thousand people

Fig. 2. Drawing of Joanga. A Joanga, an Iranun warship of the eighteenth century with three banks of oars under full sail. Upward of 100 feet long, these vessels were equipped with large bamboo outriggers. Both sides were rowed and paddled by more than 190 men. Source: Rafael Mouleon, *Construccion Navales: bajo un aspecto artestico por el restagador del Museo Naval, Catalogo Descriptivo dos Tomos.* 3 vols. (Madrid: Museo Naval, 1890).

to be trained to work alongside the Samal Bajau Laut in the *tripang* and mother-of-pearl fisheries.[19] More than anything else it was this source and use of labor power that was to give Sulu its distinctive predatory character in the eyes of Europeans in the nineteenth century as a "pirate and slave state (fig. 2)."[20]

Tausug *datus* became "silent partners" in large scale raiding endeavors and advanced the outfits and, equally important, provided a market at Jolo for all spoils taken—including the captives. In 1762, James Rennell, an East India Company factor, mentions that "some Illanians are now settled at Basselan and Sooloo from whence they fit out their privateers."[21] Twelve years later, Forrest described many of Jolo's inhabitants as Orang Illano, "who live in a quarter by themselves."[22] In the second half of the eighteenth century, Iranun slaving activity intensified with a series of terrifying raids on the coasts and shipping of the Philippines, the Straits of Malacca, and the islands beyond Sulawesi. Their primary targets were unprotected coastal settlements and sailing boats that traveled throughout Southeast Asia bringing valuable commodities from

China and the West. Many of these marauders were sponsored with the authority of their Muslim rulers from the trading states of Sulu, Cotabato, Siak, and Sambas.

The Raids, Networks, and Bases

As the last quarter of the eighteenth century advanced the length and duration of Iranun slaving voyages made the establishment of satellite settlements necessary and convenient, staging points (bases) were set up within the target areas or at crossroads along their routes. The Iranun established semipermanent settlements on the coasts of Borneo, Celebes, and Sumatra. The distance both in space and time between Sulu and Mindanao and the isolated islands and little-frequented corners of Southeast Asia were shrunk by this network of distant settlements based on ramified kinship, mobility, and alliances for purposes of slave raiding and obtaining plunder. At these outlying fortified bases captives could be put to work temporarily, or transferred as chattel in local markets; raiding vessels were safely careened and repaired; and premeditated attacks launched with impunity (fig. 3).[23]

During the period 1768–1800, the islands surrounding southern Luzon were one massive raiding base. A ring of satellite stations stretched out around the provinces of Tayabas, Camarines, and Albay and Iranun raiding *prahus* sailed from all of them, including Burias, Masbate, Capul, and Catanduanes. Hence, few coastal towns and villages in southern Luzon were spared the onslaught of the saltwater slave raiders.[24] In an effort to avoid enslavement, some villagers built stone forts and churches with lookout towers. Some people went to live in larger villages; some looked for new village sites, often on elevated ground; others abandoned the coast altogether for an equally harsh life in the mountain vastness of the interior.

Captives and the Arms Trade

With the gunpowder weapons of the English, readily obtained at Jolo, the Iranun descended on coastal settlements throughout the central and northern parts of the Philippines and even ventured inland to pillage and burn churches and towns. From one end of the archipelago to the other they carried out "a pattern of tragedy so recurrent as to become almost tedious."[25] One Spanish writer described the wholesale misery inflicted by the Iranun on the inhabitants of the archipelago, as a chapter

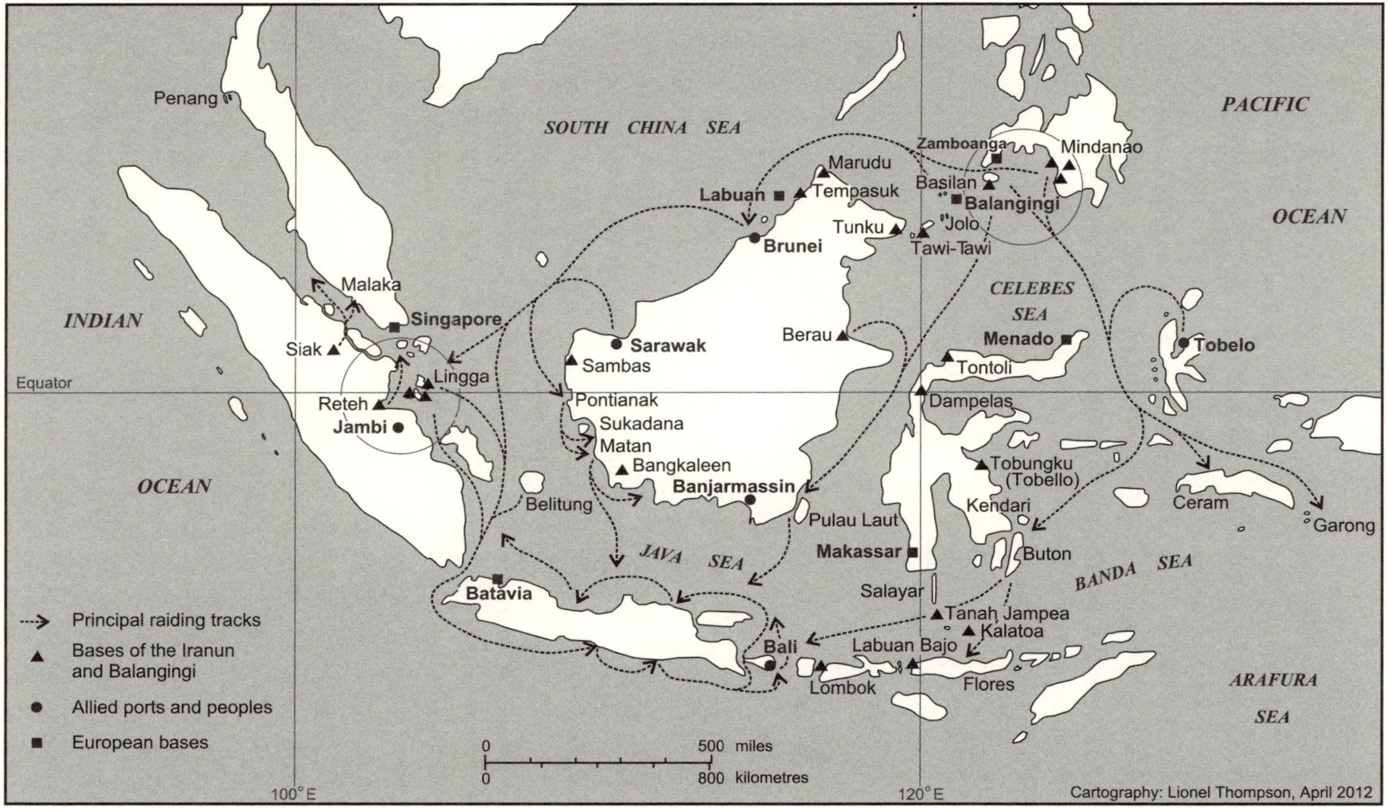

Fig. 3. Map of Iranun Raiding Patterns of Southeast Asia. Iranun-Balangingi Maritime Raiding and the Malay Archipelago in the First Half of the Nineteenth Century. Map courtesy of the author.

in the history of Spain in the Philippines "written in blood and tears and nourished in pain and suffering" (fig. 4).[26]

Indeed the post-1780 era saw slave raiding more widespread and devastating than at any earlier time as the Iranun and Balangingi borrowed both knowledge and technology from European and Chinese traders: Chinese compasses, European firearms, mariner's charts, and brass telescopes were all widely used to great advantage as "weapons of war" by these sea raiders.[27]

In 1832, Gamaliel Ward, captain of the Spanish brig *Leonidas,* reveals the extent of the arms trade at Jolo. A leading *datu's* request included "1,000 25 pound kegs of gunpowder, 6 swivel guns, 6 large cannon preferably brass, 600 muskets, 100 pistols, 4 bags of shot of varying sizes, gun flints, 2 dozen boxes of percussion caps and 8 dozen matchlets."[28] By 1835, gunpowder and muskets were the principal items desired by the Taosug *datus.*

Fear and Memory

Nearly two centuries ago, the name Lanun struck fear into the hearts and minds of riverine and coastal populations across Southeast Asia. The terrors associated with the sudden harsh presence of these well-armed raiders lives on in the oral recollections, reminiscences, popular folk epics, and drama of the victims' descendants in the Philippines, Indonesia, and Malaysia, to this day.[29] To the Europeans of the late eighteenth and nineteenth centuries, the Iranun were like the spawn from hell (fig. 5).

Few homes in Southeast Asia's vast patchwork of coastal settlements were left unscathed by the "Illanun." Many people lost their kin and were forever haunted by their loss. The Iranun and Balangingi in the nineteenth century lay in wait along paths or creeks or on isolated stretches of coast near small settlements. They captured silently and retreated swiftly. Time has not obliterated the memory of their once-fearful presence.[30]

On Camiguin Island,[31] lingering evidence of these attacks is provided by the name of the principal coastal town, Catarman, which means "place of fear."[32] The fate of many lowland Christianized inhabitants was a house of cards existence. They knew any strong wind or sudden pealing of the church bell could bring serious danger that could destroy their community. Maritime raids also had a profound impact on Southeast Asia. The Iranun and Balangingi have been rightly blamed for demographic collapse, loss of agricultural productivity, and economic

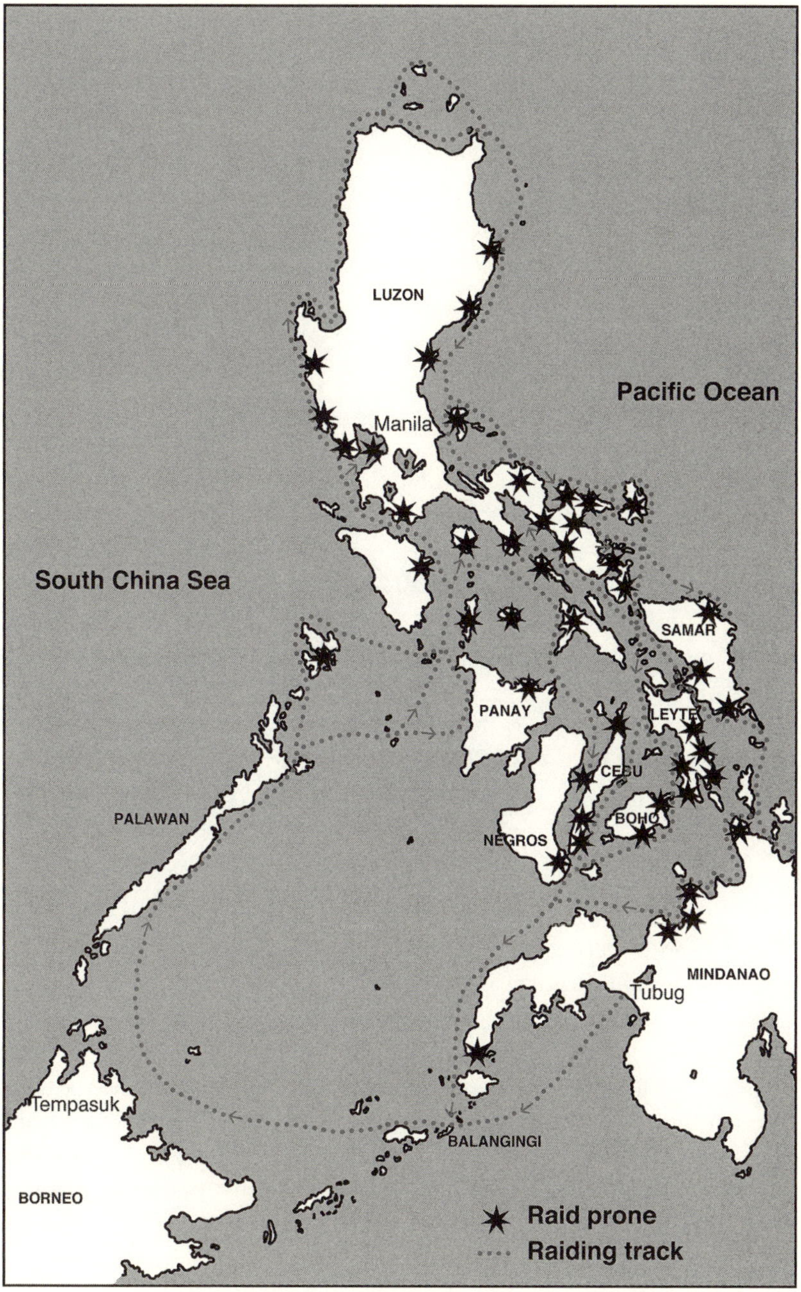

Fig. 4. Map of Raiding Patterns in the Philippines. Slave Raiding in the Philippines, 1768–1868. Courtesy of Han Vu.

Fig. 5. Drawing of Iranun Warrior. An Iranun maritime warrior attired in the distinctive thick cotton (*kapok*) quilted blood-red vest, and armed with a kris, long spear, and the *kampilan*, a long, heavy "lanoon sword" ornamented with human hair. Source: Frank Marryat, *Borneo and the Indian Archipelago* (London: Longman, Brown, Green & Longman, 1848).

decline, as well as the breakup of the Dutch stranglehold in the Straits of Malacca and Eastern Indonesia.[33]

Recent ethnohistorical research has shown that the agony and extreme fear associated with these alien slave raiders is still present in the social memory of coastal and island communities in Indonesia, the Philippines, and Malaysia.[34] Beyond the collective memory of such a traumatic period, the very landscape itself in much of the region constitutes a persistent feature of this officially sponsored violence, bearing silent witness to the suffering of its victims. For example, the Iranun sailing from Sulu in the 1790s raided and settled in Berau, where a low green hill, Gunung Illanun, still bears their name.[35]

Until quite recently, villages in many parts of eastern Indonesia, particularly on Buton, were either situated well inland or, if on the coast, on steep cliffs with extremely difficult access; the historical legacy of defense against the threat of Iranun and Balangingi marauders and slave raiders.[36] Robert Barnes, in his classic study of Lamalera—a remote community on the south coast of the island of Lembata, near the eastern end of Flores—notes the village is really a "twin settlement," with a lower one (Lamalera Bawah) on the beach and an upper one (Lamalera Atas) on a nearby cliff top for protection from earlier maritime slave raids. Evidence has also emerged supporting the widespread fear and dread of the Iranun in the Java Sea. Kurt Stenross came across people with terrifying memories of the Iranun in Tamberu. He found evidence of centuries-old Madurese oral traditions about the "Lanun" that are indicative of cultural confrontations and conflicts.

In their reconstruction of the regional demographic history of Cebu, Michael Cullinane and Peter Xenos stress that the memory and fear of "Moro depredations" are embedded in the legends and folk histories of many municipalities and parishes of Cebu to this day.[37] "Moro" came to symbolize all that was dangerous, dark, and cruel about the tragic confrontation, and the Iranun and Balangingi's adherence to Islam.

The Sea Is Opportunity

Stratified prestige in Taosug society was perpetuated through a system of ranked social categories. The aristocratic *datus* formed the most significant group at the top of the estate system. They were principally royal *datus*—the sultan's kindred and those who claimed descent from the first sultan—and others recruited through appointment by the sultan.[38] High ascribed status assured social prestige in Taosug political culture, but not power. For the scions of "royal blood," wealth and leadership talents were essential to attain political office. They established themselves in the political sphere, attracted and retained followers through vigorous participation in the commodity-procurement trade, and in the promotion of slave raiding.

Slaving and slaveholding were among the principal means of enlarging and consolidating the political influence and wealth of upwardly mobile Taosug, Magindanao, and Samal chiefs. Slaveholding provided labor power, prestige, and more importantly, differential access to force and authority in the sociopolitical hierarchy of Taosug and Magindanao society. The key to the high status of particular *datus,* who held a great deal of power and authority at this time, was control of global-local trade and the possession of large retinues of slaves and those "freemen" who were neither *datus* nor had wealth, prestige, or followers, and were not recently captured slaves.

By the beginning of the nineteenth century, the Jolo market offered British-manufactured brassware, glassware, Chinese earthenware and ceramics, fine muslins, silk and satin garments, Spanish tobacco and wines, and opium from India. There was a constant increase not only in the variety but in the volume and quality of these objects of trade. These luxury goods for personal adornment and pleasure and for the household were translated into power and prestige symbols by the aristocracy to form the material basis of their social superiority. It is in this sense that the global trade became a vital element in the overall functioning of the Taosug social system.[39]

The Sea Is Change

The boundaries between aspects of both Chinese and western culture and Malayo-Muslim culture, practice and belief were porous in the late eighteenth-century zone. Indeed, Jolo was the site of repeated complex cultural exchanges linked to the China tea trade that highlighted the ways cultural difference and diversity were increasingly blurred. From trade documents, journals, and illustrations left by traders and travelers we can learn more about how certain commodities bridged two worlds and in the process changed people's cultural attitudes and practices toward daily life.[40]

My emphasis here, as we follow various trade objects with their readily defined characteristics of "modern" and singular "magicalities and enchantments"[41] from Europe and India to the Sulu Zone and China and back, is on the complicated role commodities of the modern world system played in creating and breaking down the "borders" of cultures across Eastern Asia—cultures, usually regarded as non-capitalist and "traditional."

Apart from opium, textiles (fine muslins, madras chintz, colored long cloth, and coarse cotton stuffs) were the single most valuable items to be obtained in Bengal by country traders bound for the Sulu Archipelago. By 1812 the dress of a person in the zone no longer was determined necessarily by their status. The poorest were still said to have gone about naked, some wore grass skirts, but it was far more common for an ordinary person or slave to have at least a Chinese jacket or sarong in coarse, white cotton cloth. In 1834, Samal villagers at Siassi were described as

Fig. 6. Drawing of Taosug Datu or Malay Chief. A "Malay Chief" of Jolo in the early 1840s, whose wealth and power was based on the strict regulation of external trade. He is dressed in high fashion—a headcloth of American or Bengal manufacture; a brightly colored Chinese silk jacket, embossed and filigreed with gold thread; satin pants; and several splendid kris, or daggers—a reflection of his high status, power, and social prestige. Source: Frank Marryat, *Borneo and the Indian Archipelago* (London: Longman, Brown, Green & Longman, 1848).

being almost naked, except for a *nankin* (cotton) skirt or a pair of loose trousers of the same fabric, cut off at the knees. While the children were stark naked in the outer islands, some *banyaga* or slaves in the capital (Jolo) managed to make their clothes out of imported silk and satin. At Jolo it was difficult for visiting Europeans to distinguish *banyaga* from Taosug because of this remarkable change in fashion behavior linked to the proliferation of new material objects, such as the handkerchief.

Some people, like the coastal *datu* depicted in Marryat's travel account, were clearly even more tied to several worlds. His costume is a flamboyant mixture: Chinese in style but also made up of material objects from Europe and the Malayo-Muslim world. Here, lifestyle and fashion, as a statement, was an extrapolation of globalizing social, political, and economic forces (fig. 6).

Commercial textiles as commodities changed traditional productive processes such as handloom weaving, altered people's sense of fashion, domesticated their bodies, and fostered a deep-seated material dependency that bound local consumers in the zone to the interventions of an expanding world capitalist economy, as flows of "modern" objects intensified in speed and volume.

The Comaroffs note in their discussion of the "history" of a commodity and its impact on social and cultural life, that from the analysis of the career of valued everyday goods, we can comprehend the evolution of complex social processes and individual intention and action.[42] Ideas about fashion, work, property, value, class, and authority all changed and were changed by the relation between particular trade commodities and "zone" life.

Ceramics: Porcelain and Earthenware

Cheap mass-produced earthenware of every conceivable shape and size (cups, saucers, kettles, bowls, platters, dishes, basins, and spoons) and exquisite porcelain along with metal utensils captured the markets and imagination of the zone and transformed daily life in the zone.

Chinese potters had not only used singular glazing techniques but they proved more efficient in mass-producing crockery on a hitherto unprecedented scale than their regional competitors or the Europeans themselves. The junks from Southeast China dominated the transport of these new material objects that were becoming essential to the maintenance of the frantic pace of life in the zone.[43] Cooking methods and utensils changed in the late eighteenth century. Fish that was formerly roasted on an open fire, or salted, was now also boiled in iron or copper kettles and served with aromatic spices and rice in ornate earthenware basins and dishes. The disproportionately large number of fragments of Chinese ceramic produced in the late eighteenth and nineteenth centuries discovered at various sites around the zone confirms the accelerated tempo of Sino-Sulu trade at this time.[44]

Opium and Addiction

Opium was widely introduced among the Taosug as a consequence of the trade dealings of John Herbert's East India Company regime.[45] It is evident from Herbert's official correspondence that he never considered the social and political consequences of the introduction of opium among various segments of Sulu's population. It was those aristocrats involved in external trade who first acquired a taste for the habit-forming drug. Within two decades, opium smoking was considered an essential part of the lifestyle of the *datus* near the coast. In the nineteenth century, it was still more widely used and few among the wealthier *datus* avoided addiction. Opium had a debilitating effect on the Jolo aristocracy and prevented some among its ranks from exercising proper leadership.

Opium was not among the articles the Taosug desired from Bengal in 1761. But, by the time of Hunt's visit in 1812, annual consumption had risen to six chests of Benares opium, and the importation of this article increased fivefold over the next twenty years. The Tau Gimba, hill Taosug, scornfully observed that the rapid spread of the "opium-eating habit" among the coastal populace was responsible for their declining martial spirit.[46] In the early decades of the nineteenth century, the Taosug were also supplied with the drug by Chinese and Bugis traders, and the Iranun obtained it from native vessels they captured among the Dutch Islands. By 1839, the annual trade in opium at Sulu averaged twenty-five to thirty chests, which weighed between 100 and 120 pounds each.[47]

Ironically, for many of the wealthy coastal Taosug, opium was now less a business than an addiction—a sinister friend and a new habit-forming way of life. The trade in opium by then had wreaked a devastating social transformation almost as significant for understanding the meaning of force, and reframing the idea of culture itself and heterodox practices, as the technological and social innovations introduced by imported firearms.

The exchange of opium for tea in China accelerated the disintegration of China at the very same time that it contributed toward Sulu's ascendancy. However, this commodity ultimately also sowed seeds of discontent among an increasing number of addicted Taosug coastal *datus* which culminated in a manifest loss of will on the part of key individuals in the "chiefly" class as the Spanish swept out of the Visayas into the Mindanao-Sulu region with steam gunboats, after 1848. No longer completely in control of either themselves or the adjacent seas, these *datus* now willingly collaborated with their arch political rivals and enemies—the Spanish—against the alleged "pirates" and "ruthless people"—the Samal Balangingi—in order to retain control over the global-local commerce and, redistributive network in the zone.[48]

Ethnicities and Shifting Identities

In the late eighteenth and nineteenth centuries, the population of Sulu was heterogeneous and changing—economically, socially, and ethnically. This was a direct result of global trade. The populating of the Sulu

Zone by captives from the Philippines and various parts of the Malay world and their role in the redistributive economy centered on Jolo cannot be underestimated.[49] The first half of the nineteenth century was an important period of transformation in the demographic history of the Sulu Sultanate because of the rapid growth of the population, stemming from the incorporation of captives and their enslaved descendants into Taosug society. The size, composition, and regional distribution of the Sultanate's population were radically altered by 1848.

Most slaves were put to work procuring marine and jungle products, but some became local traders, while others became saltwater raiders and preyed upon the far-flung villages of their origin. Some though never gave up the hope of escape, and returning to their home villages, but most came to regard their new lives in the Sulu Zone as no worse than those they had left behind. Some never forgot their ancestry, but many, particularly those captured when young, readily adopted new ethnic identities.[50]

By arguing for a broader interconnected global economic perspective interesting complex questions are raised about what constitutes our conception of "culture" and ethnicity. While thousands of captive people were allocated as slaves throughout the zone each year in the period under consideration, the borderlines of race, "culture," and ethnicity were increasingly blurred by the more inclusive practices of incorporation and pluralism in a traditional Muslim social system. In *The Sulu Zone* I have maintained that the Taosug, Iranun, and Samal not only lived in an increasingly interdependent world, but that they also lived in an emergent multi-ethnic society and polity—the multi-cultural inhabitants of which came from many parts of Asia and elsewhere in the world.[51]

The Sea Is History

Framing and Trade with China

My framing and interpretation of the "Zone" as a spatial system rested on the axiom that it was inherently unstable and generally dynamic and that it was thrust on the global stage at a specific moment or era in "regional time."

My understanding and discussion of global economic-cultural interconnections and interdependencies linking the Sulu Zone and the China trade was based on the premise that these intersections were governed by particular economic systems and set in a specific era and locality. The Taosug lived in a singular time and time meant change. They also lived in a singular place and geography meant destiny. The zone was a place where borders were becoming ever more porous, less bounded, less fixed, stimulated in large measure by global-regional flows of commodities, people and ideas; a kind of powerful magnet whose force European and Chinese traders were drawn to because that was where much of the exotica for Chinese cuisine and medicine, as well as other commodities destined for the Canton market were being procured and processed. What then is the importance of *The Sulu Zone*'s thesis, about the China tea trade's complicated place within its "borderless" history? It is a central argument of this book that we cannot think of societies and cultures in isolation, as self-maintaining, autonomous, enduring systems.[52]

A key factor in the Taosug's ascendancy was Europe's globalizing trade with China. The West's search for suitable local commodities to exchange for Chinese tea is certainly the most convincing explanation of the origin of the Sulu Sultanate's startling regional expansion to the west and south.[53] Since the British primarily wanted *tripang*, shark's fin, pearls, and bird's nest for the trade in China tea, the issue of the nature of productive relations in Sulu—slavery—suddenly became primary. The demand for certain local commodities in return for imports affected the allocation of labor and the demand for fresh captives throughout the Sulu Zone. In this globalizing context, tea was more than simply the crucial commodity in the development of trade between China and Britain, it was also an evergreen shrub or small tree that was instrumental in the stunning systematic development of commerce, power, and population in the Sulu Zone—a transformation which changed the regional face and history of insular Southeast Asia.[54]

The Iranun Age—Expansion and Decline

Whether as warriors, traders, or pioneering settlers, the Iranun reached almost every part of Southeast Asia. The regularity of their raiding sweeps was as predictable as the winds which carried the Iranun and Samal Balangingi boats to their target areas. Each year, on the approach of the "pirate wind" in August, September, and October that brought these lords of the eastern seas to the Southern Visayas, Borneo, and the Straits of Malacca, the Dutch, Spanish, and English issued customary warnings to coastal towns and small craft.[55] Scattered along the coastline of the Philippine Archipelago one is still able to find the remnants of the century-long terrifying presence of these raiders—an old stone watchtower, a crumbling church-cum-fortress, or the ruins of a Spanish fort and cemetery—decaying monuments to the export of *tripang* and bird's nest and a host of other commodities and the import of firearms and gunpowder from Europe and the United States along with numerous Chinese goods. The remains of such neglected sites—primarily concentrated along the coasts of Ilocos, Catanduanes, Albay, Cebu, Leyte, and Samar—bear silent witness to the advent of sudden affluence and new patterns of consumption in the Sulu Sultanate and deep despair, displacement, and dispersion of people throughout the Philippines.

The end of the Iranun age is usually set in the years following 1848. In that year the able and liberal-minded governor general of the Philippines, Narciso Clavería, secured from England the first prefabricated steam gunboats for the defence of the islands. The arrival of the steamers marked a turning point in the ninety-year sea war (1758–1848) against Iranun and Balangingi raiding and slaving. The beginning of the end of their wind-driven way of life was inextricably tied to Clavería's decision to discipline and punish them with the *kapal api,* or "fire ships," which marked the start of a new era of bitter conflict with the Iranun and Balangingi. This sea war would signal the end of their way of life in less than twenty-five years and was the defining moment when the slave hunters became the hunted.[56]

In the context of the world capitalist economy and the advent of the China trade, it should be understood that the slaving of the Iranun and Balangingi, so readily condemned in blanket terms as "piracy" by European colonial powers and later historians, was a means of consolidating the economic base and political power of the Sultan and coastal chiefs of Sulu, and which functioned as an integral, albeit critical, part of the emergent statecraft and sociopolitical structure(s) of the zone.[57]

Thus, the history of slaving and the slave trade and the rise of the Iranun and Balangingi must be framed as part of a unitary historical process, which explains the major factors contributing to the formation and maintenance of their ethnic identity. Maritime raiding, or what the Spanish, Dutch, and British labeled "piracy," was not a manifestation of decay and dependence, but rather it was the result of remarkable economic growth and strength. The central focus of these cultures in conflict had always been in the sea—the sea which, in more ways than one, was "discovered" by Spain and Britain and functioned as a political instrument, a commodity, a national prerogative, and aspiration. The Iranun and Balangingi were defined by it, measured by their domination and use of it, and were to be dispossessed of it.[58]

The point is that history consists of the interaction of variously structured and geographically distributed social entities which mutually reshape each other. The transformations of the West and China and the rise of the Iranun and Balangingi in modern Southeast Asian history cannot be separated: each is the other's history.

The Accomplishment of Ethnicity and the Meaning of "Culture"

As Roger Keesing noted, there is no part of Eastern Asia where both the production and reproduction of "culture" and cultural meaning can be characterized as unproblematic, without glossing over or disguising radical changes in relation to ethnicity, power, and hierarchy that have differentially affected states like the Sulu Sultanate and marginal settings like the zone.[59]

Ethno-historians, like Jean and John Comaroff, Edmund Leach, Roger Keesing, Eric Wolf, and myself, have stressed that the genesis, persistence, and transformation of ethnicities must be understood within a set of fluid self-defining systems, embedded in economic and political relations and contingent upon specific historical forces and events. We are now acutely aware of the "origins" of Balangingi and Iranun ethnicity to their present configuration, and of the dynamic and ecological aspects of "culture" and place, as well as how ethnic identities have been constructed and reconstructed in local-regional history on a large scale.

Yet, perhaps, the most striking aspect of the untidy history of the Iranun and Balangingi since the late eighteenth century, which argues against bounded homogeneous ethnic groups, is how little attention it has received among most Filipinos—both Muslim and Christian alike. Its forces and characteristics are historical and mythical, global, and local, and Islamic and Christian. The relationship(s) is a dialogic one between varying periods of Southeast Asian and Philippine history. The historical representations, certain phases, and traditions are inseparable from one another and flow through each other.[60]

Historians and ethnographers of the region need to locate the emergence, maintenance, and abrogation of populations and the "cultures" they encompass within the framework of a series of historically changing, imperfectly bounded, multiple and branching integrated sets of local, regional, and global social and economic alignments. Here, some of the questions posed about the birth and accomplishment of ethnicity and ethnic identities in *The Sulu Zone* for understanding both the recent and more distant past, especially at the margins of states, and beyond their borders in Southeast Asia are far-reaching, particularly if one considers the contemporary complex theoretical cultural implications of the nature of "ethnicity" (often associated with economic and political conflict in developing societies) as a key factor for unraveling the development and history of the terms "Indonesian," "Malay," "Thai," "Burmese," and "Vietnamese." These labels have been successfully fostered as part of a national imaginary by modern states in the interests of forging national unity and to mythologize history.[61]

Commodities, Seas, and Global Connections

The world was changed through the intersections of the global economy centered on the Sulu and Celebes Seas, as well as the sultanate's critical place within it. Here, ordinary Southeast Asian farmers and fishers were captured, traumatically uprooted and forced to live in a distant economic region—the zone. European traders joined with Taosug *datus* to spark one of the largest population movements in recent Southeast Asian history with hundreds of thousands of individuals sent into slavery across the zone. Turnover in Iranun-Samal slave trafficking was in excess of several million dollars a year; human cargo and Chinese tea then were as profitable as drugs and guns. Hence, all these commodities became inextricably linked with one another in a deadly global trade.

The individuals in this economic arena changed one another's lives by means of tens of thousands of transactions conducted over nearly a century, involving artifacts and goods that reflected the people who manufactured them, and the global economic imperatives that bridged two worlds to facilitate trade in them. When they traded commodities, Europeans and Chinese exchanged some of the signs and practices of their societies as well, which, in turn could have complex, devastating social consequences in the world of the zone, as in the case of opium addiction. Increasingly, particular artifacts played a role, albeit a critical one, as a bridge between their respective cultures and worlds. The impact of these unanticipated "marrying" processes could be discerned in a myriad of everyday choices over such things as appropriate language and speech, fashion, interior decoration, and the characteristics of food and eating habits. Stratification symbols brought luxuries from Europe and China like Waterford crystal, bone china, and porcelain to trading enclaves in the tropical wilderness of the zone. The sultan of Sulu and leading *datus* had their own stores of wine, chocolate, brandy, and cigars and served their guests on Chinese porcelain. In terms of what Mann calls a "history of power," many of these commodities and artifacts were overt symbols of hierarchy and social distinctions among the Taosug, and critically important to the self-maintenance of the trading world of the zone.[62]

The mingling of commodities served not only as motors of change but as realized signs, signifying that two or more worlds met as well. This meeting of commodities and peoples highlighted in different ways the interconnectedness of the modern world. These commodities led to a continuous redefinition of belonging to a place as either "here" or "there" and/or as markers of social identity. Lives and cultures blended wherever commodities changed hands in the zone. Individuals traded, and their worlds and lives were altered in the process, sometimes irrevocably so. After 1768, a consequence of this "globalization" accelerated by the world capitalist economy was that areas, encompassing remote maritime villages and tribal long houses in the zone to entire continents, were "caught up in processes which linked them to events that, though geographically distant, (were) culturally, economically, politically, strategically, and ecologically quite near (and) the distinctions between 'here' and 'there' broke down."[63]

The sea-based history of the Sulu Zone demonstrated clearly the links between large economic and cultural systems and social mechanisms and institutions, on the one hand, and on the other, singular patterns of consumption and desire and the making of collective worlds of more localized smaller communities. In short, as Kenneth Prewitt, head of the Social Science Research Council puts it: "The global-local notion is not a metaphor invented by social theorist"[64]; rather, it was the lived experience of millions of people in the zone and on several continents, inextricably bound to one another as product and fate. Part of the challenge for me had been to identify and link broad patterns and variations in interactions of the global economy and macro-historical trends with the autonomous local history of a barely recognized economic region in Southeast Asia. The long-term changes that occurred in these patterns and trends, based on economic interconnections, trajectories of tastes, and the imperatives of the world capitalist economy, could only be perceived through their interdependent effects on the environment, on ideas, on events, and on the social and cultural transformations in the making of the world of the zone.

NOTES

1. See James Francis Warren, *The North Borneo Chartered Company's Administration of the Bajau, 1878–1909* (Athens: Ohio University Press, 1971). On the

radical nature of social and cultural change experienced by the Samal Bajau Laut see Carol Warren, *Ideology, Identity and Change: The Experience of the Bajau Laut of East Malaysia, 1969–1975* (Townsville: James Cook University, 1983), and Clifford Sather, *The Bajau Laut: Adaptation, History, and Fate in a Maritime Fishing Society of South-Eastern Sabah* (Kuala Lumpur: Oxford University Press, 1977).

2. My initial thinking on how the late eighteenth-century global economy and inter-regional cross-cultural trade created a "borderless world," or "zone," both geographically and historically, in the area of the Sulu and Celebes Seas owes much to the influence of John Smail's thinking, who in turn, had been strongly influenced by the hemispheric, cross-regional historical orientation of Marshall Hodgson. Templates for the study were provided by the classic works of E. R. Leach and Fernand Braudel. See E. R. Leach, *Political Systems of Highland Burma: A Study of Kachin Social Structure* (London: London School of Economics and Political Science, 1954); Fernand Braudel, *The Mediterranean and the Mediterranean World in the Age of Philip II*, 2 vols. (New York: Harper and Row, 1972).
3. See James Francis Warren, "The Sulu Zone, the world capitalist economy and the historical imagination: problematizing global-local interconnections and interdependencies," *Southeast Asian Studies* 35, no. 2 (September 1997): 177–222, 181–83.
4. James Francis Warren, *The Sulu Zone 1768–1898: The Dynamics of External Trade, Slavery and Ethnicity in the Transformation of a Southeast Asian Maritime State* (Singapore: NUS Press, 2007), xxvii.
5. See Nicholas Tarling, *Piracy and Politics in the Malay World: A Study of British Imperialism in Nineteenth Century Southeast Asia* (Singapore: Donald Moore, 1963), and *Sulu and Sabah: A Study of British Policy towards the Philippines and North Borneo from the Late Eighteenth Century* (Kuala Lumpur: Oxford University Press, 1978); L. A. Mills, *British Malaya, 1824–1867* (Kuala Lumpur: Oxford University Press, 1966).
6. On the important orientation of historians engaging in fieldwork akin to anthropologists, see James Francis Warren, *At the Edge of Southeast Asian History* (Quezon City: New Day Publishers, 1987), xiv–xv.
7. On the meaning that people attribute to certain commodities and how those things are used and circulated, see Arjun Appadurai, ed., *The Social Value of Things: Commodities in Cultural Perspective* (Cambridge: Cambridge University Press, 1986); Alan Warde, *Consumption, Food and Taste* (London: Sage Publications, 1997). On commodity chain methodology and circuits see Jennifer Bair, "Analyzing economic organisation: embedded networks and global chains compared," *Economy and Society* 37, no. 3 (2008): 339–64; Edward R. T. Challies, "Commodity chains, rural development and the global agri-food system," *Geography Compass* 2, no. 2 (2008): 375–94; and William H. Friedland, "Reprise on commodity systems methodology," *International Journal of Sociology of Agriculture and Food* 9 (2001): 82–103.
8. See Igor Kopytoff, "The cultural biography of things: commoditization as process," in Appadurai, ed., *The Social Value of Things* (Cambridge: Cam-

bridge University Press, 1986), 64–95; and Florence Palpacuer, "Bringing the social contexts back in: governance and wealth distribution in global commodity chains," *Economy and Society* 37, no. 3 (2008): 392–413.

9. *The Penguin English Dictionary* (Harmondsworth, England: Penguin Books Ltd., 1985), 519.
10. James Francis Warren, *Iranun and Balangingi: Globalization, Maritime Raiding and the Birth of Ethnicity* (Quezon City: New Day Publishers, 2002), vi.
11. Wu Ching-hong, *A Study of References to the Philippines in Chinese Sources from Earliest Time to the Ming Dynasty* (Quezon City: University of the Philippines, 1959), 110.
12. Alexander Dalrymple, *Oriental Repertory,* vol. 2 (London, 1808), 525.
13. Alexander Dalrymple, *Oriental Repertory,* 13–14; "A Memoir on the Sooloogannan Dominion and Commerce," 26 February, 1761, Public Records Office, Egremont Papers, 30/47/20/1.
14. Alexander Dalrymple, "A Memoir on the Sooloogannan Dominion and Commerce," 26 February 1761, Public Records Office, Egremont Papers, 30/47/20/1.
15. T. C. Lai, *At The Chinese Table* (Hong Kong: Oxford University Press, 1984), 6.
16. Lai, *At The Chinese Table,* 11.
17. Hunt, "Some particulars relating to Sulo in the archipelago of Felicia," 54.
18. Warren, *The Sulu Zone 1768–1898,* xxxvi.
19. Warren, *The Sulu Zone 1768–1898,* 74.
20. Warren, *Iranun and Balangingi,* 53.
21. Rennell, *Journal of a Voyage to the Sooloo Islands and the Northwest Coast of Borneo,* 38.
22. Thomas Forrest, *A Voyage to New Guinea and the Moluccas from Balambangan: Including an Account of Magindano, Sooloo and Other Islands* (London: G. Scott, 1979), 322.
23. Forrest, *A Voyage to New Guinea from the Moluccas from Balambangan,* 125.
24. No. 482, Gobernador Capitan General a Senor Secretario de Estado, 31 March 1775, Archivo General de Indias, Filipinas 360; No. 7, Gobernador Capitan General a Senor Secretario de Estado, Filipinas 510, 61, 95; Tomas de Comyn, *State of the Philippines in 1810, Being an Historical, Statistical and Descriptive Account of the Interesting Portion on the Indian Archipelago,* trans. William Walton (Manila: Filipiniana Book Guild, 1969), 119.
25. Norman G. Owen, *Prosperity Without Progress Manila Hemp and Material life in the Colonial Philippines* (Berkeley and Los Angeles: University of California Press, 1984), 27.
26. Pablo Fernandez, O. P., *History of the Church in the Philippines, 1521–1898* (Manila: National Book Store, 1979), 203.
27. James Francis Warren, *The Global Economy and the Sulu Zone: Connections, Commodities and Culture* (Quezon City: New Day Publishers, 1998), 34.
28. Ships' log 656/1833, *Albree.* Abstract log from Boston toward Calcutta, 16 July 1833 to 11 September 1833, kept by Gamaliel E. Ward. Included also: Memo of goods obtainable at Sooloo and a list of goods for the Sooloo market. Philips Library, Salem Peabody Museum.

29. See Charles Frake, "Abu Sayyaf displays of violence and the proliferation of contested identities among Philippine Muslims," *American Anthropologist* 100, no. 1 (1998): 41–54; Benedict Sandin, *The Sea Dayaks of Borneo Before White Rajah Rule* (London: Macmillan, 1967), 63–65, 127; Esther Velthoen, "Wanderers, robbers and bad folk: the politics of violence, protection and trade in Eastern Sulawesi 1750–1850," in *The Last Stand of Autonomous States, 1750–1870 Responses to Modernity in the Diverse Worlds of Southeast Asia and Korea,* ed. Anthony Reid (London: Macmillan 1997), 367–88; James F. Warren, *The Sulu Zone, the World Capitalist Economy and the Historical Imagination* (Amsterdam: VU University Press/CASA, 1998).
30. Warren, *Iranun and Balangingi,* vi.
31. No. 226, II, Archivo General de Indias. Filipinas 492, 10.
32. Historical Data on Catarman, Historical Data Papers Misamis Oriental, National Library of the Philippines, 45.
33. Historical Data on Catarman, Historical Data Papers Misamis Oriental, National Library of the Philippines, 6.
34. See Frake, "Abu Sayyaf displays of violence"; Sandin, *The Sea Dayaks of Borneo Before White Rajah Rule,* 63–65, 127; Esther Velthoen, "Wanderers, robbers and bad folk"; Warren, *The Sulu Zone The World Capitalist Economy and the Historical Imagination.*
35. Gavin Young, *In Search of Conrad* (London: Penguin, 1991), 254; Warren, *The Sulu Zone 1768–1898,* 85–86.
36. Michael Southon, *The Navel of the Perahu: Meaning and Values in the Maritime Trading Economy of a Butonese Village* (Canberra, Australia: Research School of Pacific and Asian Studies, 1995), 22.
37. Michael Cullinane and Peter Xenos, "The growth of population in Cebu during the Spanish era: constructing a regional demography from local sources," in *Population and History the Demographic Origins of the Modern Philippines,* ed. Daniel F. Doeppers and Peter Xenos (Madison: University of Wisconsin Press, Center for Southeast Asian Studies, 1998), 71–138, 89.
38. Cesar A. Majul, "Political and historical notes on the old Sulu Sultanate," *Journal of the Malaysian Branch Royal Asiatic Society* 38, no. 1 (1965): 23–43, 28.
39. Warren, *The Sulu Zone 1768–1898,* 252.
40. Warren, *The Global Economy and the Sulu Zone,* 25.
41. John Comaroff and Jean Comaroff, *Ethnography and the Historical Imagination* (Boulder, CO: Westview Press, 1992), 6.
42. Comaroff and Comaroff, *Ethnography and the Historical Imagination,* 14.
43. For a list of the ceramic and earthenware that comprised the cargoes of Amoy junks circa 1776 and 1814 see Appendix B in Warren, *The Sulu Zone 1768–1898,* 259–60.
44. Alexander Spoehr, *Zamboanga and Sulu: An Archeological Approach to Ethnic Diversity,* Ethnology Monograph No. 1, Dept. of Anthropology, University of Pittsburgh, 1973, 214.
45. Warren, *The Sulu Zone 1768–1898,* 19.
46. Jules Sebastion Cesar Dumont D'Urville, *Voyages au pole sud et dans L' Oceanie sur les corvettes L'Astrolobe et la Zelee pendant Les Annees 1837–1838–1839–1840,* vol. 7 (Paris: Gide et J. Baudry, 1841–1854), 198.

47. Thomas Forrest, *A Voyage from Calcutta to the Mergui Archipelago lying on the East Side of the Bay of Bengal* (London: 1792), 83.
48. See Margarita de Los Reyes Cojuangco, *Kris of Valor: The Samal Balangingi's Defiance and Diaspora* (Manila: Manisan Publishing Ltd., 1993), and Warren, *The Sulu Zone 1768–1898,* 104–25.
49. Warren, *The Sulu Zone 1768–1898,* xix.
50. Warren, *The Sulu Zone 1768–1898,* xv.
51. Warren, *Iranun and Balangingi,* xxix.
52. Wolf, *Europe and the People without History,* 390.
53. Warren, *The Global Economy and the Sulu Zone,* 9–10.
54. Warren, *The Global Economy and the Sulu Zone,* 9–10; Peter Burke, *History and Social Theory* (Oxford: Polity Press, 1992), 152.
55. Warren, *The Sulu Zone 1768–1898,* 154; Owen Rutter, *The Pirate Wind Tales of the Sea Robbers of Malaya* (Singapore: Oxford University Press, 1986), 28.
56. Warren, *Iranun and Balangingi,* 123.
57. Warren, *Iranun and Balangingi,* 23.
58. Warren, *Iranun and Balangingi,* 388.
59. Roger M. Keesing, "Asian cultures?" *Southeast Asian Review* 15, no. 2 (1991): 43–50, 46.
60. Warren, *Iranun and Balangingi,* 412.
61. See Benedict Anderson, *Imagined Communities: Reflections on the Origin and Spread of Nationalism* (Manila: Anvil Press, 2003).
62. Michael Mann, *The Sources of Social Power* (Cambridge: Cambridge University Press, 1986), 490.
63. Kenneth Prewitt, "Presidential Items," *Items Social Science Research Council* 50, no. 1 (March 1996): 15–18, 16.
64. Prewitt, "Presidential Items," 16.

SIX

Connecting Maritime and Continental History: The Black Sea Region at the Time of the Mongol Empire

Nicola Di Cosmo

The Athenian, the Venetian, the Portuguese: three thalassocracies of different scales that became political Leviathans of their own age, and as such established their power onto "their" seas and oceans.[1] But maritime history, as David Armitage reminds us, is not just the struggle between states and empires. In his register, maritime history is about the life, economy, and culture of seafaring peoples; the networks they created; and the connections they established between discrete places, whose history can be traced back into the remotest prehistory.[2] As Peter Miller acutely points out in this book's introduction, the phrase "maritime history" glosses over the critical distinction between seas and oceans, which, in keeping with the dualism between waters and lands, would be similar to speaking of "terrestrial history" whether we look at continents or peninsulas. Control exercised by states over oceans or seas is not just a matter of scale, and the relative historiography needs to be cognizant of it. Thalassography retained, however, a distinct sea-based meaning before the discovery of Atlantic navigation and the advent of ocean-faring ships. The contact points between land and sea form specific "dots" which, once they are connected, become historically and geographically dense nexuses. Several world regions whose seas have been explored and frequented since prehistory—Europe, North Africa, India, Southeast and East Asia—could not be understood without the networking apparatus developed and subsumed by the communities traveling across the connecting routes. The thalassography of the seas represents, then, the prehistory of the thalassography of the oceans; the stages of development of maritime history that precede the qualitatively evolved "invention" of oceanic historiography, but at the same time provide the blueprint for some of its organizing principles.

Historians have studied these networks, at least to a degree, in terms of systems of communication, of economic exchange, and also of political and administrative control. Once the stability, density, and frequency of the relations among distant places created by maritime "agents" are taken into consideration, it is natural to question whether such relations amount to a series of loosely concatenated networks, to multiple circuits with a high degree of interconnection, or to several subsystems subsumed under some overarching integrated system. When wrought into larger temporal frameworks, the ebbs and flows of maritime history have been inscribed in periodizations based on evolving networks created in response to technological advances, competition among states, geographic discoveries, imperial expansions, population movements, and so forth. At the macrohistorical level, sea- and ocean-borne communication has been indispensable to the development of theories of world history, from ancient to modern times.[3] In large part, one may even claim, the world from around 1500 has to be conceived as modern because it began to be shaped as a growing mesh of networks of which the maritime routes were the supporting scaffolding.

Before 1500, sea navigation was limited to the internal or coastal spaces of some bodies of water (the Mediterranean, the Indian Ocean, the southeast Asian and China Seas) with relatively minimal communication between them. While a certain degree of interpenetration between more or less contiguous maritime areas certainly thrived in history (as, for instance, between the Red Sea and the Indian Ocean), and various places performed important roles as "joints" between separate areas and "circuits," the overall connectivity remained tenuous at best and vulnerable to political upheaval. The conceptual leap from regional to world history appears therefore to depend upon the breaking out of these regional systems into oceanic connections that offered entirely new configurations.

A persistent problem in pre-early modern history has been the absence of an effort to define the existence of connectivity between and among regional systems that appear to be both separate and fundamentally different. At least two questions are in order: first, what degree of permeability can we establish between apparently separate systems? and, second, what are the forms of connectivity and the forces (or agents) that made such connections possible? When it comes to connections between the Mediterranean and the Indian Ocean from the ancient to the medieval period it is of course in the Arabian Peninsula, Red Sea, Persian Gulf, and Indian Ocean coastal areas that one must investigate.[4] When it comes, on the other hand, to fleshing out connections and networks between Europe and "High Asia," most efforts have been directed to giving historical dignity to that singularly slippery concept that is the Silk Road. The results cannot be said to have been, to date, especially promising, not because the efforts have not been valiant, but because the sources are so scant that interpretations stand on extremely thin ice, which tends to crack under more weighty scrutiny.

Iconographic, material, and textual sources that document movement across the continental expanse of Asia still do not provide sufficient evidence to trace the continuous existence of commercial or other (such as religious) networks over time, gauge their consistency, or establish their relevance in the societies that participated in them. Moreover, our sources do not tell us how the putative existence and growth of such continent-wide networks may be connected with networks in other regional circuits. Therefore, the manner in which the Mediterranean economy may be linked to China by overland routes from early Roman times to the end of the Byzantine Empire, or, more specifically, how silk inscribed with Chinese characters ended up in Palmyra, remains, for all practical purposes, a mystery.[5] More promising may be the study of coastal navigation through the southern routes of India and southeast Asia, but this is not the focus of my interest here nor is it indeed the main focus of Silk Road studies.[6]

What I am concerned with is the existence of long-distance contacts between the Mediterranean and Asia through the "steppe route" that connects northern China with central Asia, reaching in the West the shores of the Black Sea, the sea known to the Greeks as *Pontus Euxinus* (the "hospitable sea"), to the north of which Herodotus placed Scythia. The East-West sequence of grassland, oases, plateaus, and low-lying mountain passes can be negotiated with relative ease, thus making it possible for mobile people to move across Eurasia from prehistoric time. The overland routes and the traffic along them form the concept of a "Silk Road," which refers of course not just to silk trade but to any goods transported over long distances.[7]

The notion of long-distance connections across continental Asia is intriguing in a thalassographic sense for two reasons. The first, and less important, is that the horse and camel caravans associated with long-distance trade across Asia have often been viewed as quasi-analogous to maritime transportation. The second is that a few times in history it has happened that the "system" constituted by continental Asian trading networks has joined, in ways that of course are different in each case but whose existence is impossible to deny, the Mediterranean networks. It is at these historical junctures that something rather different and special seems to happen in the relations between two regions—the Mediterranean and continental Asia—that appear at other times far less permeable, if not entirely closed off. Whether these connections have always been operative and simply come to the surface at times when social, cultural, or political transformations produce more—or more transparent—evidence, or whether it is political and economic changes that actively create such connections, is a matter of the greatest historical interest, requiring a separate investigation in each case.[8] Many valuable studies exist, but there lacks a rationalization or sufficient conceptualization of this phenomenon as a probe into macrohistory, or into the history of world systems before 1500.

Aligning the goods-laden Bactrian camel with the seafaring ship has a certain romantic appeal, and as such has been endorsed by writers who have imagined the steppe as a sea, and the nomads as sailors. It was most famously Arnold Toynbee who established an analogy between steppes and seas as performing similar roles in world history.[9] Other historians have seen the oceanic transcontinental navigation of the post-1500s as the replacement of those caravans that crossed the steppe and deserts of Inner Asia from an unfathomed antiquity.[10] But what are the bases for the analogy? In part it has to do with the way in which the region itself (Inner Asia) has been understood. It was Alexander von Humboldt who first gave a scientific definition of this region as the drainage basin of inland-flowing rivers comprised between, roughly, the Ural and the Tianshan Mountains.[11] That notion of the Inner Asian "heartland," surrounded by "marginal lands," or regions of established and continuous civilizations, could easily be compared, after all, to a body of water such as the Mediterranean.[12] Second, as previously mentioned, the presence of an uninterrupted grassland belt connecting Hungary with Mongolia made the region as vast and as crossable as an ocean. Third, the presence of oases made it possible to find the occasional harbor. Fourth, historians have pointed to the cultural cohesion of the steppe zone, where the nomads' mobility favored interregional connections and long-range contacts. Their technology, from metallurgy to chariots, and the advances in horse domestication and pastoral production were critical to the formation of complex nomadic societies, large political entities, and even empires. In this, their agency in cultural transmission can be compared to that which was occurring between maritime societies. Fifth, the nomads' economic specialization stimulated commercial exchange as well as less peaceful relations with city dwellers, not unlike those traders and raiders that in every historical time have crossed the seas. Sixth, like the Mediterranean, Inner Asia was at times incorporated into overarching imperial structures, but it was more commonly politically divided. Within it, subsystems can also be easily identified, such as the oases in the Tarim basin (modern Xinjiang, in northwest China), the Transoxiana region between the Amu Darya and the Syr Darya, or the Pontic region north of the Black Sea. These were open to interactions with other regional empires and civilizations such as the Chinese, the Indian, the Iranian, and the Greco-Roman. For most of its history the Mediterranean was not much different. On its shores flourished different religions, empires, and civilizations, and the only time it was effectively unified politically was under Rome.

But in order for the analogy between steppe and sea to have some analytical value, one has to go beyond superficial similarities, aside from the fact that camels, after all, are not ships. In particular, what seems to have been missing in world historical studies is a systematic attention to those rare moments when a special historical conjuncture makes it possible for the system of networks that developed within the steppe region and the system of networks that developed within the Mediterranean to become more closely integrated; namely, do special things happen, in the greater Eurasian region, when the maritime system of Mediterranean networks and the continental system of central Asian networks are connected?

The intuition that a higher plane of historical (and thalassographic) analysis needs to be engaged when Asia and Europe are more closely connected is not entirely new. One of the earliest attempts can be perhaps traced back to Frederick Teggart's influential study that mutual stimuli occurred between Rome and Han China—a theory that has shaped to a degree our current notion of the Silk Road as a conduit of communication between East and West.[13] The realization that a degree of connectivity existed between Han China and the Roman Mediterranean rested on the theory that the Han advance into central Asia was the spark that ignited the Silk Road engine as the Chinese exported silk to Greek Bactria (and further west to the Roman world) and imported horses from Ferghana, among other things. Frederick Teggart placed much emphasis on the role of empires and their conflicts with barbarians, focusing on correspondences and correlations rather than on what we might call "connections." To him, correspondences suggested causal nexuses that needed to be included in a systematic approach to historical correlations. It itself, Teggart's search for correlations between Asian and European history was certainly not a novelty. In the eighteenth century, Edward Gibbon and the Orientalist Joseph de Guignes endorsed the idea that the barbarians in the East had a direct and powerful impact on the more familiar barbarians who invaded Rome.[14] The late nineteenth-century pronouncement by Charles de Ujfalvy, according to which Rome fell because China built a wall, is one of several links, in an ideal discourse of Eurasian connectivity, between eighteenth-century Gibbon and de Guignes with twentieth-century Teggart.[15] But the space between the two empires—the connecting tissue, so to speak—has since been left essentially unexplored on the historical level, and while archaeological work has shed some light, this is not sufficient to allow for a better understanding of connections.

More recent scholarly approaches have been quite different. The dominant paradigm of Sino-Roman studies is one of comparisons between empires rather than one of interactions and encounters across Eurasia.[16] This approach may yield valuable insights into the forms and formations of ancient empires, but what matters to the historian of networks of the Silk Road is the broader picture outside the cone of light shed by empires, and in particular, the fate of systems of circulation of goods, of people, and the mechanisms that governed the many loci of interaction, exchange, and transmission. In other words, beyond the correlations or comparisons that one may establish between China and Europe, the questions that matter to the macrohistorian are if and how the connections were established, their origin and nature, and the people who activated and controlled them. This process can be represented, perhaps, as a Rube Goldberg machine in which between cause and effect are the most intricate and convoluted intermediate steps whose form or shape cannot be predicted. The historian can see, at best, a beginning and an end, but all that happens in between is most often barely visible, if at all.

This is a type of investigation that has so far eluded rigorous study at the macroscopic level, but that can be pursued by focusing on those periods in which Inner Asian history is more penetrable and connections are less tenuous. In the premodern history of the central Asian steppes we can recognize, roughly speaking, two or three major areas where strong polities, and sometimes empires, emerged. In the West, the Scythians, and later the Parthians, Avars, Khazars, Cumans, and Pechenegs played historically relevant roles. In the East, the inhospitable steppes and deserts of Mongolia were the birthplace of a number of empires, two of which unified the whole of its territory, albeit for a relatively short time; namely, the Türks in the sixth century and the Mongols in the twelfth and thirteenth centuries. The central segment of the Silk Road—what is today the region of Xinjiang in northwestern China—was since prehistory a critical node along the transcontinental routes, and historically dominated, from ancient to early modern times, by Xiongnu, Türks, Tibetans, Tanguts, Uighurs, Kara-Kitai, Mongols, Kazaks, and Zungars, until it was finally conquered by the Qing dynasty in 1758. Much of the history of this region is about trade routes and who controlled them.

At the same time, we know precious little about the actual extent of communication, movement, migration, and exchange: this is a history, like much of nomadic history, that has been recorded very poorly. Yet, our sources yield sufficient clues to assume that, at least intermittently, trading networks spread between China and the Mediterranean achieved a degree of internal coherence, and may perhaps be regarded as forming a system that allowed a higher level of connectivity.

We can clearly identify three historical periods in which a system of trading routes appear to flourish. The first such favorable conjuncture, sometime hailed as the beginning of the Silk Road, occurs in the second–first century BCE, when Xiongnu imperial expansion and then the Han dynasty's leap into central Asia allowed China and Greek Bactria to connect. The second moment of expansion takes place in the sixth century CE with the establishment of the Türkish empire and of direct contacts between Byzantium and the Türkish Khaganate, and exchange of diplomatic and commercial missions.[17] Around this time, trade—as well as religious and cultural contacts—grew in intensity, with epochal repercussions on China, India, Iran, and of course, Byzantium.[18] The third moment of integration is generally attributed to the Mongol conquest, more about which shall be discussed shortly.

I do not see these periods as corresponding to cycles; that is to say, as resulting from the cyclical appearance of powerful nomadic empires. Nor do I see them as discrete stages in a linear process of creating gradually more complex systems. I see them, rather, as periods in which given circumstances allowed for a reconfiguration of some parts internal to previously existing networks so that two separate systems were able to come into closer contact. Here my understanding of system is based on Edgar Morin's definition, following Philippe Beaujard, according to which a system is: (1) a unit of the entire complex of relations between the whole and its parts, (2) is made of cumulative interactions, and moreover, (3) these interactions constitute the organization of the system.[19] In other words, a system is organized through the interactions between the whole and its parts, between the individual parts, and within each part. This definition of a system is not in contradiction with the Mediterranean Paradigm described by Alain Bresson, and in fact allows us to distinguish, as Bresson does, between macro and micro interactions, in what Bresson calls "forms of connectivity."[20] The whole question of connectivity is tied to several ideas and views advanced in Nicholas Purcell and Peregrine Horden's celebrated *Corrupting Sea*, and to the debate that has followed, to which the essays included in *Rethinking the Mediterranean* constitutes an especially notable contribution.[21] Connectivity, as well as interaction, implies communication, and within a regional system various forms of communication can be identified. Environment, trade and exchange, technology, and demographic movements, as well as political action, are all closely linked to the definition of the interactions that in turn generate a system. In these general and theoretical terms the network system operating in the steppe region of Inner Asia was not necessarily different from any other, and displayed both macro and micro interactions.

Arguably, it was during the century-long Mongol domination of Eurasia (c. 1250–1350) that the mechanism of interlocking network systems between continental Asia and the Mediterranean achieved its highest level of integration in the premodern world.[22] But there is one additional element that makes the Mongol period different from the previous ones. Namely, for the first time in history a clear, deliberate construction of institutions, social bodies, and political apparatus allowed the linkage between the continent-based overland trade networks and the Mediterranean exchange and transportation system to become operative.

Around the end of the thirteenth century and especially in the first half if the fourteenth, the Mediterranean maritime powers of Genoa and Venice established, with Mongol political consent and logistical support, their *comptoirs* (emporia, colonies, or commercial settlements), on the Black Sea. The main point of interest to Mediterranean historians, as well as to world historians, is whether we can draw any significant conclusion from an analysis of the motives that were behind the connections deliberately sought and created between the steppe system and the Mediterranean system. In other words, was this connection one that gave rise to special phenomena, greater than its parts, and can this be taken as an example, or a type-case of "amphibian" connectivity between maritime and continental history?

The Black Sea Region and the Mongol Conquest

At the time of Mongol domination the Black Sea became, in the words of Gheorghe Bratianu—the Romanian historian who was the first to investigate its role in relation to both Asia and Europe—the *plaque tournante* of international trade.[23] The concept of a *plaque tournante* fits well with the distributive mechanism among different, otherwise disconnected areas of which the Black Sea became the central hub at the time of the Mongol conquest. The Black Sea was neither an extension of the Mediterranean system, nor a system in its own right, but rather a conveyor belt that allowed two fully separate ecological and economic systems to become integrated into a common market of exchange and circulation whose capillaries potentially extended to China in the East and Spain in the West. The central question that has exercised generations of historians is, of course, how such a system came into existence.

Historians have often looked at the Venetian and Genoese expansion from a contemporary European perspective or from the perspective of the later European oceanic voyages and colonial-maritime empires. J. R. S. Phillips, for instance, stated that the "activities of European merchants in central Asia and the Far East were not simply the result of opportunities created by the Mongol conquest in the thirteenth century, but grew directly out of the expansion of the European economy . . ."[24] Philip Curtin has credited the Venetian and Genoese system of trading-posts empires of the Mediterranean (and of course, of the Black Sea) as a model for the post-1500 Portuguese colonial expansion.[25] Looking at the way in which maritime and continental systems became integrated across and through the Black Sea, it is essential to understand what forces acted upon the various systems, what mechanisms were activated to connect them to each other, and how (and why) such an Eurasian amphibian macrosystem eventually failed and ceased to exist, effectively opening the road to other and more daring attempts that favored Oceanic exploration and colonization. Herein lies the fundamental importance of this experiment in macro-system creation: it was the product of forces that—albeit rather weak and immature, and eventually inadequate to the task—were conscious of what they were attempting to do. Has their role been fully appreciated within the field of world history and the history of globalization?

In a recent work, the Black Sea has been presented as an independent unit of analysis, thus different from Braudel's notion of it as the "backyard of the Mediterranean," and instead leaning toward Bratianu's synthesis of the Black Sea as a region that historically alternated between being an extension of the Mediterranean and a separate system in its own right, with ramifications into the Asian steppes and other regions.[26] If we can regard the Black Sea as a self-supporting system, this is only because of its role as the mechanism that allowed the establishment of a linkage between separate and larger network systems. It became the meeting point of Eurasian continental routes and European maritime routes, but it was also the terminus coming from the West, beyond which urban, agrarian, and seaborne civilizations were gradually replaced by pastoral and nomadic cultures interspersed with sedentary enclaves, oases, and a few cities. Its eastern and northern coasts were the departure point of trade routes that would eventually reach the Caspian Sea, Transoxiana, the Tarim Basin, and China. The ship and the camel met on its shores.

For the ancients, the Black Sea was a place to avoid—rather remote and extraneous to the Mediterranean space.[27] In antiquity, Persians, Scythians, and Greeks had inhabited the regions it bounded, crossed its treacherous waters, and established (not always amicable) contacts among each other. It was only under Byzantine rule, and specifically from the sixth century, that the Black Sea acquired a clearer transcontinental vocation, as a bridge between a Roman world that gravitated on the eastern part of the Mediterranean and the world of the steppes that had been placed under the domination of the Türks, the horseriding warriors hailing from Mongolia whose nomadic empire stretched from China to central Asia. The diplomatic relations between Byzantium and the Türks followed and were possibly a direct consequence of the commercial networks that had already emerged in central Asia at an earlier time. Controlled by central Asian merchants (especially Sogdians), these networks—possibly in close symbiosis with the political and military might of Türks and other powers—were responsible for phenomena such as the vaunted cosmopolitanism of the early phase of the Tang Dynasty (618–907), while silk production spread to the West.[28]

The Black Sea itself came under the domination of the Khazar Khaganate, a Turkic kingdom that appreciated trade and knew how to protect it. Among the things traded along the Khazar routes were silks, furs, candle wax, honey, jewelry, silverware, coins, and spices. Jewish, Sogdian, and other traders profited from the open and tolerant rule of the Khazars.[29] Generally the Khazars protected the trade routes from the predatory activities of the Pechenegs and other loose nomadic peoples, while trying to maintain peaceful relations with Byzantium and the expanding Islamic empire through careful diplomacy. From the tenth century onwards, the power of the Khazars waned. They lost control of the trade routes to the nomadic Pechenegs and Volga Bulgars. Eventually, the expansion of Rus' in the north and of the Seljuk Turks in the Anatolian region, as well as the Islamic advance in central Asia, rearranged the political boundaries in ways that appear to have hindered long-distance trade. At the same time, Tang China was shaken at the foundations by the disastrous An Lushan rebellion (755–758), which was followed by the disruption and disappearance of the Sogdian commercial networks in the eighth and ninth centuries. Later, the Song dynasty (960–1274) was surrounded by powers that attempted to establish their hegemony over various segments of the transcontinental trade network.[30] Yet, the period from the tenth to about the twelfth century was defined by a less integrated system, even though by no means one in which all long-distance trade disappeared. Still, a clear and organic linkage between the Mediterranean and China has yet to be documented, nor does the Black Sea appear to have had a central position in transcontinental trade before the Mongol conquest.

The *Pax Mongolica*

In the thirteenth century, the Mongol conquest changed everything. One after another, Asian and European powers fell before the onslaught of the Mongol armies. The early Mongol military operations in central Asia, which foreshadowed the entire campaign for the conquest of Russia and Persia, began allegedly to seek revenge for the outrageous cold-blooded assassination of Mongol envoys by the Khwarizm shah. The shah had presumably (none of this can be said for certain) perceived the Mongol mission as an attempt to extend Mongol control over the trade routes that connected central Asia with China. If this were true, then he sorely miscalculated, by offering the perfect *casus belli* to Chinggis Khan. The Mongols proceeded then to conquer central Asia and to establish their hegemony over transcontinental trade routes. The mutual support between Mongol rulers and representatives of the merchant and administrative classes of Khwarezm is well known and testifies to the broader views and attitudes of the Mongols in relation to commerce. Once the conquest was completed and the Mongols achieved their military and political supremacy, they also dominated the trade networks, but how this control was exercised has to be seen in light of social, cultural, and administrative practices. Under Mongol rule the roads were secure, and men and goods crossed the lands with few or no risks of being robbed. According to Mongol regulations, moreover, if merchants met with misfortune and died while traveling, their goods and property could be recovered by relatives.

If in general we can say that the Mongol conquest was favorable to trade, the Mongols' role has been more often interpreted as one of enabling it, than one of direct implementation of specific trade-friendly policies. This is true to a point, and we need to bear in mind that the political cultures of the different realms (*ulus*) into which the Mongol empire was soon to be divided evolved in different directions. But it is difficult to deny that the Mongols were active participants in a culture of trade and exchange that extended across their entire domain.

It is at this point that we must consider two specific features of the Mongol Empire: the *jam* and the *ortaq*. The *jam* (or *yam* in its Turkic form) was a system of postal stations used by Mongol military messengers and other officials which extended over the lands they had conquered and provided logistic support to whomever had government permission to access it. The *jam* system, which functioned across the whole Mongol Empire, allowed merchants to accurately calculate travel times and costs, while commercial traffic was at the same time subject to scrupulous government supervision.[31]

Ortaq (a Turkic word meaning "partner") was the term used to indicate the officially licensed merchants that formed an association with members of the Mongol elites.[32] Aristocrats and powerful military leaders allowed such merchants to organize the provisioning of Mongol courts, cities, and their personal estates. Karakorum, the capital of Mongolia in the mid-thirteenth century, had to be supplied daily with five hundred camel loads of sundry staples and merchandise. The remoteness of the place and the quantity of provisions that it required suggest the existence of a sophisticated logistic network. The *ortaq* system is also mentioned in the *Codex Cumanicus* (c. 1330), in which the word is translated in Latin as *socium*.[33] Some of these merchants amassed enormous fortunes. They also, eventually, drew popular opposition in Yuan China, as foreigners in the service of a foreign dynasty, abuses of power, alienation from the local population, and their frequent employment as tax collectors and government agents made them odious to the Chinese people. However, feelings of resentment and resistance were mostly confined to the central and southern regions of China, and therefore to areas that were geographically peripheral to the overall Eurasian commercial networks controlled by the Mongols. The *ortaq* and the *jam* remained in different forms institutions integral to the various realms of the Mongol Empire until its end.

What emerges from these cursory considerations is that the Mongols had both the political culture and the administrative and military means to support a vast commercial network within the territories under their control. These means were the necessary conditions that allowed the Mongol continental commercial network to effectively link to the Mediterranean maritime one. Historians have concentrated, by and large, on trying to document the presence of Italian merchants on the Black Sea and beyond, and especially on the reasons that made such far-flung colonies so successful and resilient, while the eastern extension of the trading network, beyond the Caspian Sea, has received relatively less attention.[34] Why would Europeans be interested in trade with the East, and what could or did they sell and buy from China? This has been an issue of debate for decades, and based on rather few pieces of solid evidence that, while they show a substantial and thriving presence of merchants on the Black Sea and a modest but still remarkable number of individuals who pushed themselves all the way to China, do not allow us to establish with any degree of certainty the characteristics of trans-Caspian Sea trade as to size, frequency, or type of merchandise.[35]

In the first place it is necessary to appreciate the fact that the establishment of the Genoese and Venetian bases on the Black Sea has little to do with an assumed desire to extend European hegemony into farther Asia. It is also questionable that the wars that marked the commercial penetration of the Black Sea were generated by a desire to expand the Mediterranean trade networks into new and profitable markets. The Treaty of Nympheus of 1261 between Genoa and Byzantium that allowed Genoa passage through the Straits and access to the Black Sea was conceived and realized as part of an overall strategy to achieve control of the Mediterranean trading routes, rather than to see it as an attempt to build bridgeheads toward Asia and China. The Black Sea acquired special relevance not so much because of the profits that could be generated, but because of the importation of rich grain supplies and agricultural products. Both Genoa and Venice were vulnerable to commercial embargos from their hinterland that could have interrupted the provisioning of vital supplies to their cities. Access to these supplies from other markets, reachable by sea, was therefore vital. Martin da Canal, the Venetian chronicler, mentions the trade that Venice was also starting to carry out with the Black Sea exactly in these terms: the high probability of famines or wars was the main incentive that spurred the Venetian government to seek an opening toward the Black Sea—far more valuable as a reservoir of wheat than as the gateway to Oriental markets of spices and silk.[36] The relevance of the strategic over the purely commercial aspects of this expansion into the Black Sea must be given due weight as we analyze the development of two major commercial bases: Caffa in the Crimean peninsula, and Tana on the Sea of Azov. Commercial ventures were at the heart of the political success of Genoa and Venice, as the Fourth Crusade (1204) amply demonstrated, but on the Black Sea, at least initially, both of them were looking for security. The search for profits was second nature to them, and the range of commercial ventures expanded as soon as the Genoese and Venetians realized that there were possibilities to be exploited. But commercial expansion should not be regarded as the prime motive for the arrival of the Italian maritime republics on the Black Sea.

Caffa and Tana

Caffa, the Genoese basis, and Tana, the main Venetian emporium, were run very differently, and we cannot enter here the details of their political machinery, but in terms of the relations with the Mongol political power we only need to point to two aspects: trade and treaties.

By trade I mean the system of exchange that was mutually beneficial to the Europeans and to their commercial and political partners in the region—first of all, the Mongols. Trade assured the Mongols tariffs and taxes, but especially allowed a very important outlet for local products that increased enormously the revenues of the native merchant class and aristocrats: the slave trade found a ready and unlimited market in Europe, the Levant, and North Africa. Other products of the local pastoral economy were much in demand—such as hides—in addition to the fish and caviar from the Black and Caspian Seas, along with wine and grain. The Mongols won twice: by increasing the volume of the exports produced in lands that they controlled, and by taxing trade. Additional income was provided by transaction costs such as sales of pack animals, security guards, passage fees, and supplies for merchants—not to mention the growth of the economy in the emporia along the Black Sea, where various professional people (artisans, interpreters, guides, servants, and so one) were employed by the Latins.

The relationship between Mongols and Italians was favored by legally granted protections and by a system of mutual responsibilities and accountability established by treaties. Technically, these were not actually bilateral treaties, but rather were edicts issued by the Mongol khans that provided certain concessions and privileges. In reality, there is no doubt that the terms of the edict were arrived at through a process of negotiation between representatives of the Genoese and Venetian governments and the khan himself. The treaties regulated tariffs, the standing of foreign authorities in Mongol lands, the procedures to be adopted in case of legal disputes, the location and size of the foreign settlements, and the security provided to foreign persons and to their goods. Both the commercial and the political environments guaranteed a fairly solid linkage between the continental networks of the Mongols and the European and Mediterranean trade controlled in the fourteenth century by the Italian maritime powers.[37]

Let us now look more closely at what this linkage actually produced. It has already been mentioned that it allowed Genoa and Venice to access markets of strategic interest to them, and it has also been said that it served as an outlet for the export of Asian goods. There were other results as well that are better measured on a longer time span. If we look at the relationship between Asia and the Mediterranean through the lens of early modern transatlantic history, and in particular what has been called the "Columbian Exchange," then we can discern a level of exchange that goes well beyond trading goods, and engages the artistic, scientific, and broadly cultural spheres. The Mongol courts became magnets of products and peoples that came from every corner of the known world, and centers of a new cosmopolitanism that surpassed or at least rivaled that of any contemporary European court. Europeans with valuable skills—from silversmiths to falconers, and from astronomers to philosophers—were not just welcomed but also actively sought after. On the artistic level, the spread of precious Islamic textiles throughout the Eurasian world at this time, especially the gold-thread silken products much praised and appreciated by Mongol rulers, was favored by the Mongols' own political culture and by the permeability of the regions under their rule.[38] Among the most significant instances of cultural transmission we should mention is the possible transfer of gunpowder from China to Europe, with the intermediation of Mongols and Muslims, sometime in the fourteenth century—a technology transfer with epochal long-term consequences for the future rise of Europe.[39]

Food and medicine circulation was unprecedented, as we can see from the *Yinshan zhengyao* ("Proper Essentials for the [Emperor's] Food and Drink"), a 1330 text about the variety and international nature of imperial Mongol cuisine and pharmacopeia.[40] This book clearly shows the Mongols as the rulers of a world empire, whose daily necessities were gathered from the whole of Eurasia. Hunting, too, whose technical knowledge and practices were shared across all the major Eurasian courts, contributed to the creation of networks of exchange, in particular of animals and animal trainers.[41] The greater availability of artisans and their products, foodstuffs, medical supplies, plants, and animals was made possible by an integrated and ideally universalistic system of exchange and circulation.

The biological exchange must also include the downside that undoubtedly was the Black Death. The actual nature of the epidemic, where it came from, and how the contagion spread to Europe have recently been the object of intense scrutiny, and recent studies have concluded that the disease was carried from Asia to Europe with the movement of peoples and armies along the steppe routes.[42]

The actual depth of this exchange has not been plumbed yet, but it certainly involved the participation of various European merchants, travelers, and missionaries, whose presence in China has been documented during the period of Mongol rule. If Marco Polo in a sense pioneered the European exploration of China, he was by no means alone.[43]

The End of the *Pax Mongolica*

The unprecedented openness to trade and volume of exchange across Eurasian continental routes that had been common under Mongol rule were not to last. The end of the *Pax Mongolica* is usually considered responsible for the closure of intercontinental routes. The story is well known. From around 1350 onwards, gradual disruption of the "Mongol peace" caused the interruption of trade routes amidst political chaos and military activity. In Persia, the Il-khanid dynasty suddenly collapsed in 1335, in the Golden Horde from the 1350s various lords began to vie for power, and the end of the Yuan dynasty in China in 1368 seemed to give the *coup de grâce* to any residual Mongol supremacy in Asia. The rise of Tamerlane toward the end of the century further disrupted trade in central Asia and threatened the very existence of the Italian bases on the Black Sea. These bases eventually survived Timur's wrath, but the connection between Mediterranean and deeper Asia did not. As the story goes, within a few decades the development of Atlantic navigation would open a new era in European and world history. If we ask ourselves, however, what happened to the European merchants who had been active in Asia during the Mongol period, they simply seem to have disappeared, even though the Italian *comptoirs* continued to operate, albeit in a reduced capacity.

The Black Sea was the terminal point beyond which Venice and Genoa could not extend their commercial and political influence as states. Beyond that, the merchants had to entrust themselves to the Mongol authorities and to their support system. The Black Sea did not represent only a *plaque tournante;* that is, a system of distribution of goods. It was also the terminus beyond which the Mongols (the quintessential continental power) and the Italian republics (the quintessential maritime powers) could *not* project their power. The Mongols depended, for their access to the Mediterranean system, on the transport ships provided by the Italians, and the Italians depended on the Mongols to guarantee the conditions that made overland trade possible and profitable.

We have evidence of this mutual dependency not just in the virtual disappearance of any further contact between China and Europe once the Mongol states collapsed, but also in the embargo that Genoa and Venice—for once united—imposed against Janibek, khan of the Golden Horde, to force a reopening of the trade routes after hostilities erupted in 1343, followed by the breakdown in relations (political and commercial) that continued for several years. The Italians were convinced that the Mongols would eventually relent, and they were right. Janibek eventually came to terms with both Venetians and Genoese, and the previous favorable conditions were resumed, albeit with slight variations.[44]

Had the Italians established an official presence in the Far East that was not simply left to individual merchants and to their private relationship with their Mongol partners and protectors, the fate of overland trade would have likely been different. But neither Genoa nor Venice, even when presented with the opportunity, had wanted to send their ambassadors to China.[45] The reason can be explained in relatively simple terms: the high cost of the mission in the absence of any obvious political or strategic interest. We do not know exactly why no accredited diplomatic representatives reached the Yuan capital in China, but we may not be too far off the mark if we assumed that such a notion would have been regarded as totally extraneous to the intentions of the two Republics: their orientation was toward the sea, not the land. The few (or the many, depending on how we look at this problem) Europeans who did go to China might have generated a network of commercial relations that could theoretically spread over both systems. Without the support of their own states, and given their total dependency on their Mongol lords, however, the survival of such networks could not endure past the end of the *Pax Mongolica.*

Concluding Remarks on Amphibian History

At the time of the Mongol conquest, an amphibian system of conjoined maritime and continental networks came into being. It lasted long enough to show that such a connection could be created. Counterfactually, we may argue that the right circumstances could have allowed for the development of a more permanent system of communication between China and the Mediterranean. This, however, would have required that the Mongols be in power long enough to allow for greater involvement and more substantial investments from the European merchants and the establishment of more permanent structures. Secondly, European governments had to provide the financial means and political will to sustain such an ambitious commercial expansion. Neither condition was obtained, and yet knowledge gathered by the people who travelled East remained a powerful stimulus for future travels. But of course, Christopher Columbus had the force of the Spanish crown behind him.

The proposition mentioned above that interesting things happen when maritime and land-based networks are joined together leads us to question how the connection was forged, what obstacles were overcome, and what positive steps and measures were taken. In other words, what were, at least in the thirteenth and fourteenth centuries, the forms that allowed a degree of connectivity to extend between the land and sea systems? If we can use the Mongol-European connection as a guide, the following points may be suggested.

The political relationships between states are surely important because the merchants appear extremely weak once deprived of political protection. Hence, connectivity increases based on the groundwork laid by states in their mutual relations to create an environment favorable to trade, which in the last instance, as we can see in the Mongol-Venetian treaties, are meant to reduce risks and increase profits.

The relationship between states and merchants is also crucial to establish especially the boundaries of the protected sphere of commercial activity. Merchants evidently established different relations with different states. They could be protected by the foreign states that had diplomatic relations with their own state or fall prey to those that were hostile. They could also forge personal allegiances with states with which their own had no relation at all, as in the case of Yuan China. In terms of the relationship of merchants with their own state, both Mongols and Latins had their own ways to institutionalize it, by way of government, military support, and financial means.

We must also consider the commercial environment created by those who actively participated in trade. It is remarkable that the main participants in the amphibian networks we have described—Europeans (Latins, Slavs, Christians), Mongols, Muslims, and Jews, all of whom lived intermingled or in close proximity in the cosmopolitan cities on the Black Sea—formed a cohesive commercial environment in which nationality, language, creed, and culture were apparently not obstacles to the formation of commercial partnerships, hiring clerical help, or resolving legal disputes.[46] The very substantial role played by notaries, interpreters, and officers who supervised commercial transactions shows also a sophisticated legal and administrative infrastructure that served a broad multicultural community.

Next we need to examine the cities, towns, and settlements used as markets, ports, and fortified outposts. The cities on the Black Sea showed enormous resilience, and their presence has been compared to the Greek colonies of the archaic and classical periods. How important were cities—as administrative centers and markets, as well as military bases and fortresses—for the consolidation of connections between the two systems? Lastly, it is necessary to pay attention to a series of other issues—such as the intrinsic value of the goods themselves, the routes, the geography, the technology, and more generally the knowledge that people had at their disposal to plan travels to distant and unknown regions and manage commercial transactions. Knowledge is deployed and knowledge is gained as long as the conditions remain favorable and experience is accumulated. What matters is that at the conjuncture of macrosystems there are necessarily a series of nexuses through which connections are made. What has been called in this essay "amphibian history," at the interface between land and sea networks, mariners and nomads, ships and caravans, is probably the closest approximation to world history connections that the pre-1500 world generated.

NOTES

1. I refer to the "Leviathan" as a maritime power in the sense attributed to it by David Armitage in "L'elefante e la balena: imperi terrestri e imperi marittimi," in *Gli imperi: Dall'antichità all'età contemporanea,* ed. Ruth Ben-Ghiat (Bologna: Il Mulino, 2009), 55–72.
2. See for instance Barry Cunliffe, *Facing the Ocean: The Atlantic and Its Peoples, 8000 BC–AD 1500* (Oxford: Oxford University Press, 2001).
3. I use the term "macrohistory" in the same sense that William H. McNeill

has propounded and defended, and that is, not only as the opposite of microhistory but also as very different from large units of historical analysis such as regions, nations, or civilizations. In this sense, macrohistory is near-synonymous with world history. Much has been accomplished in this respect since McNeill wrote "A Defence of World History," *Transactions of the Royal Historical Society* 32 (1982): 75–89, and new definitions of macrohistory have been proposed, but in this essay I only mean by it, essentially, an historical outlook that can encompass multiple regions and multiple time scales.

4. An important and groundbreaking work in this respect is Roxani Eleni Margariti, *Aden & the Indian Ocean Trade: 150 Years in the Life of a Medieval Arabian Port* (Chapel Hill: University of North Carolina Press, 2007).
5. Lothar von Falkenhausen, "Die Seiden mit chinesischen Inschriften," in *Die Textilien aus Palmyra: Neue und alte Funde,* ed. Andreas Schmidt-Colinet, Anne-Marie Stauffer, and Khaled Al As'ad (Mainz: Philipp von Zabern, 2000), 58–81.
6. An example is the pioneer work by Xinru Liu, *Ancient India and Ancient China: Trade and Religious Exchanges AD 1–600* (Delhi: Oxford University Press, 1988).
7. For a general overview of the Silk Road as a historical concept see Xinru Liu, "The Silk Road: Overland Trade and Cultural Interactions in Eurasia," in *Agricultural and Pastoral Societies in Ancient and Classical History,* ed. Michael Adas (Philadelphia: Temple University Press, 2001), 151–79. See also David Christian, "Silk Roads or Steppe Roads? The Silk Roads in World History," *Journal of World History* 11, no. 1 (2000): 1–26.
8. This problem was acutely noted by Philip Curtin some time ago, as he mentioned that overland Asia trade is not given much relevance in historical works not because it did not exist but because "evidence about these commercial practices is not as rich as other evidence about other times and places." See Philip D. Curtin, *Cross-Cultural Trade in World History* (Cambridge: Cambridge University Press. 1984), ix.
9. For Toynbee, the concept of conductivity was central to the comparison between steppes and seas or oceans. See for instance Arnold J. Toynbee and D. C. Somervell, *A Study of History. Abridgment of Volumes VII–X* (Oxford: Oxford University Press, 1957), 3.
10. This thesis is commonly associated to the influential work by Niels Steensgaard, *Carracks, Caravans and Companies: The Structural Crisis in the European-Asian Trade in the Early 17th Century* (Odense: Studentlitteratur, 1973). See also Morris Rossabi, "The 'Decline' of the Central Asian Caravan Trade." In *The Rise of Merchant Empires: Long-Distance Trade in the Early Modern World 1350–1750,* ed. James D. Tracy (Cambridge: Cambridge University Press, 1990), 351–70.
11. On Humboldt's notion of central Asia see V. M. Masson, "The Environment," in *History of Civilizations of Central Asia. Vol 1: The Dawn of Civilization, Earliest Times to 700 BC,* ed. A. H. Dani and V. M. Masson (Paris: UNESCO, 1996), 29.
12. We owe the distinction between heartland and rimland to the seminal essay by Harfold J. Mackinder, "The Geographical Pivot of History." Originally

published in 1904 in *The Geographical Journal* XXIII.4, it has been reprinted in *The Geographical Journal* 170, no. 4 (2004): 298–321; the entire issue was specially dedicated to a retrospective of Mackinder's influential theory. On the interpretation of Inner Asia as a cultural area (mostly made of nomadic tribes) surrounded by agrarian civilizations see Denis Sinor "Introduction: the concept of Inner Asia," in *Cambridge History of Early Inner Asia* (Cambridge: Cambridge University Press, 1990), 1–18. The dichotomies between steppe and sown and "barbarians" and "civilized" that inform this influential view, however, are surely overdetermined.

13. Frederick J. Teggart. *Rome and China. A Study of Correlations in Historical Events* (Berkeley and Los Angeles: University of California Press, 1939). For the use of Teggart's theory in explaining transcontinental cultural as well as commercial traffic along the Silk Road see Joseph M. Kitagawa, "Paradigm Change in Japanese Buddhism," *Japanese Journal of Religious Studies* 11, no. 2–3 (1984): 120.
14. See John G. A. Pocock, *Barbarism and Religion, Vol. IV: Barbarians, Savages and Empires* (New York: Cambridge University Press, 2005), 135–38.
15. The exact quote is: "The building of the great wall of China was an event fraught with the greatest consequences, and one may say without exaggeration, that it contributed powerfully to the premature downfall of the Roman Empire." See C. de Ujfalvey, *Les Aryens au nord et au sud de l'Indou-kouch, 1896,* 24.
16. Recent and important examples of this orientation are: *Rome and China: Comparative Perspectives on Ancient World Empires,* ed. Walter Scheidel (New York: Oxford University Press, 2009); and *Conceiving the Empire : China and Rome Compared,* ed. Fritz-Heiner Mutschler and Achim Mittag (New York: Oxford University Press, 2008).
17. For the contacts between Byzantium and the Turk Khaganate see the summary by Peter B. Golden, "Turks," in *Late Antiquity: A Guide to the Postclassical World,* ed. G. W. Bowersock, Peter Brown, and Oleg Grabar (Cambridge, MA: Harvard University Press, 1999), 735–36.
18. Both Byzantine and Sasanian coins have been found in China, in particular northern China. How the presence of Byzantine coins may or may not reflect trade is discussed in François Thierry and Cécile Morrisson, "Sur les monnaies byzantines trouvées en Chine," in *Revue numismatique,* 6e série 36 (1994): 109–45.
19. Philippe Beaujard "The Indian Ocean in Eurasian and African World-Systems before the Sixteenth Century," *Journal of World History* 16, no. 4 (2005): 412.
20. Alain Bresson, "Ecology and Beyond: The Mediterranean Paradigm," in *Rethinking the Mediterranean,* ed. William V. Harris (Oxford: Oxford University Press, 2005), 94–114, esp. 104 ff.
21. Peregrine Horden and Nicholas Purcell, *The Corrupting Sea: A Study of Mediterranean History* (Oxford: Blackwell, 2000). "Connectivity" is discussed in Chapter V.
22. The work credited with the creation of a pre-thirteenth-century system of

interlocking trading circuits is Janet Abu-Lughod, *Before European Hegemony: The World System A.D. 1250–1350* (Oxford: Oxford University Press, 1989). See in particular the map on page 34. While the work has met with some justifiable criticism on account of insufficient data or overreaching interpretation, I feel that the notion that a greater degree of "connectedness" (my word) was achieved exactly at the time of the Mongol conquest cannot be challenged.

23. G. I. Bratianu, "La Mer Noire, plaque tournante du trafic international à la fin du Moyen Age," *Revue Historique du Sud-est Européen* 21 (1944): 36–69.
24. J.R.S. Phillips, *The Medieval Expansion of Europe* (Oxford: Oxford University Press, 1998), 96.
25. Philip D. Curtin, *Cross-Cultural Trade in World History,* 138.
26. Eyüp Özveren, "The Black Sea World as a Unit of Analysis," in *Politics of the Black Sea. Dynamics of Cooperation and Conflict,* ed. Tunç Aybak (London and New York: I. B. Tauris, 2001): 72–73 [61–84].
27. On the Black Sea not being regarded in antiquity as an extension of the Mediterranean see G. W. Bowersock, "The East-West Orientation of Mediterranean Studies and the Meaning of North and South in Antiquity," in *Rethinking the Mediterranean,* 175.
28. On the Sogdian networks on the Silk Road see Jonathan K. Skaff, "The Sogdian Trade Diaspora In East Turkestan During The Seventh And Eighth Centuries," *Journal of the Economic and Social History of the Orient* 46, no. 4 (2003): 475–524. According to what Procopius reports it was during the reign of Justinian (527–565) that the Byzantines acquired the secret of silk production, but imports and trade of silk continued under the later emperors. See Nikolas Oikonomidès, "Silk Trade and Production in Byzantium from the Sixth to the Ninth Century: The Seals of the Kommerkiarioi," *Dumbarton Oaks Papers* 40 (1986): 33–53.
29. To my knowledge, the best essay on the economy of the Khazar khaganate is Thomas S. Noonan, "The Economy of the Khazar Khaganate," in *The World of the Khazars: New Perspectives,* ed. Peter B. Golden, Haggai Ben-Shammai, and András Róna-Tas (Leiden: Brill, 2007): 207–44.
30. On the foreign trade of the Song dynasty and its international connections see Yoshinobu Shiba, "Song Foreign Trade: Its Scope and Organization," in *China Among Equals: The Middle Kingdom and Its Neighbors 10th–14th Centuries,* ed. Morris Rossabi (Berkeley and Los Angeles: University of California Press, 1983), 89–115.
31. On the system of postal stations in China see Peter Olbricht, *Das Postwesen in China unter der Mongolenherrschaft im 13. und 14. Jahrhundert* (Wiesbaden: Harrassowitz, 1954). The system itself was adopted from the imperial predecessors of the Mongols—the Khitans. On this see Michal Biran, "The Mongol Transformation: From the Steppe to Eurasian Empire," *Medieval Encounters* 10, no. 1–3 (2004): 344 [339–61].
32. On the *ortaq* in the early Mongol empire see the classic work, see Thomas T. Allsen, "Mongolian Princes and their Merchant Partners 1200–1260," *Asia Major* 2 (1989): 83–126.

33. Peter Jackson, *The Mongols of the West 1221–1410* (Harlow: Pearson Education Ltd., 2005), 291.
34. There is an immense historical literature on the Black Sea and the Crimea from the thirteenth to the fifteenth centuries, which would be tedious to repeat here.
35. Research into the activities of Italian merchants in central Asia and the Far East include: Luciano Petech, "Les marchands italiens dans l'empire mongol," *Journal Asiatique* 250 (1962): 549–74; Roberto S. Lopez, "Les méthodes commerciales des marchands occidentaux en Asie du onzième au quatrième siècle," in *Sociétés et compagnies de commerce en Orient et dans l'Océan Indien,* ed. M. Mollat (Paris: S. E. V. P. E. N., 1970): 343–48; Roberto S. Lopez, "L'importance de la mer Noire dans l'histoire de Gênes," in *Colloquio Romeno-Italiano, I Genovesi nel Mar Nero durante i Secoli XIII e XIV—Colocviul Romano-Italian, Genovezii la Marea Neagră in secolele XIII–XIV,* ed. Ş. Pascu (Bucarest: Editura Academiei Republicii Socialiste România, 1977), 13–33.
36. Martin da Canal, *Les Estoires de Venise. Cronaca Veneziana in lingua francese dalle origini al 1275,* ed. A. Limentani (Firenze: Olschki, 1972), 324–25.
37. Nicola Di Cosmo, "Mongols and Merchants on the Black Sea Frontier in the Thirteenth and Fourteenth Centuries: Convergences and Conflicts," in *Turco-Mongol Nomads and Sedentary Societies,* eds. Reuven Amitai and Michal Biran (Leiden: Brill 2005), 406–12 [391–424].
38. On the spread of Islamic textiles and the artisans who produced them, see Thimas A. Allsen, *Commodity and Exchange in the Mongol Empire* (Cambridge: Cambridge University Press, 1997). This study is especially valuable for the close attention to mechanisms of cultural transmission (see especially pages 71–106).
39. I base this assumption on the role of the Mongols in spreading this technology to India in the fourteenth century. See Iqtidar Alam Khan, "The Role of the Mongols in the Introduction of Gunpowder and Firearms in South Asia," in *Gunpowder: The History of an International Technology,* ed. Brenda J. Buchanan (Bath, England: Bath University Press, 1996): 33–44; and Iqtidar Alam Khan, "The Coming of Gunpowder to the Islamic World and North India: Spotlight on the Role of the Mongols," *Journal of Asian History* 30, no. 1 (1996): 26–45.
40. This text has been translated and edited in Paul D. Buell and Eugene N. Anderson, *A Soup for the Qan: Chinese Dietary Medicine of the Mongol Era as Seen in Hu Szu-Hui's Yin-shan Cheng-yao* (New York, Kegan Paul International, 2000).
41. On the exchange and traffic of animals and animal knowledge on a Eurasian scale see Thomas T. Allsen, *The Royal Hunt in Eurasian History* (Philadelphia: University of Pennsylvania Press, 2006). Information about the use of animals in hunting in Mongolian courts is interspersed through the book. For information about traffic in animals and animal trainers see especially chapter 12.
42. The literature on this topic is vast, and ranges from the highly specialized to the highly popular. For an overview of the medical issues and historical

implications of the fourteenth-century plague and other epidemics see Samuel K. Cohn Jr., "Epidemiology of the Black Death and Successive Waves of Plagues," *Medical History Supplement* 27 (2008): 74–100. For a recent theory that links specifically the spread of the plague with trade routes see Brian H. Bossak and Mark R. Welford, "Did medieval trade activity and a viral etiology control the spatial extent and seasonal distribution of Black Death mortality?" *Medical Hypotheses* 72 (2009): 749–52.

43. The presence of Marco Polo in Catay and the Far East has been the subject of some scrutiny recently. For an exhaustive and instructive study of Marco Polo's travels and connections with China under the Mongols see David Jacoby, "Marco Polo, His Close Relatives, and His Travel Account: Some New Insights," *Mediterranean Historical Review* 21, no. 2 (2006): 193–218.
44. On this crisis see S. P. Karpov, "Génois et Byzantins face à la crise de Tana de 1343 d'après les documents d'archives inédits," *Byzantinische Forschungen* 22 (1996): 33–51.
45. On this point see Nicola Di Cosmo, "Black Sea Emporia and the Mongol Empire: A Reassessment of the Pax Mongolica," *Journal of the Economic and Social History of the Orient* 53 (2010): 83–108, especially 100–106.
46. On the cosmopolitan and multiethnic population of the Italian colonies on the Black Sea see Michel Balard, "Les Génois en Crimée aux treizième-quinzième siècles," *Archeion Pontou* (1979): 210–11.

SEVEN

An Ocean of Islands: Islands, Insularity, and Historiography of the Indian Ocean

Roxani Margariti

In his marvelous philosophical novella starring self-enlightened hero Ḥayy b. Yaqẓān (Alive son of Awake), the Andalusian philosopher Abū Bakr Ibn Ṭufayl (d. 1185/6) situates Ḥayy's home in an equatorial island "lying off the coast of India below the equator," and characterized by "the most tempered climate on earth."[1] It is precisely thanks to ideal insular conditions that Ḥayy comes about, through spontaneous genesis "without father or mother,"[2] and it is perhaps partly thanks to the distraction-free universe of his island that he attains ultimate spiritual clarity. Ibn Ṭufayl allows for an alternative story of Ḥayy's origins, presenting us with an additional insular construct: if not born of the island's soil, the sage-to-be may have arrived at this same perfectly temperate predator-free world as an infant castaway from a more worldly kind of island, one that is "rich and spacious," densely populated, and ruled by "a proud and possessive . . . king."[3] Finally, the story ends with an episode that involves yet a third island, a small world inhabited by a diverse population that teaches Ḥayy and the audience a fundamental lesson: full-blown spiritual enlightenment is possible only for the exceptional few.[4] At the end of the story, Ḥayy returns to his insular island.

The three islands are, of course, imaginary, and there is nothing in Ibn Ṭufayl's metaphorical geography and idealized topography to suggest that he was giving any serious thought to real Indian Ocean archipelagoes. But, the choice of archipelagic setting for Ḥayy's development "off the coast of India" and below the equator, that is, at the other end of the known world from the author's native Maghrib, is noteworthy. It lifts the tale to the "bounded, isolated, self-sufficient, and temporally distanced"[5] space of conceptual islands, thus signaling its metaphorical quality. Yet at the same time it echoes the very literal conception of the eastern sea as a world of islands similar in this trait to the Mediterranean. This same conception appears graphically in the sea maps of the celebrated geographical compendium known as the "Book of Curiosities": both the Mediterranean's and the Indian Ocean's green surfaces are dotted with circular and oval figures of islands, some nameless beyond the marking "island" (*jazīra*).[6] Island shapes are the graphic equivalent of the color green: they evoke the sea.

This chapter argues that islands and insularity are crucial subjects of study for Indian Ocean history and historiography—as they are for the Mediterranean—and that examinations of island lives enrich our understanding of Indian Ocean polities and communities and offer a new view of their participation in an interconnected world—as they do of the Mediterranean. I suggest that for certain chronological periods the Indian Ocean world can best be described as "an ocean of islands," that is, a geohistorical entity of which the constituent parts were geographical and geopolitical islands. In particular, the era of ever-expanding maritime commerce from the tenth century on was characterized by the formation of independent, autonomous or semi-autonomous communities that played a significant role in the articulation of trading and other networks. Of these communities, many were based on geographical islands, so investigating their development, use of land and sea spaces, and connections with other island and non-island communities substantially fleshes out our portrait of the region in these early and little understood times. Furthermore, by exploring the connection between geographical insularity and geopolitical autonomy we arrive at a better understanding of the mechanisms that facilitated this autonomy. Moreover, insularity shapes the conduct of the long-distance trade, the flow of human migration, the formation of networks of knowledge, and the practices of pilgrimage; that is, all the major modes of connectivity that characterized the Indian Ocean world. In advancing this argument and then presenting the outline of a specific island case study, this chapter offers a rehabilitated concept of insularity, one that is only intelligible within a conceptual framework of interaction, and that is indeed crucial to understanding the articulation of the Indian Ocean world in the period that interests me, between the tenth and the thirteenth centuries.

Indian Ocean Historiography, Thalassography, and Nesiology

Ḥayy's island metageography suggests conceptual schemes about the nature of insularity that have informed ancient thought and—perhaps not surprisingly—also modern historiographical approaches to Indian Ocean islands. First, fabulous islands abound in the Indian Ocean in both outsiders' and insiders' descriptive imaginations; whether of the kind that gave birth to Ḥayy's body and intellect or others producing single-gender societies,[7] these utopias stem from the intersection of medieval geographical experience and conceptual geographies and as such constitute a particularly fertile subject of investigation.[8] Second, the three kinds of islands that function as settings to Ḥayy's story sketch out a basic island typology—oceanic, coastal, archipelagic—and single islands have been and perhaps should continue to be made sense of through a typological approach in Indian Ocean historiography. Third, the three islands that Ibn Ṭufayl conjures are microcosms, complete natural and/or social worlds that we can easily and conveniently wrap our minds around; historical and literary studies of Indian Ocean connectivities use a similar microcosmic conception of islands and as such Indian Ocean islands have been thought to encapsulate the dynamics of Indian Ocean history. Fourth, and related to this concept of the microcosm, Ibn Ṭufayl compellingly describes Ḥayy's island as something of a biological and intellectual laboratory—a place whose sheltered geography and environment allow for a physical and mental experiment to reach its full expression in the figure of Ḥayy; although the application of the laboratory assumption to either physical or societal models has been proven problematic, the concept affected the colonial perception and exploitation of the more remote Indian Ocean islands, and recent work has exposed those interactions in and with the oceanic realm. Finally, and most importantly for the kind of island history that this chapter is concerned with, Ḥayy's story reminds us of the interplay between isolation and interconnectedness that characterizes the life of every island; this interplay defined the economic, social, and political phenomena that conferred unity to the Indian Ocean realm in the medieval period.

Thalassography writes into history human interactions with and across the sea. In the Mediterranean—the birthplace of thalassography—islands play a dramatic role in this particular endeavor, and the same applies in historical and anthropological approaches of the archipelagoes of the Pacific, of the Atlantic, the Baltic Sea, and the Caribbean. The critical mass of island studies in these historiographical contexts amounts to a set of "nesiologies"; that is, a corpus of studies of island histories and cultures within the broader framework of maritime history. These studies have provided new impetus and method to the efforts to describe the worlds constituted through wide, watery spaces.[9] Indian Ocean islands, by comparison, have been largely neglected and the roles they have played historically and culturally in the creation of Indian Ocean unity have received concerted attention only very recently and quite patchily; what work has been done, moreover, has focused primarily on the colonial period.[10] Following up on the recent work of Edward Alpers, Shawkat Toorawa, Megan Vaughan, Aparna Vaidik, and a few others, the goal of this section is to summarize the advances made in island studies since the beginning of Indian Ocean history.

Islands and insularity play a remarkably consistent and pervasive role in Braudelian and post-Braudelian definitions of Mediterranean unity. Islands are many things in and for the Mediterranean, and Braudel gives us an evocative few pages on the importance of islands and the roles they have played in the history of the sea. Microcosms and miniature continents, stepping stones, entrepôts, and navigational landmarks, sources of commercially recognizable "brand" goods and of people on the move, that is, of materials, groups, and individuals that circulate throughout the realm, islands are central to the study of the sea that surrounds them. Island archaeology has become a sophisticated subfield of the discipline and Mediterraneanists have been forging the way.[11] And in their provocative opus on the history and historiography of the same sea, Horden and Purcell argue for the connectedness of islands, ascribing them a greater degree of connectivity than isolation. On another level, we might say that *The Corrupting Sea* revisits the concept of insularity of islands and island habitats with dramatic effect; for what are the authors' "microregions" that make up the puzzle of the Mediterranean world if not islands, selectively isolated, circumscribed, and at the same time defined by the modes of connection with other places of the same ilk?

Indian Ocean historiography has been inspired and animated by Braudel's work, *The Mediterranean*, and by the challenge it posed for non-Mediterranean history. In his seminal synthetic survey of Indian Ocean history, at that formative stage of Indian Ocean historiography, Auguste Toussaint highlighted the significance of the ocean's islands. A native of Mauritius, the island's archivist, and an accomplished scholar of its history, Toussaint eagerly seized on the Braudelian allusion to the islands' historiographical importance, but as opposed to Braudel who found that islands are often "both far ahead and far behind" the region's historical trends, Toussaint boldly asserted that the pulse of the ocean can be best felt at the remotest among them.[12]

Although Edward Alpers has rightly pointed out the limitations of Toussaint's narrow—and as Alpers puts it "patriotic"—focus on small, remote, oceanic, and only recently populated islands,[13] it is worth returning to the Mauritian historian's justification of the insular lens. For the purposes of historical analysis, the oceanic realm, according to Toussaint, should be divided four ways: the watery realm itself, the coast of Africa, the coast of Asia, and the islands scattered across the whole ocean. Of the latter, he argues, the Mascarenes, Seychelles, and Chagos archipelagoes, that is, groups of small remote islands, are the most profoundly maritime. The inhabitants of the ocean's larger islands are not necessarily sailors. Coastal islands partake too much of the neighboring mainlands, and life there is not as entirely dictated by the ocean's season. By contrast, the islands in the middle of the ocean are neither African nor Asian, and are entirely dependent on and shaped by the sea, "nées de la mer et pour la mer."[14] These definitions of the particular islands' *sui generis* nature and encapsulated historical experience would seem to exclude the majority of the ocean's islands, but a closer reading reveals that Toussaint's call was for something more inclusive: along with the Mascarenes, Seychelles, and Chagos, "the Comoros, the Maldives, the Laccadives, the islands of the Arabian coasts, and some others" merit scholarly attention.[15]

Toussaint's work thus lays the foundation for writing histories of Indian Ocean through histories of its islands primarily by raising the question of what constitutes the Indian Ocean and its unity, and answering that water, shores, and islands are its most basic, and by extension ubiquitous, elements. In this scheme, one might say that islands emerge as a salient combination of the other two constituent parts. This enduring conceptualization of islands as part of the oceanic frame's "deep structure"[16] (i.e., of the Indian Ocean's Braudelian *longue durée* and as essentially maritime), render them particularly relevant to the current thalassographic quest for a deeper understanding of maritime lives and the maritime experience. Arguably, it is these lives and experiences that the thalassographic lens can most clearly illuminate, and for Indian Ocean historiography, this quest has a strong tradition.[17] In the 2007 volume of the periodical *Azania*, devoted specifically to definitions of the Indian Ocean as unified cultural space, John Mack tellingly entitles his introductory essay "The Land viewed from the Sea," and calls for an effort to understand maritime lives while cautioning that "it is very difficult to de-center our thinking from its habitual terrestrial location and reposition it off-shore."[18] Michael Pearson's contribution to this quest is the concept of the littoral society,[19] which has consequences for the way we view islands. In Pearson's scheme, islands are where one is most likely to find "littoral societies." Small islands in particular host exclusively littoral societies while the people of very small islands can be called amphibious and even aquatic; that is, living primarily or even wholly from and on the sea.[20] Pearson thus contributes to a typology of islands, each type playing a different set of roles in the definition and articulation of the maritime. In other words, Pearson's formulation helps us think of islands as the locus for a concentrated study of what Purcell, elsewhere in this volume, characterizes in terms of the trinity of "beach, tide, and backwash."

Yet, while islands thus emerge from this historiography as quintessentially maritime worlds, we have no detailed and comprehensive account of the range of roles that different Indian Ocean islands and island groups have played in the study of the region as a whole. Syntheses exclusively focused on Indian Ocean islands are few. The closest thing to a survey is the work of geographer Jean-Louis Guébourg on "small islands and archipelagos of the Indian Ocean."[21] In his preface to this volume, geographer Roger Brunet praises the author's indefatigable exploration of "microcosms that make of the Indian Ocean an immense archipelago, and render that anti-continent into the world's eighth continent."[22] Although his is not a historical work, Guébourg highlights some of the common threads in island histories including exploration, settlement, colonization and, for the present day, the tourist gaze. The one decidedly historical chapter focuses on the discoveries and mapping of the Indian Ocean's island world and argues that the islands of the Indian Ocean provide an important illustration of the process and progress of geographical exploration, discovery, and definition.[23]

A collection entitled *The Western Indian Ocean: Essays on Islands and Islanders* is another, more recent, work that draws attention to the potential for comprehensive historical inquiry on Indian Ocean islands and for understanding the ocean's history through its islands.[24] The argument that this volume makes is that islands constitute powerful entries into the study of oceanic history and society as a whole. The essays here fall under two very broad themes of inquiry: the representation of islands on the one hand and the economic, social, and political dimensions of island lives on the other. Under the first theme, islands emerge as the exotic realm where everything is topsy-turvy, as shown in Arabist Shawkat

Toorawa's examination of how the medieval geographical imaginary produces meaningful mythologies about the Waqwaq "(is)lands." His work, along with Michael Laffan's recent essay on the complex history of a set of Arabic terms for the Indonesian archipelago,[25] show us how outsider perspectives, even ones that are reasonably well informed, produce miraculous simulacra to populate faraway lands.[26] In an earlier piece, Aziz al-Azmeh highlighted some of the more extreme representations of Indian Ocean peoples and especially of islanders in the Arabic literary tradition of "wonders"; by producing the "other," these accounts of deviation from or inversion of normal reality function to "sustain a sense of normality, continuity, and affinity" among the audience.[27] Distance seems to generally amplify the degree of the inversion, even in cases where at least some contact existed with the (is)lands in question. These groundbreaking studies remind us more generally that the majority of written sources for the medieval Indian Ocean convey the perspectives of outsiders and landlubbers, who are all too easily drawn to the default representation of the Indian Ocean world as slightly strange, especially at its perceived edges. Travelers and those using firsthand reports by people who may have had contact with the islands in question do not necessarily provide a corrective to these visions. Maritime lore itself thrives on exaggerations or fabrications of the exotic that function, at least partly, as examples of what happens outside the bounds of normalcy that the audience inhabits.

On the second issue that the island volume raises, that is, the participation of the islands in all that makes the distinctly patterned economic and social Indian Ocean world as we know it, East Africanist Edward Alpers writes of the "island factor" in the integration of Africa into the Indian Ocean world.[28] This study provides an excellent historiographical introduction to Indian Ocean island studies, offers the most substantial bibliography for the topic to its date of publication, and reiterates the sense that syntheses on island histories are still missing. Alpers outlines the role of African seaboard islands in commercial expansion and integration of different parts of the ocean, both as stepping stones where traders' communities installed themselves or found temporary shelter, and as independent island polities; in the process of Islamization of coastal East Africa; in the deployment of slave trading networks; and finally, in the economic and political development of colonial and postcolonial East Africa. In a later article, Peter Mitchell applies a similarly Afrocentric perspective to the archaeological study of what he calls "Africa's islands," including those of the Indian Ocean.[29] With a focus on the material cultures that reflect island roles in the integration of continental and oceanic systems, Mitchell's enumeration of themes pertinent to island studies adds to the list established by Alpers two important issues: the development of what he and other archaeologists see as distinct island cultures, and the implications of island colonization for maritime technology at any given historical period.[30] Taken together, these two pieces present us with a valuable thematic framework of inquiry that can be adapted and tested to explore island histories in other parts of the ocean, as we shall soon see.

The bulk of recent studies on Indian Ocean islands focus on the colonial encounter. In his pioneering monograph, *Green Imperialism: Colonial Expansion, Tropical Island Edens, and the Origins of Environmentalism* (1995), Richard Grove explored the "Edenic island discourse" and the roots of Western environmentalism as it took shape in colonized islands; the Mascarenes, and particularly Mauritius, featured prominently in this story. The scope of Grove's work extended beyond the Indian Ocean, but it opened up space for the consideration of Indian Ocean islands in the framework of environmental history and literary utopias, and most importantly, colonial history. For the same period and under the same broad rubric of the colonial encounter, new studies have recently appeared and provide fresh and exciting views on individual islands and archipelagoes and their participation in ocean-wide phenomena, including imperial expansion, slavery, and penal history. Megan Vaughan's *Creating the Creole Island: Slavery in Eighteenth-Century Mauritius* (2005) continues and amplifies the tradition of scholarship on the Mascarenes initiated by Toussaint, but emphasizes the social dynamics that obtained not only here but at many other places across the ocean. Finally, Aparna Vaidik's *Imperial Andamans: Colonial Encounter and Island History* (2010) rescues the history of one of these places construed as peripheral from the perspective of the colonial metropoles. Problematizing the view of the ocean as fully integrated and showing how the local and regional perspectives fit with the broader transregional oceanic frame, she offers a methodological blueprint for the writing of similar histories of neglected parts of the ocean not only in the early modern and modern eras, but also in earlier times.

To sum up, the existing literature on Indian Ocean island history relates to four major themes: first, the historiographical exploration of colonial encounters; second, questions of representation, mapping, and the imaginary; third, the integration of discrete regions into the Indian Ocean realm (especially East Africa and South Asia) and, fourth, the links between these regional historiographies, whether oceanic or continental.[31] A perusal of this work reveals that the coastal islands of the Red Sea, the Persian Gulf, and even the Arabian Sea are generally left out of the discussion. From the geographical perspective this omission is perhaps less problematic: Guébourg clearly defines his "small islands and archipelaga of the Indian Ocean" as tropical or equatorial lying between the latitudes of 13° 45′ north (North Andaman islands) and 21° 20′ south (Reunion), with a hot and humid climate, with the exception of the island of Socotra.[32] But climate aside, some of the common parameters in the histories and modern experiences of the islands he studies—endemic features and resources, degrees of isolation and distances from continental structures, the non-insular imaginary and its effects—obtain as important forces in the trajectories of all islands in the region. The same applies to other parameters that emerge from the historical studies surveyed above, especially Islamization, trade, maritime technology, and slavery networks.

There is also a chronological gap in the existing nesiology of the Indian Ocean. With the exception of the few aforementioned efforts to explore the role of islands in the political, economic, and religious integration of Africa into the Indian Ocean realm and in the medieval geographical imaginary, the experience and meaning of island worlds in the premodern Indian Ocean remains underdeveloped. The effort to understand the increasing integration of the world of trade in the Indian Ocean in this period, as well as the rise of small independent maritime polities in the region from the eleventh century onward, some of them on islands, must include the study of islands and islanders.

This historiographical marginalization of some islands of the Indian Ocean may partly result from a bias of time and preservation. The apparent blind spot may also reflect the bias of sources—it reproduces the sense that these places were at the margins of empires, or at least of large territorial states that produced the texts we read. Somewhat related is the effect of modern national boundaries: islands and archipelagoes that constitute discrete nations function as "necessary" historiographical foci more readily so than islands that are simply parts, often marginal parts, of non-island territorial states.[33] Inspired by Aparna Vaidik's exemplary effort to recenter and de-marginalize the story of one Indian Ocean archipelago and its people, in the final section of this chapter I will use the example of a traditionally marginalized group of islands to show how its history partakes in themes that are central to Indian Ocean thalassography as a whole.

Land-and-Sea Realms of the "Medieval" Indian Ocean

The eleventh, twelfth, and thirteenth centuries constitute the period of expanded and increasing contacts between the Mediterranean and the Indian Ocean worlds, and of encounters between religious, ethnic, linguistic, and vocational groups and identities along the western Indian Ocean littoral. These encounters once seemed straightforward and unidirectional (Persian and later Arab commercial expansion across the ocean; Islamization and the directional "spread" of Islam from an Arabian center southward to Africa and eastward to Southeast Asia), but have now been shown to be much more complex and intricate, as we learn more about the agency of a number of historical actors, to the point that "indigenous" and "foreign" become murky categories in the context of the Indian Ocean rim.[34] This period also saw the rise of independent port city states, most well defined in the case of the Swahili coast, but also along the coasts of the Red Sea, the Persian Gulf, and the Arabian seaboard—the part of the ocean that my own research explores. Aden, Sharma, and Mirbāṭ along the South Arabian coast; the island Kish in the Persian Gulf; and the Red Sea islands of Bāḍiʿ, Sawākin, and Dahlak (which will be discussed in more detail later) all show signs of independent political organization between the eleventh and the thirteenth centuries.

In this period, the histories of these discrete parts of the Indian Ocean are best told through the prism of their participation in and constitution of an archipelago of port cities,[35] connected to and simultaneously separate from each other and their hinterlands. Many of these places, along and across the western Indian Ocean rim, are mostly invisible from a continental, terra firma perspective, and at best they emerge to view only as marginal and inconsequential. Some, like Sharma and Bāḍiʿ, disappear from view by the thirteenth century when more seaward-oriented larger territorial states reroute their traffic and subsume their functions; others, like Aden, Kish, and Dahlak, continue to function as the ports or dependents of these larger states. Losing sight of these places and times results in losing sight of economic mechanisms, social articulations, and ideological affiliations—both in the places themselves and in their historiographically more visible continental neighbors.

Two concepts offer the keys to an analysis of the dynamic processes that constituted and dissolved polities in the world of the Indian Ocean in medieval times.[36] The first of these is the notion that the polities in questions comprised a land-and-sea realm, that they filled an expansive and hybrid geographical space. The second concept is the insularity of small city states, especially ports, whether on terra firma or on geographical islands; thus to be added to the geographical islands are social and geopolitical islands which also seem to float, because they are not anchored, or at least not firmly anchored, to large territorial states.

The notion that islanders think of the sea and their forelands as an extension of their living space finds ample support in the anthropological literature on maritime peoples. An important contribution to this subject is that of ethnologist and anthropological writer Epeli Hau'ofa. His seminal essay, "Our Sea of Islands," takes issue with the concept of the smallness and concomitant economic helplessness of the island states and territories of the Pacific.[37] He shatters that prevailing stereotype by countering that for the islanders their territory does not stop at each island's end. According to Hau'ofa, Oceania is "a sea of islands with their inhabitants," and the peoples of Oceania:

> did not conceive their world in such microscopic proportions. Their universe comprised not only land surfaces, but the surrounding ocean as far as they could traverse and exploit it, the underworld with its fire-controlling and earth-shaking denizens, and the heavens above with their hierarchies of powerful gods and named stars and constellations that people could count on to guide their ways across the seas. Their world was anything but tiny.[38]

The sea, whether surrounding an island or bordering a mainland littoral, may at times constitute a connecting corridor or an insurmountable barrier, depending partly on the state of maritime technology and competence of the littoral populations and on the permeability of other geographical/environmental boundaries present. For example, work in the Alaskan archipelagoes has shown that internal features of islands present less permeable boundaries whereas the surrounding sea is used as a series saltwater passages from one island to the next.[39] The delineation of the sea and its resources as part of the maritime realm of a community or polity, the carving of maritime turf, follows from such observations of the permeability of watery boundaries and the subsistence practices of maritime peoples. A recent episode on the island of Matinicus off the coast of Maine, in what was described as "lobster wars" by the media, illustrates well one mode of informal territorializing of water. A Matinicus islander shot a fellow Matinican when the latter tried to prevent him from sharing informal territorial waters, reserved for islanders only, with his mainland son-in-law. Interviewed about the incident, a woman from the same island declared that "the only difference between this place and the family farms that were settled out West is our land is covered with water."[40] It is an interesting story about islands, but also about the relationships of maritime people to the sea in general; far from being empty, or even simply a "common property resource" (as suggests Pearson, quoting anthropologist Estellie Smith),[41] the sea is or can be marked through the practices of littoral or "aquatic" peoples. Anthropological work on the spatial dimension of fishing practices by Maine lobstermen provides fascinating evidence of such territorializing tendencies as informally sanctioned practices of maritime communities.[42] In the period that we are concerned with here, Emily Tai describes "marking water" in the medieval Mediterranean in a more formalized and state-centered, rather than community-driven, mode.[43]

In his essay on the new thalassology and Indian Ocean studies, Markus Vink argues that "it is imperative for 'the new thalassology' to disentangle the complex strand of spatial categorizations and explore the permeable inner and outer boundaries of the Indian Ocean world(s)."[44] The history of island and insular polities in the medieval period, and particularly the way that their land-and-sea realm was construed by the participants in these polities and contested by outsiders, may well reveal a great deal about those "permeable inner" boundaries of the Indian Ocean system. There is also a temporal dimension to the new history of islands. In a 2009 workshop on the material culture and overseas connections of the medieval site of Sharma, Yemen, Eric Vallet called for the return, after a long and by now venerable corpus of Indian Ocean historiography pursuing the *longue durée* and the broad structures of the Indian Ocean world, to *histoire événementielle*—the shorter scale that is sensitive to change and particularity.[45] Expansion or contraction of the land-and-sea realms of individual polities can be shown to correlate with political phenomena that affect broader swaths of the ocean.

Insularity and Indian Ocean Ports

If geographical islands are for the most part connected, and "bounded, isolated, and temporally distanced" islands exist primarily in the imagination and only occasionally in reality, then we are forced to rethink the concept of insularity. Instead of bespeaking isolation, insularity becomes the essential quality of worlds both connected and isolated to varying degrees by that which surrounds them—sea, or land, or both. Indeed the distinction between insularity and isolation is vital and, in geographical terms, has been made strongly by Guébourg, who speaks of isolation as a variable of insularity.[46] Moreover, a variety of scholars of island cultures in other seas have recently engaged with the assessment of the complex relationship of insularity to isolation.[47] The multiple etymologies and frequent ambiguity or open-endedness of the terms for islands across languages helps support the case for a multivalent definition of the concept. Take for instance the Latin root of the English word for island, *insula,* and juxtapose it with the Greek root for the same meaning, *nesos.* The former signifies the inwardness of terra firma, while the latter, according to one etymological theory, shares something with the words for ducks and boats;[48] in Greek, islands are floaters, whereas in Latin they are self-contained parcels of land. More linguistically pertinent for the Indian Ocean, the Arabic root *j-z-r* can be equally derived from to the sense of severing of parts from a whole—thus according to some lexicographers, islands are "so called because cut off from the main land"—or from the sense of emerging after the withdrawal of waters—thus islands are "so called because of the retiring of the water from it."[49] In Sanskrit-derived languages, on the other hand, *dvīpá* is used for island, peninsula, and sandbank, as well as a "division of the terrestrial world."[50]

Etymologies and linguistic usage thus allow the application of the island concept to entities that are not necessarily surrounded by sea. Similarly, the port cities of the Indian Ocean can be said to be islands whether or not they were located on geographical islands. There is something intuitive and obvious in applying the concept of insularity to the study of ports. For one thing, islands and ports share the sea and its many possibilities. This locational and functional affinity of ports and islands is intuited by remarkable Book of Curiosities (*Kitāb Gharā'ib al-funūn wa-mulaḥ al-'uyūn*), which devotes one of its chapters to "the depiction of the seas and their islands and havens."[51] In the case of the western Indian Ocean, moreover, the affinity is also topological; that is, ports' physical geography and topography share to various degrees in the insularity of islands. Several ports are located on islands close to continental shores: Manda, Shanga (on Pate), Lamu, Kilwa, and others down the Swahili coast; Sawākin, Bāḍiʿ, and Dahlak down the Red Sea; Kish and Hormūz in the Persian Gulf. Several others have what we might agree on calling an insular topography: archaeological and textual evidence shows such ports to have natural and manmade barriers separating them from their umland and hinterland. An excellent example of this is Aden, the peninsular port *par excellence,* nested in a volcanic crater on a peninsula that according to legend was once an island, and that is connected to the mainland with what can properly be called a land bridge. In addition, in their description of the medieval port of Jeddah, a traveler and a native author alike describe how early settlers built a wall around the entire city, and later reinforced the landward defenses with a perimeter seawater moat running from both ends of the city's Red Sea waterfront, thus rendering the city a literal island. At the site of al-Balīd, identified as the Dhofari port of Ẓafār, archaeologists discovered that similar moat-like structures extended from the waterfront flanking and surrounding the town. The recently excavated Hadrami port of Sharma, destroyed and abandoned in the mid-twelfth century, features a landward wall. Walls and moats defended these towns from potentially hostile hinterlands, or at the very least demarcated the space of the port as separate, thus rendering it insular.[52]

Pearson cautions that including port cities in littoral society complicates the definition of the latter. While islands are mostly littoral, cities on the sea are somewhat ambiguously so; islanders are, by his lights, too cosmopolitan to qualify as fully "of the coast."[53] But if occupation and range of connections divide major port cities of a region from the more humble littoral, and port city cosmopolitans from harbor and seashore people of the sea, the more broadly defined concept of insularity that includes not only literal islands but also island-like geopolitical entities provides a different and more unifying lens for the Indian Ocean littoral. Thinking about the functions of insularity helps us understand the world of small port-city states that propelled the conduct of long-distance cross-cultural trade linking two maritime systems from the tenth century onwards—especially the period between the eleventh and the thirteenth centuries.

Dahlak Archipelago: A Case Study

The history of Dahlak Kebir (Big Dahlak), in the homonymous archipelago of more than 200 islands, exemplifies the main themes and dynamics of island and insular political geography and economy in the Indian Ocean between the eleventh and thirteenth centuries.[54] An extensive and largely unexplored settlement site lies at the southwestern end of this big island, bearing witness to a once-thriving medieval port town. A large number of water reservoirs suggest either a large permanent population, or perhaps more plausibly, the onetime passage of many transients through this port. One aspect of the archaeological record of this forgotten port that has been extensively studied is a corpus of funerary stelae. Put together through the efforts of epigraphists Madeleine Schneider and Giovanni Oman, this corpus consists of about 100 monuments still *in situ* on the island and more than 170 others dispersed in collections from New York (through London, Bar-le-Duc, Treviso, Modena, Athens, Cairo, Khartoum, and Mumbai) to Calcutta. The corpus tells the story of a diverse community with far flung connections; a number of the funerary inscriptions, moreover, reveal that between the eleventh and the early thirteenth centuries, and then again around the time of the Portuguese forays in the Red Sea, local potentates had assumed the title of sultan. I argue that at least in the case of the more robustly attested eleventh–thirteenth century sultans, this assertive titulature and the concomitant claims to authority and legitimacy accompany a moment of expansion of the island polity and an assertion of control, if not sovereignty, over a land-and-sea realm, extending throughout the Dahlak archipelago and including the sea routes that traversed it.

Like the islands of the Swahili coast, Dahlak Kebir lies relatively close to the mainland, at about 50 kilometers from the port of Massawa, and about eighteen miles from the closest mainland shore on the Buri peninsula. It is part of an archipelago, which not unlike more remote, oceanic archipelagoes of the Maldives, the Laccadives, and even the Andamans, exhibits an inherent unity or boundedness; described in another context as a "dispersed city," archipelagic coherence lends itself naturally to the creation of a geopolitically configured land-and-sea realm.[55] The Red Sea, although an arm of the Indian Ocean rather than an open oceanic space, lies along intersecting networks of routes, north-south and east-west, regional and transregional, and as such partakes both of the local and regional Red Sea networks, of oceanic Indian Ocean networks, and of transregional networks connecting the Indian Ocean with the Mediterranean. There are several ways in which the Dahlak polity fits into the prevailing patterns of Indian Ocean connections in the formative period of the transregional system between the eleventh and the thirteenth centuries. Insularity and a subtly visible bid for control over its maritime realm are key factors in this participation, and result in five phenomena: the early use and subsequent conceptualization of the island as a place of imprisonment, the equation by outsiders of local bids for control and definition of turf with piratical activity, the development of the island as a node in a commercial network of ports connecting the Mediterranean and the Indian Ocean, its participation in the slaving network connecting the African and Arabian shores of the Red Sea, and its incorporation in larger territorial states.

The Island Imprisons

In introducing his description of the port city of Aden, thirteenth-century traveler Ibn al-Mujāwir (fl. early thirteenth century) goes off on a tangent to enumerate the places that ostensibly functioned as prisons in the Islamic and pre-Islamic past: Aden, a virtual island, was first among them and so were other important ports of the Arabian seaboard including the island of Dahlak. The early Muslim historian al-Ṭabarī (d. 923) testifies to the use of the island as a place of exile during the Umayyad and Abbasid caliphates, and Ibn al-Mujāwir simply repeats this piece of information at a time when the fragmentation of the Islamic world had given the island a different role. Ibn al-Mujāwir's contemporary, geographer Yāqūt al-Ḥamawī (d. 1229), conveys fragments of poetry condemning the place for its isolation and desolation.

The concept and reality of islands as places of imprisonment or exile is a recurring theme in the study of maritime places, from Robben Island in Cape Town to Gyaros and Makronesos in the Aegean—and significantly, the perceived suitability of islands for incarceration factored into the use of the Dahlak island of Nokra as an Italian colonial prison in the mid-twentieth century, and in the maintenance of a notorious penitentiary on Dahlak Kebir itself in our days.[56] But although penal history is a productive framework of inquiry for islands that have been imagined as natural prisons and used as such in practice, it is important to note here what Vaidik's recent work on the Andamans invites us to consider that the topos of the insular prison, of islands as natural and ideal places of incarceration, artificially and forcibly amplifies the dimension of insular isolation. The prison image, in turn, contributes to the construction of certain places as marginal. While political marginality may be a historical fact at certain times—when these places lay at the peripheries of large territorial empires—it should obscure neither the other factors that shaped the human experience of these places and periods, nor the transformations that took place with the expiration of the marginalizing geopolitical realities. The application of a variety of geospatial frameworks of inquiry—local, regional, transregional—will produce a more nuanced account of both historical trajectory and lived experience of islands and other insular entities.

The Insular Pirates' Lair

As Braudel shows, islands in certain times and certain parts of the Mediterranean came to be synonymous with pirates, so that to the viceroys of Sicily in the early modern period "clearing the islands" signified eliminating piracy.[57] A correlate of Dahlak's once (during the early Islamic, Umayyad, and Abbasid unified state) real and later (in post-Abbasid times) enduringly perceived marginality is the representation of the island's rulers—as happens with the sovereigns of other island and insular ports in this region—as predators. Medieval travelers, from the twelfth-century Spanish Muslim pilgrim Ibn Jubayr, to the anonymous early thirteenth-century Jewish trader whose report of a first foray down the Red Sea was preserved in the Cairo Geniza, complain about the rapacity of local people, subjects, and rulers, and the extortionism of port city customs houses. Modern historiography picks up this perspective somewhat too quickly, and the ruler of Dahlak— one of those who claimed the title of sultan and advertised their defense of Islam on their tombstones—becomes "a petty ruler" or a "dangerous pirate." Similar judgments apply to the ruler of the island of Kish in the Persian Gulf, a state that in the twelfth century was explicitly said by geographers to have commanded a land-and-sea realm beyond the immediate confines of its small island base. I argue that these designations of piracy elide the dynamics of a geopolitical landscape in which small island and insular polities grew along the busy sea lanes, laid claim on their immediate maritime realm, and eventually came into conflict with larger states with vested interests in these routes.

Stepping Stones, "Routes des Îles," and Maritime Corridors

How did these islands attract mercantile traffic in the first place? In his powerful evocation of Mediterranean islands as stepping stones, Braudel paints vivid images of insular "stretches of relatively calm waters" and islands as "stationary fleets."[58] The navigators and cartographers of the medieval Islamic world evince a similar notion of navigational islands in their writings and visual representations respectively. Aḥmad Ibn Mājid, the renowned and much-studied Arab navigator, writing just before the turn of the sixteenth century, includes islands among the basic elements of seafaring knowledge—he devotes an entire chapter to islands,[59] and in a summary of navigational fundamentals he recommends thorough knowledge of "all the coasts and their landfalls and their various guides, such as mud, or grass, animals or fish, sea-snakes and winds, the alterations of the tides and the sea currents and the islands on every route."[60] The recently discovered eleventh-century geographical treatise mentioned earlier includes maps of the Mediterranean and the Indian Ocean, the former replete with islands, the latter a bit more sparingly littered with them. Emilie Savage-Smith interprets maps like this one as designed to be "an aid to memory and a means of imposing order on new and complex material and not as a visual model of physical reality."[61] Reproducing physical forms was not as important as reminding viewers the fundamentals of knowledge acquired while sailing and trading at sea, and islands were clearly central to that knowledge.

Dahlak was one of these islands, a link in maritime itinerary chains (but not stationary fleets as there was no one dominating power) leading up or down the Red Sea. Medieval Jewish traders' letters preserved in the Cairo Geniza testify to routine stops on Dahlak; in one case, the somberly portrayed itinerary mentioned earlier, links a number of ports down the Red Sea (some of them on islands) into a well-defined chain of landings. Moreover, this and other Geniza traders' letters attest to a thriving market: a mysterious but clearly much sought-after product of the archipelago linked local producers and traders to overseas buyers. Mediterranean and other merchandise deposited here, ranging from Egyptian textiles and dyeing chemicals recorded by the Jewish traders, to Chinese pottery and glass bangles retrieved in a recent archaeological surface survey of the medieval site, must have made their way to the Abyssinian mainland across the water and beyond.

The Island and the Slave Trade

The link with that nearby littoral (the coast of modern-day Eritrea, eighteen miles to Buri peninsula, about fifty miles to the Massawa area) forged one of the island's most robust relationships across the water, with principalities in Yemen, through the transfer of slaves either as merchandise or as tribute. Markus Vink ascribes "the 'Afrocentric' focus of Indian Ocean historiography on slavery and the slave trade" to the legacy of Atlantic slavery studies and the concomitant focus on plantation slavery in the Mascarenes and the Swahili coast.[62] The same historiographical forces, coupled with the dispersed and diverse nature of the pertinent sources, are probably at the root of a chronological bias that has left premodern slavery networks relatively unexplored. Here, with Dahlak (and possibly other island and insular ports of the western Indian Ocean), is a chance to look at slavery from a greater variety of angles in a precolonial period: the African hinterland networks, the agency of the island intermediaries, and the impact on the receiving societies.

The complex and little-known story of the rise and fall of the Najahid princes of Zabīd (1020s–1150s), a city that was one of the early recipients of Abyssinian slaves through Dahlak, illustrates the closeness of ties between the Yemeni coast and its opposite shores across the Red Sea, as well as the multifaceted impact of slavery networks in the region. Dahlak Kebir and its archipelago play a pivotal role in this connection and the production of a fascinatingly hybrid littoral society. On losing their city to the rising Sulayhid power of the Yemeni highlands, the defeated Najahid rulers, who were of Abyssinian slave origin, took refuge in Dahlak, where they plotted their return. In preparation for storming the Najahid city, the Sulayhid leader, al-Mukarram, instructed his troops to refrain from killing black Africans in Zabīd, and instead to subject them first to a linguistic test; if when asked to pronounced the Arabic phoneme "ẓ," they produced a "z," then they were fair game, their accent having just betrayed them as pure Abyssinians and presumably part of what was perceived as a foreign Abyssinian cadre ruling the city; but if they pronounced the phoneme in the standard peninsular Arabic way, they were to be considered Arabs and spared, because "Arab men in these coastal regions have children with black slaves and black skin is shared by free and slave alike."[63] Work on the implications of the cosmopolitan and hybrid nature of medieval port and littoral societies, on islands and beyond, such as the ones on Dahlak and in Zabīd, is only just beginning.[64]

Islands Integrated: Territorial States with an Oceanic Outlook

There is a significant temporal component in the pursuit of nesiology. The fragmentary record—textual and material—of a number of small islands and port cities of the western Indian Ocean suggests the emergence of independent polities at these places in the course of the eleventh century. In the absence of an active and sustained concern with the mechanics of overseas networks on the part of powerful regional states at that time, the same period that saw the development of the medieval world system also saw the rise of independent or semi-independent insular polities in the northwestern corner of the Indian Ocean, in the Red Sea, the eastern Arabian seaboard, the Persian Gulf, and along the African coast. In the second half of the twelfth century, a new dynasty founded by Salāḥ al-Dīn b. Ayyūb took over the remnants of the Fatimid state in Egypt and Syria. Fighting the crusaders in the Levant and Egypt, the Ayyubids also expanded southwards to subsume the sphere of waning Fatimid influence in Yemen. They thus found themselves in charge of the transregional trade routes connecting the Mediterranean with the Indian Ocean. Their hold on the region was tenuous, however, and soon a clan of deputies, charged by the Ayyubids with the administration of Yemen, took over the government, and through conquest and alliances unified this fractious land. Eric Vallet has recently posited that this new state in the western reaches of the Indian Ocean forged a distinct oceanic policy. He points to the preoccupation of the Rasulid state with shipping and maritime transport as evidenced in remarkable Rasulid administrative documents. In these documents and in the abundant chronicles for this period he also discerns Rasulid patronage of Muslim communities and religious authorities overseas, notably along the Indian littoral and in the Persian Gulf.[65]

What happened to the city-states that had been part of the geopolitical landscape up to that time? Some of them, like Aden, became firmly incorporated into the Rasulid realm, and continued to function as nodes of transoceanic networks, albeit with strengthened connections to the hinterland. Others, like Dahlak, appear to have lost their significance and possibly their autonomy and to have come under direct Rasulid influence, and at the periphery of the Rasulid state. The exact nature of Rasulid control over Dahlak remains to be fully explored and its elucidation will add to our understanding of the extent of Rasulid influence over its oceanic foreland. What we can say with some certainty is that Dahlak was drawn more closely into the orbit of a powerful Yemeni state once again, and that it perhaps became a piece of that state's extraterritorial, oceanic realm. The insularity that characterized it in the earlier period changed, and the island-based principality was not to reassert itself autonomously until more than two centuries later, when the demise of the powerful Rasulid state and the arrival of outsiders onto the western Indian Ocean scene appears to have offered the opportunity to a local potentate to exercise once again a kind of localized power.

Concluding Remarks

Dahlak's slipping in and out of a larger, more powerful state's sphere of influence is indicative of more generalized trends in Indian Ocean power structures and provides an example of how the focus on insular entities illuminates the geopolitical dynamics of the region through time. A similar lesson is taught dramatically by another island, one that appears to have resisted firm control throughout its long history. Ibn al-Mujāwir tells the story of Socotra's resistance to outside control very eloquently. He relates that when the Ayyubids conquered Yemen, they sent five galleys to Socotra in order to conquer the island. The islanders, however—all Christians and sorcerers according to the author—had other plans. "When the enemy approached the island, it disappeared from sight. They patrolled up and down, up and down, night and day for several days and nights, but found no sign of the island, nor had any news of it at all."[66]

What is this story about? Sitting strategically on the way between Africa and the northwestern part of the Indian Ocean and one of the first islands of the Indian Ocean to be inhabited, Socotra was a known quantity in the thirteenth century when Ibn al-Mujāwir was writing. It was visible from several north-south and east-west navigational routes, even if not visited very often, perhaps because of the treacherous maritime approaches to its landmass.[67] The island was known for its Christian community, and it appears that its closest links were not with the relatively nearby opposite shore of Yemen, but rather with the Persian Gulf and its Nestorian Christian communities.[68] Religious polemics and literary roots aside, Ibn al-Mujāwir's motif of the disappearing island hints at Socotra's disconnection from Yemen, and thereby at the fact that Socotra formed part of different networks at different times in its history. As Zoltan Biedermann has shown, in the early modern period, when the Ottomans and Portuguese fought for control and influence over Indian Ocean routes and littorals, Socotra drifted decidedly away from western Yemen's tenuous sphere of influence and instead became part of an axis that connected the Portuguese-controlled Persian Gulf with East Africa.[69] Far from being a pure construct, a product of the obvious inversions and polemics that bristle at the surface, Ibn al-Mujāwir's story may be hinting at different lines drawn across the water, as his Socotra partakes of the real Socotra: an insular form both connected and separated from the rest of the region by the Indian Ocean.

In his important work on island archaeology, Paul Rainbird has advanced the argument that islands "form only a part of a much more complex story, the story of maritime communities."[70] Taken to its extreme, this suggests that treating islands and their histories as uniquely meaningful is misleading. This chapter has argued something slightly different. The histories of Indian Ocean islands like Dahlak and Socotra are indeed only a part of the complex story of the Indian Ocean, but a part that provides dramatic illustrations of the dynamics that obtain in the oceanic world as a whole. Focus on islands and on the concept of insularity as a major characteristic of the nodes of the Indian Ocean world enables us to fine-tune the periodization of Indian Ocean history, and decipher the ways in which historical events and political change over time affected lives and livelihoods in the region as a whole. Looking at islands exposes the mechanisms through which the constituent parts of the Indian Ocean world functioned and related to one another.

But we can look from an Indian Ocean derived nesiology to the question of a future thalassography. For the formation of land-and-sea realms centered on islands or island-like port cities does not only illuminate the possibilities and reach of maritime life in these land-and-sea realms. Islands connect. Studying how they connect—nesiology—models one of the forms thalassography can take. Islands may be dependent or independent, large or small, but as long as they are inhabited their study will always link water, matter, and people.

NOTES

1. Ibn Ṭufayl, *Ibn Tufayl's* Ḥayy Ibn Yaqẓān: *A Philosophical Tale* (Chicago: University of Chicago Press, 2009), 103. For the Arabic text, see Leon Gauthier, *Hayy ben Yaqdhan: Roman philosophique d'Ibn Thofail* (Beirut: Imprimerie Catholique, 1936), 20, 24.
2. Ibn Ṭufayl, *Ḥayy Ibn Yaqẓān,* 103.
3. Ibn Ṭufayl, *Hayy Ibn Yaqẓān,* 105, 109.
4. Ibn Ṭufayl, *Hayy Ibn Yaqẓān,* 164–65.
5. This is Aparna Vaidik's eloquent summary of attributes of islands according to an isolationist conceptualization of their essence. See Vaidik, *Imperial Andamans: Colonial Encounter and Island History* (London: Palgrave Macmillan, 2010), 10.
6. See Emilie Savage-Smith and Yossef Rapoport, eds., *The Book of Curiosities: A Critical Edition* (Internet publication, March 2007), www.bodley.ox.ac.uk/bookofcuriosities.
7. See Shawkat Toorawa, "Waq al-Waq: Fabulous, Fabular, Indian Ocean (?) Islands," *Emergences: The Journal of Media and Composite Cultures* 10 (2000): 387–402. This essay was republished under the title "The Medieval Waqwaq Islands and the Mascarenes" in a volume edited by the same author and entitled *The Western Indian Ocean: Essays on Islands and Islanders* (Port Louis: Hassam Toorawa Trust, 2007), 49–65.
8. In addition to the seminal critical engagement with "fabular" islands by Shawkat Toorawa, it is also worth noting here that in one of the earliest twentieth-century treatments of the Indian Ocean as a unified maritime

and cultural space, Alan Villiers captures—though he does not deliberately explore—this essential dimension of island history and historiography by entitling his second chapter "Fabulous Islands"; see his *Monsoon Seas: The Story of the Indian Ocean* (London: Museum Press, 1952), 12–35.

9. For the coinage of a term to designate the mapping of such studies, see the fascinating work on the poetics of islands by David Punter, "Nesiographics: A Faulty Alphabet of Islands," *Emergences* 10 (2000). For the Greek *nesologion* or Latin *isolarium* as a term that designates a late medieval genre of geographical writing, see Yorgos Tolias, *Ta Nesologia: He monaksia kai he syntrophia tōn nesiōn* (Athens: Olkos 2002).

10. In the earliest serious comprehensive historical survey of Indian Ocean history, Auguste Toussaint contrasts the absence at the time (1961) of comprehensive work on Indian Ocean islands with the undertaking of such work for the Pacific. This remains the case to this day, and it is so that Aparna Vaidik, in the most recently published work on an Indian Ocean island theme, can write of Indian Ocean historiography's mostly "apathetic attitude towards the study of island histories" and draw a similar comparison as Toussaint, this time with Pacific and Atlantic historiography; see Toussaint, *Histoire de l'océan Indien* (Paris: Presses Universitaires de France, 1961), 5, and note 1; and Vaidik, *Imperial Andamans: Colonial Encounters and Island History* (New York: Palgrave Macmillan, 2010), 9, 12. For further statements on the historiographical neglect of Indian Ocean islands especially with regard to African history, see Edward Alpers, "Indian Ocean Africa: the Island Factor," *Emergences* 10 (2000): 373–74; and Peter Mitchell, "Towards a Comparative Archaeology of Africa's Islands," *Journal of African Archaeology* 2 (2004): 230.

11. See, for example, the Cyprian Broodbank, *An Island Archaeology of the Early Cyclades* (Cambridge: Cambridge University Press, 2000), 6, as well as the reception of Broodbank's work itself.

12. Toussaint, *Histoire de l'ocean Indien,* 3–5. See Braudel, *La Méditerréanée et le monde méditerranéen à l'époque de Philippe II* (Paris: Armand Colin 1949), 116–18. The significance of islands is also emphasized in two earlier works on the Indian Ocean. Writing from the vantage point of the large, continental island of Madagascar, Jacques Auber argues that islands "marked the historical destiny of the Ocean" by playing significant roles as stepping stones, fulcrums, and bases for trade and conquest; see his *Histoire de l'océan Indien* (Antananarivo: Société Lilloise d'imprimerie de Tananarive, 1955), 16–29. And as mentioned in note 8, Alan Villiers devoted a whole chapter to "Fabulous Islands."

13. Alpers, "The Island Factor," 374.

14. Toussaint, *Histoire de l'océan Indien,* 4. To quote precursor and sympathizer of Indian Ocean-centered histories, Holden Furber, Toussaint wrote his "reflective history" of the region "while living on the island where, in the age of sail, nearly all the trends of the region's maritime history met, mingled, and parted to go their several ways." Holden Furber, Review of "Histoire de l'océan Indien" by Auguste Toussaint, *American Historical Review* 67 (1962): 422–23.

15. Toussaint *Histoire de l'océan Indien,* 5.

16. In his 2003 survey of Indian Ocean history, Pearson includes islands as one of several "deep structural topological characteristics" of the ocean, under the general rubric of "deep structure"; see Pearson, *The Indian Ocean* (London: Routledge, 2003), 17–19. Earlier, Kenneth McPherson gave islands shorter shrift in a section on the region's geographical setting, but returned to island roles later in the book, as for example when he describes the coastal waters of East Africa as "shallow, island-protected waterways" that "provided an ideal communication route for scattered coastal agricultural communities"; see his *The Indian Ocean: A History of People and the Sea* (Delhi and New York: Oxford University Press, 1993), 14, 23.
17. In the second major survey of Indian Ocean history to appear after Toussaint, Kenneth McPherson makes maritime lives central to his inquiry, as evidenced by his subtitle "A History of People and the Sea"; see McPherson, *Indian Ocean.*
18. John Mack, "The Land Viewed from the Sea," *Azania* 42 (2007): 1–14, esp. 3.
19. Pearson, "Littoral Society: the Concept and the Problems," *Journal of World History* 17 (2006): 353–73. The littoral—the zone that lies between the sea and the land and partakes of both—is everywhere across the Indian Ocean similarly porous, and as such produces a unity in Indian Ocean lives. Location at the permeable boundary zone between land and sea, occupation directly related to marine resources, and culture that reflects both aforementioned factors serve as Pearson's criteria for a working definition of littoral peoples.
20. Pearson, "Littoral Society," 358, 561–62.
21. Jean-Louis Guébourg, *Petites îles et archipels de l'océan Indien* (Paris: Karthala 1999). A second edition was published in 2006.
22. Guébourg, *Petites îles et archipels de l'océan indien,* 7.
23. Guébourg, *Petites îles et archipels de l'océan indien,* 43: "Les îles de l'océan Indien sont un exemple significatif de cette évolution des approches spatio-temporelles des découvreurs qui décrivent, des astronomes qui observent, des arpenteurs qui dessinent, des administrateurs qui nomment et des usagers qui choisissent et pérennisent des toponymes de première grandeur, chemins, parcelles, et lieux-dits, aboutissant à un pavage complet des dénominations spatiales."
24. Shawkat Toorawa, ed., *The Western Indian Ocean: Essays on Islands and Islanders* (Port Louis: Hassam Toorawa Trust, 2007). Two of the six essays in this volume, edited by Shawkat Toorawa and Edward Alpers, were first published in an issue of the periodical *Emergences* devoted to island studies.
25. Michael Laffan, "Finding Java: Muslim Nomenclature of Insular Southeast Asia from Srivijaya to Snouck Hurgronje," in *Southeast Asia and the Middle East: Islam, Movement, and the Longue Durée,* ed. Eric Tagliacozzo (Stanford, CA: Stanford University Press, 2009), esp. 24–25, 29–30, 33–34, 43, 48–49.
26. Toorawa leaves open the possibility that Arab sailors knew of the existence of the actual Mascarenes in the southern reaches of the ocean, while Laffan's account involves periods when Middle Eastern traders were present in Southeast Asia.

27. Aziz al-Azmeh, "Barbarians in Arab Eyes," *Past and Present* 134 (1992): esp. 3, 13–15.
28. Alpers, "Indian Ocean Africa: the Island Factor." This essay was reprinted with few changes and under the title "The Islands of Indian Ocean Africa" in Shawkat Toorawa, ed., *The Western Indian Ocean: Essays on Islands and Islanders.*
29. Mitchell, "Towards a Comparative Archaeology."
30. Mitchell, "Towards a Comparative Archaeology," 230.
31. For examples of an approach that stems from a more nation-centered sensibility, see *The Indian Ocean and its Islands: Strategic, Scientific and Historical Perspectives,* eds. Satish Chandra, B. Arunachalam, and V. Suryanarayan (New Delhi and London: Sage Publications 1993); and Hooshang Amirahmadi, *Small Islands, Big Politics: The Tonbs and Abu Musa in the Persian Gulf* (New York: Palgrave Macmillan, 1996).
32. Guébourg, *Petites îles et archipels de l'océan indien,* 12.
33. In connection to this effect, it is interesting to note that for purposes of an environmental assessment the United Nations' definition of "Indian Ocean Islands" region comprises only the island states of the Comoros, Seychelles, Madagascar, and Mauritius. This grouping makes sense in terms of administrative division and the practical parameters of assessment and implementation of environmental strategies, but it does produce a misleadingly named category. See Payet, Soogun et al., *Indian Ocean Islands: GIWA Regional Assessment 45b* (Kalmar: United Nations Environmental Programme, 2005), 13.
34. An example of this is the productive debate about the rise of Swahili states along the East African seaboard and the many ways in which a variety of rooted forces shaped these places, once seen as foreign Arab establishments; see Mark Horton, *Shanga: The Archaeology of a Muslim Trading Community on the Coast of East Africa* (London: The British Institute in Eastern Africa, 1996), 1–7; Chapurukha M. Kusimba, *The Rise and Fall of the Swahili States* (Walnut Creek, CA: Altamira Press, 1999), 43–66. See Erik Gilbert, "Coastal East Africa and the Western Indian Ocean: Long-Distance Trade, Empire, Migration and Regional Unity, 1750–1970," *History Teacher* 36 (2002), http://www.historycooperative.org/journals/ht/36.1/gilbert.html. Another example from the other side of the western Indian Ocean seaboard: the building traditions, aesthetics, and religious ideologies encapsulated in the production of the earliest mosques other Islamic religious buildings in Gujarat from the eleventh century on; there, what would have earlier been described as imported forms of religious architecture—with all the historiographical and present-day political ramifications that such description entails—now emerge from Alka Patel's meticulous examination of forms and craftsmanship as the result of complex interactions between Muslim patrons and craftsmen trained in indigenous building traditions; patrons, craftsmen, and the public who used the edifices in question formed a social continuum in the sense that they all participated in what Patel calls a number of different "axes of belonging"—religious, vocational, ethnic,

linguistic; see Patel, *Building Communities in Gujarat: Architecture and Society during the Twelfth Through Fourteenth Centuries* (Leiden: Brill, 2004).

35. The concept of archipelago of cities is Janet Abu-Lughod's, who deployed it to describe the geography of the economic system of the thirteen and fourteenth centuries. I am borrowing and elaborating her concept of "an archipelago of 'world cities' elevated above a sea of relatively isolated rural areas and open stretches." See her *Before European Hegemony: The World System, A.D. 1250–1350.* (Oxford and New York: Oxford University Press, 1991), 13–14, 348, 353.
36. I have briefly discussed these two ideas in the following papers: "Mercantile Networks, Port Cities, and 'Pirate' States: Conflict and Competition in the Indian Ocean World of Trade Before the Sixteenth Century," *Journal of Economic and Social History of the Orient* 51 (2008); and "Maritime Cityscapes: Lessons from Real and Imagined Topographies of Western Indian Ocean Ports," in *Histories of the Middle East: Studies in Middle Eastern Economy Society and Law in Honor of A. L. Udovitch,* eds. Roxani Margariti, Adam Sabra, and Petra Sijpesteijn (Leiden: Brill, 2010), 101–26.
37. Epeli Hau'ofa, "Our Sea of Islands," *Contemporary Pacific* (1993): 4–5. Hau'ofa's work resonates strongly with island studies across regions, and especially with material culture debates of insularity and islandness; see especially Rainbird, *The Archaeology of Islands* (Cambridge: Cambridge University Press, 2007), 39–41.
38. Hau'ofa, "Our Sea of Islands," 7–8. For the importance of Hau'ofa's ideas in the postcolonial framework of inquiry that interrogates the necessity and inevitability of the boundaries, see Rainbird, *The Archaeology of Islands,* 39–41.
39. Rainbird, *The Archaeology of Islands,* 41. Rainbird is quoting the work of Madonna Moss.
40. "Lobster Wars Rock Remote Maine Island," MSNBC, September 27, 2009, http://www.msnbc.msn.com/id/32700866/ns/us_news-life/.
41. Pearson, "Littoral Society," 359–60.
42. James M. Acheson, *The Lobster Gangs of Maine* (Hanover, NH: University Press of New England, 1988), 71–83. On the question of how boundaries are delineated, Acheson writes (71–73) that "Fishermen talk about boundaries in terms of such key elements as river mouths, major peninsulas, and islands, but the actual dividing lines between lobstering territories are relatively small features familiar only to people intimately acquainted with the area."
43. See Margariti, "Mercantile Networks, Port Cities, and 'Pirate' States," 545–46, based on Emily Tai, "Marking Water: Piracy and Property in the Pre-Modern West." Paper presented at Seascapes, Littoral Cultures, and Trans-Oceanic Exchanges, Library of Congress, Washington DC, February 12–15, 2003. http://www.historycooperative.org/proceedings/seascapes/tai.html.
44. Vink, "Indian Ocean Studies and the 'New Thalassology'," *Journal of Global History* (2007): 60.
45. Comments and unpublished paper entitled "Tribus et oligarchies mar-

chandes, puissances navales et États continentaux: Les pouvoirs cotiers de l'Arabie méridionale au temps de Sharma," presented at "L'horizon Sharma: Mutations des réseaux commerciaux de l'océan Indien, ca. 980–1150," Paris, Friday, June 5, 2009.

46. Guébourg, *Petites îles et archipels de l'océan Indien,* 21–42.
47. See Eriksen, "In What Sense Do Cultural Islands Exist?," 135; Rainbird, *Archaeology of Islands,* esp. 1–3, 39–42; Elias Kolovos, "Insularity and Island Society in the Ottoman Context: the Case of the Aegean Island of Andros (Sixteenth to Eighteenth Centuries)," *Turcica* 39 (2007): 50; Cyprian Broodbank, "Not Waving but Drowning," *Journal of Island and Coastal Archaeology* 3 (2008), 72–74.
48. Christos Doumas, "Archaeology in the Aegean Islands," in *Archaeology: Aegean Islands,* ed. Andreas Vlachopoulos (Athens: Melissa, 2010): 14, 28.
49. Edward Lane, *Arabic-English Lexicon,* Book 1, part 2 (London: Williams and Norgate, 1865), 419.
50. Monier Monier-Williams, *Sanskrit-English Dictionary,* http://www.sanskrit-lexicon.uni-koeln.de/monier/indexcaller.php.
51. Book of Curiosities; see Jeremy Johns and Emilie Savage-Smith, "The Book of Curiosities: A Newly Discovered Series of Islamic Maps," *Imago Mundi* 55 (2003): 10. For an earlier temporal context in the Mediterranean, Maria Konstantakatou makes a case for the connected insularity of the city of Athens in a chapter entitled "The Island of Athens"; see her *The Dance of the Islands: Insularity, Networks, the Athenian Empire, and the Aegean World* (Oxford: Oxford University Press, 2007), 137–75.
52. For references to each of these examples and an earlier statement of the argument on the insular topography of Indian Ocean ports, see Margariti, "Maritime Cityscapes," 117–20.
53. Pearson, "Littoral Society," 356.
54. This section is based on information compiled and conclusions reached in two articles where I first presented my research on port city states in general and Dahlak in particular. My aim in providing this condensed reading of much of the same information is to explore how the history of this single island exemplifies many of the themes of Indian Ocean island history, and in turn, how a thalassographic and nesiographic lens exposes several important themes of Indian Ocean history as a whole. For references to the relevant sources and full bibliographical details, refer to these two articles: Margariti, "Thieves or Sultans? Dahlak and the Rulers and Merchants of Indian Ocean Port Cities, 11th to 13th centuries AD," in *Connected Hinterlands: Proceedings of the Red Sea Project IV held at the University of Southampton, September 2008,* ed. Lucy Blue, John Cooper, Julian Whiteright, and Thomas Ross (Oxford: Archaeopress BAR, 2009); and "Mercantile Networks, Port Cities, and 'Pirate' States."
55. In his examination of Ming accounts of the Maldives and Laccadives, Roderich Ptak unearths an eloquent expression of the internal coherence of archipelagoes in two descriptions of the Maldives by outsiders from the East and the West, respectively. Maghrebi traveller Ibn Baṭṭūṭa and Ming records describe archipelagic physical space by comparing its layout to a port and

a city, respectively; see Roderich Ptak, "The Maldive and Laccadive Islands (liu-shan 溜山) in Ming Records," *Journal of the American Oriental Society* 107 (1987): 682. The notion of the archipelago as a "dispersed city" appears in modern historiographical terms in the work of Greek historian Spyros Asdrahas, "To helleniko archipelagos mia diasparte pole," in *Chartes kai Chartographoi tou Aigaiou Pelagous* (Athens 1985): 235–48; quoted in Kolovos, "Insularity and Island Society."

56. Alberto Sbacchi, "Italy and the Treatment of the Ethiopian Aristocracy, 1937–1940," *International Journal of African Historical Studies* 10 (1977): 218; James Walston, "History and Memory of the Italian Concentration Camps," *Historical Journal* 40 (March 1997): 174. Human Rights Watch, *Service for Life: State Repression and Indefinite Conscription in Eritrea* (New York: Human Rights Watch, 2009), 36, 40, 94.
57. Braudel, *The Mediterranean,* 149.
58. Braudel, *The Mediterranean,* 144.
59. Tibbetts, *Arab Navigation In the Indian Ocean Before the Coming of the Portuguese* (London: The Royal Society of Great Britain and Ireland, 1971), 217–47.
60. Tibbetts, *Arab Navigation,* 77.
61. Savage-Smith, "Memory and Maps," in *Culture and Memory in Medieval Islam: Essays in Honour of Wilferd Madelung* (London and New York: Institute of Ismaili Studies, 2003), 109.
62. Markus Vink, "'The World's Oldest Trade': Dutch Slavery and Slave Trade in the Indian Ocean in the Seventeenth Century," *Journal of World History* 14 (2003): 133.
63. 'Umāra al-Ḥakami, *Yaman: Its Early Mediaeval History* (London: Arnold, 1892), 34–36 (English), 25–26 (Arabic).
64. The Najahids are a particularly fascinating and little-known dynasty on whom numismatic evidence has recently shed new light. See R. Strothmann and G. Rex Smith, "Nadjāḥids," in *Encyclopaedia of Islam, Second Edition,* vol. 7 (Leiden: Brill, 1993), 861–62. For the new numismatic evidence, see Audrey Peli, "A History of the Ziyadids Through Their Coinage," *Proceedings of the Seminar of Arabian Studies* 38 (2008): 251–64.
65. See Eric Vallet, "Yemeni Oceanic Policy at the End of the Thirteenth Century," *Proceedings of the Seminar of Arabian Studies* 36 (2006): 289–96; "Les Sultans Rasulides du Yémen protecteurs des communautés musulmanes," *Annales Islamologiques* 41 (2007): 149–76.
66. G. Rex Smith, *A Traveller in Thirteenth Century Arabia: Ibn al-Mujāwir's Tārīkh al-Mustabṣir* (London: Hakluyt Society, 2008), 264. Also see Smith, "Ibn al-Mujāwir on Dhofar and Socotra," *Proceedings of the Eighteenth Seminar for Arabian Studies* (1985): 79–91.
67. Alan Villiers comments on the visibility but poor accessibility of Socotra from the maritime trade routes and notes that "its roadsteads are open, and it can be a menace to navigation, particularly when the southwestern monsoon obscures the sky"; see *Monsoon Seas,* 18.
68. Zoltan Biedermann, "An Island under the Influence: Soqotra at the Crossroads of Egypt, Persia and India from Antiquity to the Early Modern Age,"

in *Aspects of the Maritime Silk Road: From the Persian Gulf to the East China Sea* (Wiesbaden: Harrassowitz, 2010), 14.

69. Zoltan Biedermann, "An Island under the Influence," 19–20. Biedermann argues that Socotra had become "a border stone of the area under Portuguese influence in the Western Indian Ocean."

70. Rainbird, *The Archaeology of Islands,* 3.

BIBLIOGRAPHY

Abu-Lughod, Janet. *Before European Hegemony: The World System A.D. 1250–1350.* Oxford and New York: Oxford University Press, 1991.

Acheson, James M. *The Lobster Gangs of Maine.* Hanover, NH: University Press of New England, 1988.

de Acosta, José. *The Natural and Moral History of the Indies.* London: Hakluyt Society, 1880.

Alpers, Edward A. "Indian Ocean Africa: The Island Factor." *Emergences* 10 (2000): 373–86.

Amirahmadi, Hooshang, *Small islands, big politics: the Tonbs and Abu Musa in the Persian Gulf.* New York: Palgrave Macmillan, 1996.

Auber, Jacques. *Histoire de l'Ocean Indien.* Antananarivo: Société Lilloise d'imprimerie de Tananarive, 1955.

al-Azmeh, Aziz. "Barbarians in Arab Eyes." *Past and Present* 134 (1992): 3–18.

Asdrachas, Spyros. "To helleniko archipelagos mia diasparte pole." In *Chartes and Chartographoi tou Aegaiou Pelagous,* edited by V. Sphyroeras, A. Abramea, and S. Asdrachas, 235–48. Athens: Olkos, 1985.

Biedermann, Zoltan. "An Island under the Influence: Soqotra at the Crossroads of Egypt, Persia and India from Antiquity to the Early Modern Age." In *Aspects of the Maritime Silk Road: From the Persian Gulf to the East China Sea,* edited by Ralph Kauz, 9–24. Wiesbaden: Harrassowitz, 2010.

Braudel, Fernand. *La Méditerranée et le monde méditerranéen à l'époque de Philippe II.* Paris: Armand Colin, 1949.

Braudel, Fernand. *The Mediterranean and the Mediterranean World in the Age of Philip II.* Vol. 1. New York: Harper and Row, 1972.

Broodbank, Cyprian. *An Island Archaeology of the Early Cyclades.* Cambridge: Cambridge University Press, 2000.

Broodbank, Cyprian. "Not Waving but Drowning." *Journal of Island and Coastal Archaeology* 3 (2008): 72–76.

Chandra, Satish, B. Arunachalam, and V. Suryanarayan. *The Indian Ocean and its Islands: Strategic, Scientific and Historical Perspectives.* New Delhi and London: Sage Publications 1993.

Collier, L. Review of *Histoire de l'océan indien,* by Auguste Toussaint. *Geographical Journal* 127 (December 1961): 511–12.

Doumas, Christos. "Archaeology in the Aegean Islands." In *Archaeology: Aegean Islands,* edited by Andreas Vlachopoulos, 14–29. Athens: Melissa, 2010.

Eriksen, Thomas Hylland. "In What Sense Do Cultural Islands Exist?" *Social Anthropology* 1 (1993): 133–47.

Furber, Holden. Review of *Histoire de l'océan indien,* by Auguste Toussaint." *American Historical Review* 67, no. 2 (January 1962): 422–23.

Gauthier, Léon. *Hayy ben Yaqdhan: Roman philosophique d'Ibn Thofail, texte arabe et traduction française.* Beirut: Imprimerie Catholique, 1936.

Gilbert, Erik. "Coastal East Africa and the Western Indian Ocean: Long-Distance Trade, Empire, Migration and Regional Unity, 1750–1970." *History Teacher* 36 (2002), http://www.historycooperative.org/journals/ht/36.1/gilbert.html.

Grove, Richard H. *Green Imperialism.* Cambridge: Cambridge University Press, 1996.

Guébourg, Jean-Louis. *Petites îles et archipels de l'océan Indien.* Second edition. Saint-Denis: Karthala, 2006. (First Edition: Paris: Karthala, 1999).

Hamilakis, Yannis. "The Other Parthenon: Antiquity and National Memory at Makronisos." *Journal of Modern Greek Studies* 20, no. 2 (2002): 307–38.

Hau'ofa, Epeli. "Our Sea of Islands." *Contemporary Pacific* (1994): 2–17.

Horton, Mark. *Shanga: The Archaeology of a Muslim Trading Community on the Coast of East Africa.* London: The British Institute in Eastern Africa, 1996.

Human Rights Watch. *Service for Life: State Repression and Indefinite Conscription in Eritrea* New York: Human Rights Watch, 2009.

Ibn Tufayl, Abu Bakr. *Ibn Tufayl's Hayy Ibn Yaqzān: A Philosophical Tale.* Edited by Lenn Evan Goodman. Chicago: University of Chicago Press, 2009.

Johns, Jeremy, and Emilie Savage-Smith. "The Book of Curiosities: A Newly Discovered Series of Islamic Maps." *Imago Mundi* 55 (2003): 7–24.

Kolovos, Elias. "Insularity and Island Society in the Ottoman Context: The Case of the Aegean Island of Andros (Sixteenth to Eighteenth Centuries)." *Turcica* 39 (2007): 49–122.

Konstantakatou, Maria. *The Dance of the Islands: Insularity, Networks, the Athenian Empire and the Aegean World.* Oxford: Oxford University Press, 2007.

Kusimba, Chapurukha M. *The Rise and Fall of Swahili States.* Walnut Creek, CA: Altamira Press, 1999.

Laffan, Michael. "Finding Java: Muslim Nomenclature of Insular Southeast Asia from Srivijaya to Snouck Hurgronje." In *Southeast Asia and the Middle East: Islam, Movement and the Longue Durée,* 17–64. Stanford, CA: Stanford University Press, 2009.

Lane, Edward. *Arabic-English Lexicon.* Book 1, part 2. London: Williams and Norgate, 1865.

"Lobster Wars Rock Remote Maine Island." MSNBC.com, September 5, 2009, http://www.msnbc.msn.com/id/32700866/ns/us_news-life/.

Mack, John. "The Land Viewed from the Sea." *Azania* 42 (2007): 1–14.

Margariti, Roxani Eleni. "Maritime Cityscapes: Lessons from Real and Imagined Topographies of Western Indian Ocean Ports." In *Histories of the Middle East: Studies in Middle Eastern Economy Society and Law in Honor of A.L. Udovitch,* edited by Roxani Margariti, Adam Sabra, and Petra Sijpesteijn, 101–26. Leiden: Brill, 2010.

"Mercantile Networks, Port Cities, and 'Pirate' States: Conflict and Competition in the Indian Ocean World of Trade Before the Sixteenth Century." *Journal of Economic and Social History of the Orient* 51 (2008): 543–77.

McPherson, Kenneth. *The Indian Ocean: a History of People and the Sea.* Delhi and New York: Oxford University Press, 1993.

Mitchell, Peter. "Towards a Comparative Archaeology of Africa's Islands." *Journal of African Archaeology* 2 (2004): 229–50.

Monier-Williams, Monier. *Sanskrit-English Dictionary.* http://www.sanskrit-lexicon.uni-koeln.de/monier/indexcaller.php.

Patel, Alka. *Building Communities in Gujarat: Architecture and Society During The Twelfth Through Fourteenth Centuries.* Leiden: Brill, 2004.

Payet, R.A. et al. *Indian Ocean Islands.* GIWA Regional Assessment 45b. Kalmar: United Nations Environment Programme, 2004.

Pearson, Michael N. "Littoral Society: The Concept and the Problems." *Journal of World History* 17, no. 4 (December 2006): 353–73.

Pearson, Michael Naylor. *The Indian Ocean.* London: Routledge, 2003.

Peli, Audrey. "A History of the Ziyadids Through Their Coinage." *Proceedings of the Seminar of Arabian Studies* 38 (2008): 251–64.

Ptak, Roderich. "The Maldive and Laccadive Islands (liu-shan 溜山) in Ming Records." *Journal of the American Oriental Society* 107 (1987): 675–94.

Punter, David. "Nesiographics: A Faulty Alphabet of Islands." *Emergences: Journal for the Study of Media & Composite Cultures* 10 (November 2000): 241–59.

Rainbird, Paul. *The Archaeology of Islands.* Topics in Contemporary Archaeology. Cambridge: Cambridge University Press, 2007.

Rapoport, Yossef, and Emilie Savage-Smith. "Medieval Islamic View of the Cosmos: the Newly Discovered Book of Curiosities." *Cartographic Journal* 41, no. 3 (2004): 253–59.

Ray, Himanshu P. "Early Seafaring Communities in the Indian Ocean." In *The Western Indian Ocean: Essays on Islands and Islanders,* edited by Shawkat M. Toorawa, 21–46. Port Louis: Hassam Toorawa Trust, 2007.

Savage-Smith, Emilie. "Memory and Maps." In *Culture and Memory in Medieval Islam: Essays in Honour of Wilfred Madelung,* 109–27. London and New York: Institute of Ismaili Studies, 2003.

Savage-Smith, Emilie, and Yossef Rapoport., eds. *The Book of Curiosities: a Critical Edition.* Intcrnct publication, March 2007, www.bodley.ox.ac.uk/bookofcuriosities.

Sbacchi, Alberto. "Italy and the Treatment of the Ethiopian Aristocracy, 1937–1940." *International Journal of African Historical Studies* 10, no. 2 (1977): 209–41.

Smith, G. Rex. *A Traveller in Thirteenth Century Arabia: Ibn al-Mujāwir's Tārīkh al-Mustabṣir.* London: Hakluyt society, 2008.

Smith, G. Rex. "Ibn al-Mujāwir on Dhofar and Socotra." *Proceedings of the Eighteenth Seminar for Arabian Studies* (1985): 79–91.

Strothmann, R., and G. Rex Smith, "Nadjāḥids." *Encyclopaedia of Islam, Second Edition.* Volume 7: 861–62. Leiden: Brill, 1993.

"Thieves or Sultans? Dahlak and the Rulers and Merchants of Indian Ocean Port Cities, 11th to 13th centuries AD," in *Connected Hinterlands: Proceedings of the Red Sea Project IV held at the University of Southampton, September 2008,* edited by Lucy Blue, John Cooper, Julian Whiteright and Thomas Ross, 155–63. Oxford: Archaeopress BAR, 2009.

Tibbetts, G. R. *Arab Navigation in the Indian Ocean Before the Coming of the Portuguese.* London: The Royal Asiatic Society of Great Britain and Ireland, 1971.

Tolias, Giorgos. *Ta Nesologia: He Monaxia Kai He Syntrophia Tōn Nesiōn.* Athens: Olkos 2002.

Toorawa Shawkat M. "Waq al-waq: Fabulous, Fabular, Indian Ocean (?) Island(s)." *Emergences: Journal for the Study of Media & Composite Cultures* 10 (November 2000): 387–402.

Toorawa Shawkat M., ed. *The Western Indian Ocean: Essays on Islands and Islanders.* Port Louis: Hassam Toorawa Trust, 2007.

Toussaint, Auguste. *Histoire de l'océan indien.* Paris: Presses Universitaires de France, 1961.

Toussaint, Auguste. *History of the Indian Ocean.* Chicago: University of Chicago Press, 1966.

Toussaint, Auguste. *La route des îles.* Paris: S. E. V. P. E. N, 1967.

Vaidik, Aparna. *Imperial Andamans: Colonial Encounter and Island History.* New York: Palgrave Macmillan, 2010.

Vallet, Eric. "Les Sultans Rasulides du Yémen protecteurs des communautés musulmanes." *Annales Islamologiques* 41 (2007): 149–76.

Vallet, Eric. "Tribus et oligarchies marchandes, puissances navales et États continentaux: Les pouvoirs cotiers de l'Arabie méridionale au temps de Sharma." Paper presented at 2nd International Study Day of the APIM Program "L'horizon Sharma: Mutations des réseaux commerciaux de l'océan Indien, ca. 980–1150," Paris, June 2009.

"Yemeni Oceanic Policy at the End of the Thirteenth Century." *Proceedings of the Seminar of Arabian Studies* 36 (2006): 289–96.

Vaughan, Megan. *Creating the Creole Island: Slavery in 18th century Mauritius.* Durham, NC: Duke University Press, 2005.

Villiers, Alan. *Monsoon Seas: The Story of the Indian Ocean.* New York: McGraw-Hill, 1952.

Vink, Markus. "Indian Ocean Studies and the 'New Thalassology.'" *Journal of Global* History 18 (2007): 41–62.

"'The World's Oldest Trade': Dutch Slavery and Slave Trade in the Indian Ocean in the Seventeenth Century." *Journal of World History* 14 (2003): 131–77.

'Umārah ibn 'Alī al-Ḥakamī. *Yaman, Its Early Mediæval History: The Original Texts.* Edited by Henry Cassels Kay. London: Edward Arnold, 1892.

Walston, James. "History and Memory of the Italian Concentration Camps." *Historical Journal* 40 (1997): 169–83.

EIGHT

Skerries, Haffs, and Icefloes: Small Seas and Maritime Histories

David Kirby

Writing in the *North American Review* at the end of the nineteenth century, Charles Minor Blackford triumphantly declared that "the progress of thalassography has been so rapid and so great during this century, that it can be sketched only in bare outline." What Blackford and his contemporaries understood by "thalassography" was, essentially, the exploration of the world's oceans.[1] The "new" thalassography (or thalassology) proclaimed a little over a hundred years later has a rather different meaning. An outgrowth of historical geography and area studies, it proposes a different kind of approach to world history by analysis of the whole by way of its components, and how they fit together.[2] Marine life is replaced by human life, and the methodology owes more to cultural anthropology than to the natural sciences. And whereas the *locus operandi* of pioneering oceanographic expeditions such as that of the *Challenger* in the 1870s was emphatically the big oceans of the world, the new thalassographers show a distinct preference for inland seas, and particularly, the interface between land and sea.

It is with that interface—the littoral—that this chapter is mainly concerned. After a few general remarks, I propose to look at one of the smaller "inland seas"—the Baltic, paying particular attention to a time of technological, political, and socioeconomic transition—the hundred years from around 1850. I have chosen this period partly to challenge some of the underlying assumptions of the Braudelian approach, but also to caution against relying too heavily on an excessively water-linked view of global history: states and their governments, big business, and the vast majority of the world's population that do not live by the shore but visit it and have views about it, cannot be left out of the equation. I shall conclude with a few remarks on the whole idea of "inland seas," and on an alternative approach to the history of the seas.

It is perhaps stating the obvious that perceptions of the sea differ widely. There is for instance a very sharp difference between those who go down to the sea in ships and those who do not. Until recent times, seafaring communities marked themselves out not only by their dress, customs, and language, but also by the very distinctive rhythms of their everyday life. Their way of life gave them the ability to read their environment in ways which non-mariners were unable to understand. For one experienced North Sea fisherman, writing his memoirs at the end of the nineteenth century, "there's nothin' in the world can be easier, when you've once learned your lesson, than to pick yer way about in the North Sea just with nothin' else to guide yer than the depth o' water an' the natur' o' the bottom."[3] Sailors not only need to have an intimate knowledge with the sea bed, they have to be able to read the margins between land and water. Before modern sonar technology, recognition of landmarks was a vital aspect of navigation. Church towers, prominent rocks, and headlands, and even tall trees, all formed part of an intimate knowledge of the contours of the coastline. In the Baltic, as in the far larger southern European sea, "an essentially visually ordering of geography was one of the earliest ways by which an individual might understand the relationship between his own sphere of movements and far broader horizons."[4] It is this knowledge, and the sustenance the sea can provide, that explains why people have chosen to live in otherwise remote and inhospitable places, like the fisherman in northern Norway who sat in the kitchen with his back to the land, looking out to sea to the horizon of the open sea.

> Since ancient times he has seen the life of the sea, not a flat, grey surface with waves, but underneath, a landscape with shallows and depths, with clay, sand, stone and vegetation, with currents and eddies and with creatures of the sea. He fished for them and especially the cod, for these riches were the reason why people chose to live on this jagged edge, bordering on the Arctic Ocean.[5]

There is another sharp contrast between our twenty-first-century view of the sea and that of our pre-eighteenth-century ancestors, for whom the sea was in many senses more immediate, and more frightening. It was immediate in that a significant proportion of the then-population of Europe lived within a dense network of waterways linking the land and sea, not only upon the shoreline, but in the marshy waterlands, and down the rivers that carried goods to and from the hinterland. It was frightening in that the sea could and did frequently wreak immense damage, and was thought to be the abode of many unknown and fearsome creatures. Fear, as Jean Delumeau has observed, can be found in all civilizations poorly equipped technically to respond to the countless assaults of a threatening environment: but the one space where the historian is sure to encounter this fear is the sea.[6] The sea was considered to be a waste area, in much the same way as mountains and heathland, yet from the very earliest times of human existence, it had been deemed essential to traverse it and to trawl its depths. Our contemporary conceptions of the sea are somewhat different. *Navigare necesse est* no longer strikes an immediate chord in a world where goods are transported in sealed containers and offloaded at dedicated sites far from centers of habitation. Only a very small minority, even of people living on the coasts, earn a living directly or indirectly from the sea. The seaside has become a place to visit for relaxation and pleasure, and this had led to a significant shift of perception, in which the prevailing view is invariably from the shore. Expensive sea defenses and an elaborate infrastructure of services—railway lines, roads, amusement parks, hotels, restaurants, and lodging houses—serve to assure the visitors' ease and comfort. The cruise liners and car ferries that are nowadays the limit of most people's experience of a sea voyage are also designed to make that encounter an extension of consumerism. The sea, in other words, is kept at a safe distance, almost out of sight, and when things do go wrong, as in the tragic sinking of the MS *Estonia* in the Baltic in September 1994, the shock is perhaps more profound because such disasters are, in comparison with early centuries, so rare.[7]

Coasts have always presented problems to those who feel the need for definition. The shoreline is a constant encounter between the elements. The ebb and flow of tides shape and reshape the relationship between land and water each day. The silt and sand carried by the waters block channels and leave once-prosperous ports to decay. Very occasionally, the elements break loose and change the coastline dramatically: new islands can be thrown up by volcanic activity, as occurred in the Vestmannaeyjar off southwest Iceland in 1963 and 1973, or an entire township can be covered in an exceptional sandstorm—the fate in 1413 of Forvie in northeastern Scotland. And coastlines everywhere in northern Europe have been radically altered over the ten thousand years or so in which human beings have settled there by the complex workings of the isostatic-eustatic relationship.

To this constant shifting of the boundaries between land and water may be added the sheer complexity of coastal morphology. This is especially the case in the northern Baltic, with its thousands of islands and inlets.[8] This is marginal territory indeed, far more difficult of control than the fertile and well-populated plains and valleys of the hinterland. Those who have settled in the past on these margins have had rather less contact with the hinterland, especially if there are few navigable waterways linking the two, than with other coastal communities. Far more than the inland peasant-farmer, seafarers lived face-to-face with the source of their livelihood, which was at the same time the daily arbiter of their fate. Their circumstances promoted cooperation, from the building and crewing of boats to the forming of trading partnerships, and because their seagoing activities were difficult for rulers or great magnates to control, they were less subjected to the kinds of impositions than were their less fortunate brethren of the fields.

The people of the coast have also caused problems for historians. In general, historians writing about the Baltic have been primarily interested either in power politics or trade. In the decades immediately following the disintegration of the Russian Empire, historians of different nationalities argued over the "struggle for supremacy" and "the Baltic question."[9] Trade was also dominated by conflict and struggle.[10] In the two decades since the collapse of the USSR, attention has turned more to questions of Baltic identity, and there has also been considerable energy devoted to the refurbishment of regional and local histories. But, with the exception of the maritime historians, the sea as such has featured more as a backdrop than as a central object of study. Even less attention has been paid to shores and shore dwellers, which have been largely left to the natural scientists and archaeologists, whose empirical studies have in recent decades added greatly to our knowledge of early maritime life and activity in the Baltic,[11] and to a number of Swedish ethnographers, who have continued the tradition of research into cultural landscapes established in the 1930s with the pioneering studies of ethnographer Åke Campbell.[12] Most of these studies, however, are limited to a specific region, such as the west coast of Sweden, or particular communities.[13] Transmarine or inter-regional studies have until recently been rare. The 1991 publication of Poul Holm's work on the maritime communities around the entrance to the Baltic was a noteworthy advance in this respect, though mention must be made of the rather neglected writings of the East German ethnographer Wolfgang Rudolph, who in 1980 identified a "zone of maritime culture of coastal villages" stretching along the southern Baltic shores from northern Schleswig to the Åland islands, which differed markedly from the culture of the agrarian hinterland—a cultural zone that overrode differences of language and religion.[14]

The liveliness and frequency of transmarine cultural exchanges is a distinguishing feature of this coastal culture for both Rudolph and Holm, though Rudolph also dwells upon other differences, such as the higher demands placed upon learning and skills in coastal seaports, and the contrast between the seasons—summer, the busiest time of the rural hinterland, was a dead time in coastal communities, where the men were at sea. The liveliest time in coastal communities was from November until March, when festivities were celebrated. Communities heavily dependent upon the sea were also distinguished in the way that labor functions and roles were apportioned, by the connections to other maritime communities rather than with inland regions, and at an everyday level by the speech, dress, and mental horizons of the shore dwellers.[15] Both writers see the nineteenth century as the apogee of these maritime regions, which they and others trace back into the Middle Ages. The vital thread is coastal trade, which occurred in a huge variety of forms, from exchange of commodities on the shore to the export of specialized goods to the towns and cities.

Much of this fits easily into the micro-regional perspective that is now emerging in studies of the seas and oceans of the world. It questions the over-emphasis upon "high" commerce, and offers a shoreline alternative to the "top-down" approach which sees technological advances and the power of the state as major determinants. It places the sea at the center of things, and it defines maritime culture as the human utilization of maritime space by boat, settlement, fishing, hunting, shipping, and their attendant subcultures such as pilotage, lighthouse, and seamark maintenance.[16] As the authors of the highly influential history of the Mediterranean in antiquity and the early Middle Ages argue, it creates a kind of inside-out geography in which the world of the sea is "normal" and the hinterland is the marginal fringe, with an inversion of distance in which neighbors on land are seen as distant and places linked by sea are close.[17] To the "connectivity of micro-regions," which is a crucial element in this theoretical framework, one might also add the necessity and ability of maritime communities to work together collectively. The building of boats, sailing, and trading, all demanded that communities work together, as did keeping the seas at bay and providing seamarks and other navigational aids.

This shift of focus has opened up exciting new perspectives, but there is also need to sound a note of caution, which is what I shall attempt to do in the following pages. My first observation is that there has been a marked reluctance on the part of historians to carry the story of the shoreline beyond the transition from sail to steam and diesel. We have become very wary of teleological progressivism, but we are less ready to challenge the Braudelian vision of the world as a passage of *longue durée,* barely affected by the doings of kings and their minions, or indeed by steam, diesel, and electricity. In seeking to rescue the seas and their shores from the margins of historiography, might we not be in danger of actually isolating them still further, by diminishing or downplaying the relationship with the deeper hinterland, both as a physical and a politico-social entity? The fact is that there is and always has been a considerable input from the landward side, which has significantly affected the human relationship with the sea. Let me try to illustrate this with a couple of examples of how intervention has brought about decisive change—one concerning the physical relationship of land and sea, the other, a striking instance of the way in which human volition can reconfigure an entire region.

The Baltic Sea is essentially the outcome of the last major period of glaciation, which came to an end some thirteen thousand years ago. Its shorelines bear all the marks of the retreating mass of ice, and are still affected by the consequences. The land is still recovering from the massive weight of the ice. Some of the ancient post-glacial coastline of modern Estonia lies over a hundred kilometers (just over sixty-two miles) from the present-day shoreline, and it is estimated that Finland gains around seven square kilometers (just over four square miles) of new land every year. Post-glacial rebound has led to marked differences in land settlement and usage. Along the low-lying Ostrobothnian coastline for instance, intensive land uplift has meant that channels are few, narrow, and rock-strewn, making navigation difficult. Harbors have had to be relocated several times. In the skerries that fringe the southern and southwestern coast of Finland, less severely affected by isostatic rebound, there are many good harbor sites—long, narrow sheltered bays and sounds, and channels which are broad and relatively rock free. Here, settlement is scattered in small hamlets and individual farms, whereas along the Ostrobothnian coast, large villages predominate, with their communal harbors, where their own specially adapted flat-bottomed boats are stored in scores of boathouses that line the shore.[18]

Ice continues to play a major part in the life of communities around the Baltic. The freezing over of the brackish waters has enabled armies on occasion to launch invasions, bus companies to run unusual winter services, and it allowed Vladimir Lenin to escape the clutches of the Russian authorities in 1907 by taking a hazardous nighttime walk across the ice in order to catch the ferry to Stockholm. For shipping and trade, however, ice poses severe problems. Before the deployment of effective icebreakers at the end of the nineteenth century, maritime trade for the ports of the northern Baltic was at a standstill for up to five months of the year. Ice floes in the southern Baltic seriously affected the harbor works and channels of major ports. The harsh winter of 1870–71, when even the harbors of Copenhagen and Hamburg were icebound for two months, prompted renewed activity to develop an icebreaker that could cope with the conditions in the Baltic. Within three decades, icebreakers were active in opening up channels to all the major ports around the Baltic. The need to be able to export goods all year round, and the competitive pressures this created, in other words, compelled intervention, which invariably assumed national dimensions.

The building and deployment of icebreakers, alongside the wholesale development of new infrastructures to accommodate larger vessels and to facilitate trade, would not have been possible without the active involvement of the state. Three of the four icebreakers operating in Finnish waters on the eve of the First World War were state owned (the fourth was owned by a municipality), and the Finnish state had stepped in to take over the winter harbor of Hanko, built in the early 1870s by a business consortium hoping to profit from the transit traffic from Russia. This was by no means the first instance of active state involvement in maritime matters. The charting of the waters of the Gulfs of Bothnia and Finland had been undertaken by the Swedish Admiralty in the seventeenth century, and it was the Swedish state that maintained an extensive postal service network across the waters of the Baltic. Maritime states were in the forefront of shipbuilding technology, with their dockyards and wharves employing several thousand peoples.[19] But the most spectacular example of state planning on the shoreline was the creation of St. Petersburg. Within fifty years of its foundation on the marshy banks of the Neva, the new Russian capital had acquired 90,000 inhabitants; by 1850, it had swelled to half a million—by far the largest city in the entire Baltic region. Built at huge human cost to accommodate the desire of

Peter the Great to establish a Russian presence on the sea, St. Petersburg was less than ideal as a port—it did not overtake Riga as the chief Russian port in the Baltic until the mid-nineteenth century—but as the capital of a major European power which expanded substantially into the Baltic region between 1710 and 1809, it exercised immense political influence over the entire Baltic region. There was a less spectacular but no less determined effort by the German Reich to establish a presence on the Baltic after 1871, with the building of a naval base at Kiel and massive harbor works at Stettin and other ports: and the newly independent Polish state after 1918 poured money into making Gdynia its gateway to the sea.

This kind of active intervention from beyond the shore has several major consequences. Firstly, it alters the spatial geography of the region. St. Petersburg acted as a magnet not only to settlers, but also traders, and this in turn disrupted or killed off long-established trading routes and connections and established new ones. It even had an impact on the cultural life of the older towns of the Baltic. Karl Rosenkranz was worried in 1842 that a new steamboat service between Lübeck and St. Petersburg would make it even less likely that waxworks, menageries, and circuses would be stopping in Königsberg in future; the cultural life of the entire region was thus dependent on the itineraries devised by the producers in St. Petersburg.[20] On a more positive note, wealthy Russians from the capital spent time and money during the summer in the spas and bathing resorts that sprang up along the coast of the Gulf of Finland.

The building of railways and canals often had a similar effect. The creation of a railway network not only strengthened the structure of the nation-state through uniformity, standardization, and regulation, it also firmly linked the coastal regions to the hinterland. The railways enabled goods to be carried more rapidly and more cheaply, offering an alternative to coastal trading. They also greatly facilitated the movement of people: workers moving into coastal areas where new industries and construction work provided employment opportunities, but also workers moving away from areas in economic decline in search of work inland, and above all, holidaymakers. In the twentieth century, roads and air traffic have added a whole new dimension to the communications network, helping create the infrastructure upon which the modern leisure industry has grown. Seaside towns have had to engage in extensive marketing, creating "brand images" which may rely heavily on a romanticized "heritage" or the attractions of the town as a festival venue. The point is that they have to face inland to do so; their lines of communication no longer flow across the waters as they once did.

The impact of early tourism upon coastal communities around the Baltic has been extensively studied by cultural anthropologists and historians. Up until the mid-twentieth century, visiting the Baltic seaside was largely confined to those with the wealth and leisure to do so, and the influx of people with money to spend, but also with very different attitudes and standards of decorum and behavior, often led to tension. In many places, the visitors provided a welcome new source of income to the sea captains and others involved in the declining coastal trade. Many built larger houses to accommodate their guests, who also provided seasonal employment, ranging from the provision of domestic services to boat hire. The presence of the visitors, dressed in light summer clothes, and bent on pleasure, created a stark contrast with the local working population—something frequently commented upon in guidebooks. Visitors frequently complained of the odor and mess around the fishermen's boats and stores, and were worried that the local street urchins might teach their children bad habits. The locals found the habits and behavior of the visitors strange and sometimes offensive, especially in areas which were strongholds of religious revivalism, such as the west coast of Sweden.

The numbers of summer visitors along the northern shores of the Baltic were rather more modest in comparison with those who flocked to the sea-bathing establishments on the flat, sandy shores of northern Germany, and in any event one must be cautious in evaluating the impact of the encounter between shoreline communities and the tourists. What in our present age is called "the seaside industry" is only one part of a process of transition, which set in from the 1860s onwards. Like all such processes, it was by no means uniform.[21] The necessity of adapting to sudden change was nothing new; what was different was the source of change. Maritime communities in the past had had to adapt to the sudden disappearance of the herring shoals, or the silting up of harbors; the threats to their livelihood came largely from the natural world. Now they were subjected to pressures generated by human forces: the intensification of global competition which drove out the under-capitalized small partnership shipping firms, and the demands of the state for regulation and control, which spawned a whole new class of seafarers—pilots, harbor masters, fisheries inspectors, and a merchant marine largely employed by the big new shipping concerns—and brought national, patriotic values to the people through the schools it built and via conscription into its armed forces.

The consequences of these economic and political pressures have been far-reaching, not least for the ecosystem of a small, shallow, virtually tideless, and landlocked sea. Communities in which a substantial part of the population earned a livelihood from maritime activities such as fishing, seafaring, or boat building, even into the middle decades of the last century, are now a memory, displayed in the artifacts and photographs of maritime museums. Substantial parts of the Finnish and Swedish archipelagos—marginal land which was subjected to fairly intensive settlement as a result of population pressure after 1750—are now almost entirely given over to tourism.[22] There are now large conurbations on either side of the Sound, which is crossed daily by thousands of rail and road commuters; the sea channels which once made a journey from Jutland across the islands to the Danish capital a hazardous venture have now been bridged, and the sea itself has been squeezed out, as shown in an advertisement celebrating the opening of the bridge across the Great Belt in 1997.[23]

The wars and subsequent divisions of territory during the twentieth century have wrought sudden, often tragic changes. Along the shores of the Baltic, from Mecklenburg to the Russo-Finnish frontier, there have been massive changes since 1945 in the ethnic composition of the population. Königsberg is now Kaliningrad—an entirely Russian city (though there are vigorous attempts to restore various aspects of its former identity).[24] The Baltic German and Jewish inhabitants of the Latvian and Estonian capitals have long since departed, the former obeying Hitler's call to return to the fatherland in 1939–40, the latter mostly destroyed in the Holocaust; Poles, mostly from the hinterland, now inhabit Gdańsk and the surrounding coastline, whilst only a minority of the present inhabitants of Mecklenburg and Pomerania can trace their ancestry back to the region. The Cashubians that Baron von Innstetten pointed out to his young wife Effi Briest, Slavs who had lived on the Pomeranian coast for a thousand years and maybe much longer, are far less noticeable one hundred years after Theodor Fontane's novel was published; like the Livs, the Swedish-speaking settlers on the western islands of the Estonian coast, and the Ingrians around the mouth of the Neva, they seem likely to follow the Abodrites, Wagrians, Rugians, Prussians, and Curonians into linguistic and cultural oblivion.

These changes have taken place against the background of fiercely conflicting national claims. As early as the 1840s, voices were being raised for a greater German presence at sea. In 1859, the writer Anton von Etzel belittled the seafaring capabilities and capacity of other peoples around the Baltic; the Germanic peoples had always played a leading role at sea and on the broad oceans. The Baltic, von Etzel confidently concluded, was "essentially a Germanic middle sea."[25] The intimate involvement of the German state in the development and direction of the Museum für Meereskunde was made explicit in the first issue of the museum's publications series: the main purpose of the museum was to bring to the attention of the people Germany's interests at sea.[26] The Polish League of the Sea (*Liga morska*) actively propagated Poland's maritime and colonial claims during the interwar years, seeking to commemorate the day in which Poland regained a Baltic coastline with a ceremony similar to the wedding of the Doge with the Adriatic. Research institutes were founded around the Baltic, and a large amount of literature was produced, to advance the interests of the national states. The political division of the entire region after the Second World War added yet another layer of national claims, this time advanced most forcefully and consistently by the Soviet Union. The enthusiasm for a common Baltic heritage in the aftermath of the collapse of the USSR was voiced largely in the nations to the west of the Iron Curtain; countries formerly in the grip of the Soviet Union were rather keener on refurbishing their own national identity.

Alongside the drawing of the coasts into the firm embrace of the national state and economy there also occurred a more subtle form of conquest. During the course of the eighteenth century, there was a significant shift in attitudes toward the sea, brilliantly charted by Alain Corbin. Corbin speaks of a "new harmony between the body and the sea," and describes the passion for sea-bathing that set in from the middle of the eighteenth century and the emergence of a literature of the picturesque, in which the beach became a stage from which to view the elements.[27] The impetus for what Corbin terms "the invention of the beach" came not from those who dwelt by the sea, but from those who visited it. The cliffs, rocks, dunes, and beaches that had for centuries been shunned and avoided were in the course of the nineteenth century opened up to the visitor, for whom the guidebook writers were only too willing to tell them how to take in the scenery. The passion for geology aroused by the findings of scientists such as Charles Lyell, and the invocation of sublime sentiments by the Romantic poets were further inducements to visit the shore.

From the outset, there was a conflict of desires between those seeking inspiration from what they took to be unspoilt nature, and those more bent on pleasure. In the Baltic, such pleasures were taken at a sea-bathing establishment by those with the leisure and wealth to do so, although there is some contemporary evidence of what one can only describe as "wild" camping and bathing on coasts within easy access of the larger cities. Guidebooks made much of the beauties of the natural surroundings of the bathing stations, and sometimes chided visitors for preferring to idle away their time on the dance floor instead of enjoying the picturesque and unspoilt coastline; however, they were also careful to stress the comforts and health-giving properties of the spa.[28] Several places around the Baltic were famed for their sublime qualities: the cliffs at Arkona on the island of Rügen, painted in early nineteenth century by Caspar David Friedrich and immortalized in verse by the poet Ludwig Gotthard Kosegarten, the panoramic view over the Sound from the terraced gardens above the town of Helsingborg, or Skagen at the tip of Jutland, which had attracted a colony of artists by the end of the century. But the remoteness and isolation of the coast was by no means to everyone's taste. Hans Christian Andersen for one found Skagen too raw, its inhabitants too impoverished, and looked forward to the day when the houses would stand neatly side-by-side, each with its own garden and with good road access to the nearest railway station, whilst a comprehensive survey of "picturesque and romantic Germany," having concluded frankly that "the character of the German Baltic coast . . . is in general neither picturesque nor romantic," noted that a great number of "naked and bare" coastal villages have nevertheless recently been turned into bathing resorts, "thereby arousing the need to seek out pretty, less desolate places."[29]

During the preindustrial era, the outside world defined the Baltic largely in terms of trade. It was the "East Sea" from which the natural products of the eastern and central European hinterland were exported to the west, in return for salt, textiles, and certain luxury goods. The autochthonous peoples along its southern and eastern shores had been subjugated by German knights, and with that process of eastward expansion, new trading centers were developed. For much of the Middle Ages, the models for urban culture around the Baltic were German—in architecture, trading standards and regulations, models of urban government, and even in the predominant status of Low German as the language of seafaring and commerce. During the sixteenth and seventeenth centuries, as the Hanse declined and the Low Countries came to dominate the Baltic grain trade, there was a significant Dutch influence, although the great territorial power in the region from 1629 to 1700 was Sweden. By 1800, however, these seaborne influences upon what we might term the "high culture" of the region were in decay. Coastal cities such as Danzig, once richer and far more cosmopolitan than either Berlin or Warsaw, were now ridiculed as old-fashioned and provincial.

Recent research has added an invaluable new dimension to the somewhat one-sided picture of the Baltic as a seaway filled with Dutch or English ships carrying off the grain and timber products of an economically backward hinterland by showing how important coastal trade was in developing interregional contacts and as a means of transmitting cultural impulses. That trade reached its apogee in the eighteenth and early nineteenth centuries, when ships from ports all around the Baltic, mostly owned by partnerships, sailed not only to harbors all along the coast, but through the Sound to the Mediterranean and Black Sea and beyond. The Black Sea grain trade in the nineteenth century was particularly lucrative: it was rumored that earnings from a good trip could pay off half the value of the ship.[30] This kind of small-scale carrying trade was hit hard by the emergence from the 1870s of the big, capital-intensive steamship companies such as DFDS and the concentration of tonnage in the big ports. Many small harbors ceased to exist, and regions such as Løjtland in southern Jutland abandoned seafaring altogether. Some communities were better able to adapt, taking advantage of a niche in the market and using their earnings to modernize their vessels, or developing new skills and techniques. But most declined, and with that decline, long-established networks withered and died.

The picture of micro-regional connectivity during the era of sail is a very compelling one. The waterfront of cities such as Stockholm swarmed with peasant-traders, who had combined to build a boat, to stock it with goods in demand—dried fish, firewood, wooden utensils—and to crew it. Coastal trade around the Sound was dominated by the *skude*—a small, locally built vessel ideally suited for the transport of goods in these waters. The customs books for Copenhagen for 1615 reveal a total of 486 small vessels, mostly *skuder,* coming to the port. Most were from northwest Scania, and some came as many as eleven times during the sailing season. Those involved in the trade were often called "peasant," but in reality were more or less full-time sailors, often landless shore dwellers. Certain regions developed their own specialisms. Peasants from the poor, stony lands around the village of Kalanti in southwest Finland exported homemade wooden drinking vessels to Stockholm and Reval; traders from the Danish island of Langeland were regular visitors to the Landskrona horse fair in Scania, buying up large quantities of horses, in all likelihood for sale in northern Germany. Many were also prepared to travel long distances. Russian fur traders travelled along the waterways from as far as the White Sea coast to the midsummer market in Tornio, at the head of the Gulf of Bothnia, and as late as the nineteenth century, some three hundred years after the glory days of the Tornio summer fair, women from the uplands of central Sweden made the journey to St. Petersburg to sell their hair.

These are but a few examples of the variety of trading activity, which generated a wide range of social and cultural contacts, from friendship and marriage to the transmission of ideas and techniques. They are nevertheless hardly known or noted in the literature on trade in the Baltic, especially that which seeks to demonstrate the degree to which the Baltic was integrated into wider networks. In general, that literature focuses upon the sixteenth and seventeenth centuries, when the grain trade from eastern Europe to the Dutch market was at its height.[31] Some four decades ago, however, Roberto Lopez argued for the establishment during the thirteen and fourteenth centuries of firm commercial networks stretching from Livorno to Lübeck and beyond. He described the North Sea and the Baltic as a "northern Mediterranean," and this notion has been recently revisited, most notably by David Abulafia, who is struck by a range of structural similarities, such as commercial penetration by town-based merchants, and the exchange of luxury goods for raw materials as the basis for trade. Abulafia even describes the two seas as "regions which constituted frontiers between Christians and non-Christians."[32]

How valid or useful are such comparisons? At first glance, there seems little merit in attempting any such task.[33] The Mediterranean is far larger, and of course has a powerful and enduring history, with all that generates, stretching back to antiquity. The Baltic is very much a threadbare relative in comparison. True, there are the Vikings, but we prefer to visualize them sailing the open North Atlantic rather than up the muddy creeks of Denmark; there is the Hanse, but that carries with it the unacceptable political baggage of *Drang nach Osten*. The former are moreover generally perceived as unruly barbarians (whose descendants nonetheless brought about a marvelous fusion of Latin and Arabic culture in Sicily), the latter as a rather inward-looking group of traders. No Baltic Herodotus or Thucydides ever lauded the thalassocracy of the Danish kings of the Middle Ages; as a seventeenth-century Swedish poet sadly admitted, how many Gothic heroes had been condemned to oblivion by the lack of a northern Homer to record their deeds. The lemon tree does not blossom in the Baltic, there are no perfumes of wild thyme and basil, only the pungent odors of tar and cured herring, and *kåldomar* (cabbage rolls) are a poor substitute for *dolmas*.

If, however, it is possible to step outside these rather stereotyped comparisons, which not infrequently reduce the Baltic region to the status of a rather peripheral adjunct, there may be something worthwhile to say about the nature and character of inland seas. Whilst mindful of the dangers of committing the cardinal sin of slipping into exceptionalism, it would seem to me that one ought to begin by singling out those features which distinguish each sea, and which crucially determine the way in which people have lived around their shores. This is perhaps easier to do for the Baltic, which is much smaller and shallower than the Mediterranean, and subject to a less varied set of climatic and geophysical influences. Although the Baltic drainage area is surprising large—four times greater than the actual sea basin, with an estimated eighty-five million people currently living there—it has no great deserts or mountain chains, no active volcanoes or great deltas surrounding it. It has a limited range of fish species, though the brackish nature of the water means fresh as well as saltwater fish are found there. The circumjacent landmass is generally ill-suited to cultivation, and subjected to periods of prolonged and intense cold during the winter months. As has already been shown, ice has played and continues to play a crucial part in the region. Outside the port towns, settlement remains sparse; the fortified hill town from which the peasant-farmer emerged daily to till the fields, familiar around the Mediterranean, is unknown here. A seemingly barren landscape with few signs of human habitation beyond the harbors at the mouths of navigable rivers, the Baltic shoreline demanded adaptability from those who did choose to settle there, ingenuity, and an ability to make the most of meager resources.

In one respect, the historian of the Baltic may be deemed fortunate in not having to grapple with the legacy of classical antiquity that seems to cause problems for those writing about the Mediterranean.[34] Does this mean "basin thinking," to apply Kären Wigen's phrase to these waters, is not a "product of high imperialism"—an invention of the post-Enlightenment mind?[35] Traditional historiography would of course see the Baltic as a contested region, in which no one imperialism ever dominated. However, in the postmodernist world in which those of us nurtured under the old dispensation often struggle fully to comprehend what terms such as "high imperialism" might mean, "the struggle for Baltic supremacy," or a phrase favored by seventeenth-century contemporaries, *dominium maris Baltici,* do not seem to have much resonance. However, if Wigen's imperialism is essentially a master narrative imposed by a dominant culture, it may be possible to discern a common thread in European "inland seas" studies. The theme of the "civilizing mission" that crops up so often in studies of the Baltic area is in no way as powerful or evident as the Greco-Roman heritage, but it has nonetheless played an important part in defining the entire region. If the *Kulturträger* of the nineteenth century and their rather more brutal adherents of the twentieth century are no longer politically acceptable, there is still a perception of a strong undercurrent flowing eastwards into the Baltic from the more "civilized" West. "Through contact with the West," argues the Danish archaeologist Klavs Randsborg, "the North became integrated in [*sic*] general European development in a far more direct way than was the case in the early periods of Antiquity, when the lines of contact stretched mainly across land or along the rivers of central Europe."[36] "Bourgeois-mercantile initiatives" were singled out by Ahasver von Brandt, the preeminent postwar historian of the Hanse, as the decisive element in permanently and securely binding the entire northeastern quarter of Europe into the social, religious, and economic structures of the West. The incorporation of the Baltic region into the Hanseatic economic system should not be explained in terms of colonization, but rather as a bridging of the seas—from the hinterland of Germany via the island of Gotland to Novgorod, from Lübeck to Scania and Norway. "This shows beyond a doubt that economic factors were decisive here, not the politics of settlement."[37] "German-style Hanseatic culture" has more recently been singled out by the archaeologist David Gaimster as forming "a major vehicle of Europeanisation in the north."[38] It need hardly be said however that for those whom progressive-minded people in the nineteenth century were wont to call "peoples without a history," coping with the notion of civilization as something brought in by those who ruled over them, often with a harsh hand, posed serious problems—not unlike those faced by the autochthonous peoples of Africa placed under colonial rule in the nineteenth century.

Approaching the history of inland sea basins from the perspective of coastal communities can be helpful in avoiding some of the pitfalls of this particular master narrative. "Contacts and connections" are the keywords here; a high level of interchange facilitated by a network of transmarine communications. Looking at the evolution of these communities over a very long period of time reveals the richness and diversity of their contacts—the lively transfer of shipbuilding know-how, the import and export of distinctive commodities, the absorption of new fashions and ideas, and constant encounters with people from other regions and lands. It also shows that "civilization" and "integration" are not one-way processes, but are shaped out of these multilayered contacts and connections. Coastal communities—and not just the major ports—were the gateways through which much of this exchange flowed. The sailor returning home to his village after a year at sea on a Dutch merchantman was as much a carrier of new goods, habits, and ideas as the skippers and merchants of the harbor, even though the port had admittedly a far greater multiplier effect.

There are, however, a number of objections that may be raised against this "connectivist" scenario. For one thing, there were many coastal settlements as isolated and cut off from the outside world as the remotest mountain hamlet. There are sizeable coastal areas with little or no direct connection to the sea; areas of reclaimed marshland, for example, or zones of industrial production. There were also quite distinct and different gender roles in coastal communities, which undoubtedly affected the ways in which men and women experienced and perceived the wider world.[39] And, as already indicated, the fine web of coastwise contacts around the Baltic was eventually torn apart by the powerful and permanent intrusions from the land. It also offers little scope for what one might term the discovery and invention of the coast—discovery in the sense of charting the waters, marking out and guarding the seaways, but also the uncovering of the natural history of the shores—invention as a creative enterprise undertaken not only by artists and writers but also by owners of bathing establishments and tour promoters such as Thomas Cook.

It will perhaps be obvious from the above that I find the French concept of *maritimité* particularly apt for the study of the peoples around Europe's northern inland sea. As defined by the University of Caen, "maritimity" is a simple way of describing the complex relationships of man and the sea, evoking *la representation, la sensibilité, la perception que l'homme a de son milieu.* This perspective of the maritime world evolves according to different epochs, ideologies, technological evolution, and *les mentalités.*[40] It is not indifferent to the passage of time.[41] It works best on a relatively small scale—most of the research produced in France has centered upon a specific port or region, or a particular aspect, such as religious beliefs and superstitions—and is less vulnerable to suspicions of high imperialist grand narratives. It seeks to reveal the variety of human experience and on the whole avoids making too easy comparisons.[42] For the Baltic, a sea which in many ways has far more in common with large Lakeland areas than with deep and salty oceans, the intimacy of *maritimité* feels more appropriate than the more ambitious claims of thalassology. Where a national culture becomes too refined, where High German drives out the old *lingua franca* of *Plattdeutsch,* or where the revendications of Great Russian chauvinism threaten autochthonous cultures, the sense of a common maritime identity evaporates, according to the Finnish diplomat and essayist Lorenz von Numers. Writing over half a century ago, in a divided Europe, von Numers saw the Baltic as a delicate, difficult to define entity, something which only small jumbled-up communities held together by vigorous sea trading, overhanging threats, and with a common *lingua franca* could have produced.[43] In the second decade of the twenty-first century, the age of globalization, there may still be a case for seeking out the distinctiveness of small seas before they all become Mediterraneans.

NOTES

1. C. M. Blackford, "The Exploration of the Sea," *North American Review* 167, no. 502 (1898): 343.
2. P. Horden and N. Purcell, "The Mediterranean and the 'New Thalassology,'" *American Historical Review* 111, no. 3 (2006): 722.
3. E. J. Mather, *Nor'ard of the Dogger* (London: Nisbet, 1887), 178.
4. P. Horden and N. Purcell, *The Corrupting Sea: A Study of Mediterranean History* (Malden, MA: Wiley-Blackwell, 2000), 165.
5. K. Kolsrud, "Fishermen and Boats," in *The Northern and Western Isles in the Viking World: Survival, Continuity and Change,* eds. A. Fenton and H. Palsson (Edinburgh: J. Donald Publishers, 1984), 116–17.
6. J. Delumeau, *La peur en Occident, XIVe–XVIIIe siècles* (Paris: Fayard, 1978), 31.
7. This point is discussed in D. Kirby and M.-L. Hinkkanen, *The Baltic and North Seas* (London: Routledge, 2000), 261–63.
8. 85 percent of Finland's coastline is made up of islands, giving an estimated total length of around 46,000 km (28,583 miles). See *Finlands kuststrategi,* published by the Finnish Ministry of the Environment, MF10sv_2006 Fin lands kuststrategi.pdf, 9.
9. O. Brandt, *Der Kampf um die Ostsee am Vorabend der französischen Revolution* (Stettin: Ostsee-Dr. u.-Verlag, 1933); W. Sobieski, *Der Kampf um die Ostsee von den ältesten Zeiten bis zur Gegenwart* (Leipzig: Markert and Petters, 1933); E. Hornborg, *Kampen om Östersjön* (Stockholm: Bonnier, 1945); W. Konopczyński, *Kwestia bałtycka do XX w.* (Gdańsk: Inst. bałtycka, 1947); W. Kirchner, *The Rise of the Baltic Question* (Newark: University of Delaware Press, 1954). The background of political uncertainty in the region is especially noteworthy, though the theme of the struggle for Baltic supremacy can trace its roots back into the nineteenth century.
10. For example, the influential study by A. Attman, *The Struggle for Baltic Markets. Powers in Conflict 1558–1618* (Göteborg: Vetenskaps- o. vitterhetssamhället, 1979), and the more recent *In Quest of Trade and Security: The Baltic in Power Politics 155–1990,* G. Rystad, K.-R. Böhme, and W. M. Carlgren, eds. (Lund: Lund University Press, 1994, 1997).

11. For recent examples, see O. Crumlin-Pedersen, *Archaeology and the Sea in Scandinavia and Britain* (Roskilde: Viking Ship Museum, 2010); and J. Harff et al., "The Baltic Sea: A Model Ocean to Study the Interrelations of Geosphere, Ecosphere and Anthroposphere in the Coastal Zone," *Journal of Coastal Research* 21, no. 3 (2005): 441–46.
12. N.-A.Bringéus, *Åke Campbell som etnolog* (Uppsala: Kungl. Gostav Adolfs Akademien för svensk folkkultur, 2008).
13. O. Hasslöf, *Svenska västkustfiskarna* (Gothenburg: Svenska Vastkustfiskarnas Centralforbund, 1949); O. Löfgren, *Fångstmän i industrisamhället. En halländsk kustbygds omvandling 1800–1970* (Lund: LiberLäromedel, 1977); S. Erixon, *Stockholms hamnarbetare före fackföreningsrörelsens genombrott* (Stockholm: Nordisk rotogravyr, 1949).
14. P. Holm, *Kystfolk: Kontakter og sammenhænge over Kattegat og Skagerrak ca. 1550–1914* (Esbjerg: Fiskeri- og Søfartsmuseet, 1991). See also O. Mortensøn, *Sejlskibssøfolk: fra det sydfynske øhav* (Rudkøbing: Langelands Museum, 1987); W. Rudolph, *Die Hafenstadt: eine maritime Kulturgeschichte* (Oldenburg: Stalling, 1980), 128. See also his "Seefahrerdörfer der südlichen Ostseeküste. Tendenzen und Perioden der Entwicklung einer regionalen Sonderkultur (16. bis 19. Jahrhundert)," *Jahrbuch für Volkskunde und Kulturgeschichte* 10 (1977): 105–30.
15. See the chapters on "Mariners," "Sailors at Sea," and "Maritime Women" in Kirby and Hinkkanen, *The Baltic and North Seas,* 186–253, which provide an overview of maritime communities in the northern seas.
16. This definition is offered by the archaeologist Christer Westerdahl in 'The Maritime Cultural Landscape," *International Journal of Nautical Archaeology* 21, no. 1 (1992): 5–14. For a useful summary of the shifting focus of research, see J. Bentley, "Sea and Ocean Basins as Frameworks of Historical Analysis," *Geographical Review* 89, no. 2 (1999), 215–24.
17. P. Horden and N. Purcell, *The Corrupting Sea: A Study of Mediterranean History* (Oxford: Blackwell, 2000), 133.
18. H. Smeds," Skärgårdens kulturlandskap," *Skärgårdsboken 1948,* 540. In the North Sea basin, the rise in sea levels that followed the final retreat of the ice cap and the subsequent renewed periods of marine transgressions have had even more profound effects upon settlement and migration. For an interesting early discussion of the possible causes and consequences of rising sea levels on the Prussian coastline, see G. Pisanski, *Einige Bemerkungen über die Ostsee insonderheit an den Küsten von Preussen* (Königsberg: Gottlieb Leberecht Hartung, 1782), 21ff. Pisanski cites a mid-seventeenth-century epithalamium by the Memel poet Simon Dach, in which the bridegroom's father shows his son the view from the town walls, saying that where there are now ships sailing there once were sand dunes, a transformation that has occurred in his lifetime ("*Das ist bey meiner Zeit geschehn/Nur inner dreissig Jahren*").
19. On states and shipbuilding, see in particular J. Glete, *Navies and Nations: Warships, Navies and State Building in Europe and America, 1500–1860,* 2 vols. (Stockholm: Almqvist and Wiksell International, 1993).
20. K. Rosenkranz, *Königsberger Skizzen* (Danzig: Gerhard, 1842), vol. 1, 151.

21. The transition from sail to steam is a very good example, discussed briefly in Kirby and Hinkkanen, *The Baltic and North Seas,* 204–6.
22. There has been a massive increase reported in the number of summer cottages, not only in the Swedish and Finnish skerries, but also on the coasts of the Baltic states. See V. Palginõmm, U. Ratas, and A. Kont, "Increasing Human Impact on Coastal Areas of Estonia in Recent Decades," *Journal of Coastal Research,* special issue 50, ICS proceedings 2007, online ISSN 0749.0208, 114–19.
23. This image is reproduced in Kirby and Hinkkanen, *The Baltic and North Seas,* 142.
24. I. Oldberg, "The Emergence of a Regional Identity in the Kaliningrad Oblast," *Cooperation and Conflict* 35, no. 3 (2000): 269–88.
25. A. von Etzel, *Die Ostsee und ihre Küstenländer* (Leipzig: Lorck, 1859), v. The need for a strong German presence at sea was particularly emphasized by the advocates of the customs union: see for example F. Klefeker, *Der Zollverein und die Küstenstaaten Norddeutschlands* (Hamburg: Perthes-Besser and Mauke, 1844).
26. *Meereskunde. Sammlung volkstümlicher Vortrage zum Verständnis der nationalen Bedeutung von Meer- und Seewesen,* vol.1 (Berlin: E. S. Mittler, 1907), 2–3.
27. A. Corbin, *The Lure of the Sea* (Berkeley and Los Angeles: University of California Press, 1994), 57–96, 121–83.
28. For instance, J. Wallin, *Beskrifning öfver badorterna å Sveriges vestra kust* (Göteborg, 1858), 88.
29. Andersen's comments, written in 1859–60, are cited in Holm, *Kystfolk,* 282. T. von Kobbe and W. Cornelius, *Wanderungen an der Nord- und Ostsee* vol. 2 (Leipzig: Wigand, 1841) 3, 53.
30. According to contemporary testimony, cited in R. Wossidlo, *'Reise, Quartier, in Gottesnaam': Das Seemannsleben auf den alten Segelschiffen im Munde alter Fahrensleute* (Rostock: Hinstorff, 1959), 1.
31. M. Bogucka, "The role of Baltic trade in European development from the XVIth to the XVIIIth centuries," *Journal of European Economic History* 9, no. 1 (1980): 5–20. M.v an Tielhof, *'The mother of all trades': the Baltic grain trade in Amsterdam from the late 16th to the early 19th century* (Leiden: Brill, 2002).
32. R. Lopez, *The Commercial Revolution of the Middle Ages, 950–1350* (Englewood Cliffs, NJ: Prentice-Hall, 1971). D. Abulafia, "Mediterraneans" in *Rethinking the Mediterranean,* ed. W. Harris (New York: Oxford University Press, 2005), 80. The crusading analogy is rather weak; unlike the Muslim Arabs, the pagans of northeastern Europe never conquered much of the circumjacent lands of the Baltic, and the religious "frontier" that did eventually come into being was between two variants of Christianity.
33. A quick search of the web does indicate an interest in cross-comparisons between the two sea basins, mostly from marine biologists, though in 1994 there did appear a brief research report in *European Planning Studies* on marginalization, specialization, and cooperation in the Mediterranean and Baltic regions; the point of departure, inevitably, was the desired objective of European integration.
34. For a discussion of this, see P. Horden and N. Purcell, *The Corrupting Sea,*

722–40. The Finnish historian Matti Klinge has however argued that the northern parts of Europe have been defined, and have indeed defined themselves throughout the historical era in relation to the Mediterranean core, in *The Baltic World* (Helsinki: Otava, 1994), 7.

35. In her "Introduction" to the forum on Oceans of History, published in *The American Historical Review* 111, no. 3 (2006): 720, Wigen concludes that on the basis of the evidence offered in the subsequent essays on the Atlantic, Pacific, and Mediterranean "that basin thinking is the product of high imperialism."
36. K. Randsborg, "Seafaring and Society—In South Scandinavian and European Perspective," in *Aspects of Maritime Scandinavia, AD200–1200,* ed. O. Crumlin-Pedersen (Roskilde: Vikingskibshallen, 1991), 11.
37. See his "Die Hanse als mittelalterliche Wirtschaftsorganisation," in A. von Brandt et al., *Die deutsche Hanse als Mittler zwischen Ost und West* (Köln-Opladen: Westdeutscher Verlag, 1963), 12.
38. D. Gaimster, "A Parallel History: The Archaeology of Hanseatic Urban Culture in the Baltic c.1200–1600," *World Archaeology* 37, no. 3 (2005), 412.
39. See the chapter by Merja-Liisa Hinkkanen in Kirby and Hinkkanen, *The Baltic and North Seas,* 231–53.
40. http://atlas-transmanche.certic.unicaen.fr.
41. Corbin, *The Lure of the Sea,* 283.
42. On this, see especially P. Holm, "Coastal Life, "Nordic Culture" and Nation State: Relections on the Formation of a Nation State and Maritime History," in *People of the Northern Seas,* eds. L. Fischer and W. Minchinton (St. John's, Newfoundland: International Maritime Economic History Association, in conjunction with the Association for the History of the Northern Seas, 1992),194–200; R. Dettigmeijer, L. Heerma van Voss, and J. Roding, eds., *The North Sea and Culture, 1550–1800* (Hilversum: Verloren, 1996).
43. In his essay, "Kring det baltiska," in *Månen är ett säl,* ed. L. von Numers (Stockholm: Wahlström and Wistrand, 1952), 114–23.

NINE

The Mediterranean and the Mediterranean World in the Age of Peiresc

Peter N. Miller

In the early 1930s, while a *lycée* teacher in Algiers, Fernand Braudel met Nicolas Fabri de Peiresc (1580–1637). He seems to have been assigned the task of making sense of the correspondence between Peiresc and Sanson Napollon, governor of the French presidio of Bastion de France between 1628 and 1633. He seems not to have understood how Peiresc managed his correspondence network, and so concluded that the results of his inquiry were "disappointing."[1] But what might the history of scholarship in the twentieth century have looked like if the young Braudel had really come to understand how Peiresc managed the practicalities of his Mediterranean-wide correspondence?[2] Could we imagine how someone with Braudel's sense for the material dimension of life might tackle the practice of a scholar? Which is another way of asking how a materially oriented historian might tackle a corpus of culturally oriented evidence, or, how economic history and cultural history intersect. Could one pay attention to the material side of history without losing sight of the fact that history is made by human beings, with all their strengths and weaknesses?[3]

Peiresc offers an especially favorable vantage point for trying to bring together the questions of the cultural historian and the historian of material culture. His literary remains, for which I will use the term *archive*, contain on the order of fifty or sixty thousand pieces of paper. And these range widely, as befits one of the great early modern polymaths. Because of what survives we are able to evaluate his own thinking through a range of media and materials, including objects (or their drawings), informal sketches, drafts of essays, reading and conversation notes, memoranda, and letters. There are thousands upon thousands of these, the vast majority unread and unpublished, and the whole not yet calendared.

Peiresc, who worked on almost everything, did not leave Provence between his return from Paris in 1623 and his death in 1637. And since his intellectual life was mostly local, as was true for nearly everyone, it means that most of his intellectual life was necessarily conducted in the Mediterranean. Additionally, he was interested in antiquities, which inevitably meant a dense Italian correspondence network, and Oriental studies, which required contact with the Levant and North Africa.[4] But this could be accidental Mediterraneanism, what Nicholas Purcell and Peregrine Horden have called "histories *in* the Mediterranean" as opposed to "*of* the Mediterranean."[5]

So let us instead focus on Peiresc historian *of* the Mediterranean. He studied its natural history, its currents, tides, mountains, fossils, winds, and tectonic fractures. He mapped it and helped remap it.[6] He also studied its human history. He paid especial attention to the House of Barcelona-Counts of Provence, the rise, almost-empire, and fall of the Angevins, and the movement of Provencaux to the Kingdom of Jerusalem on crusade, and their landholding and ownership there.[7] This is the kind of history *of* the Mediterranean, on its axes of animate and inanimate, that was not supposed to exist before the middle of the twentieth century. Partly, this work is unknown because Peiresc did not publish his researches and, partly, because research and history have had such a tense relationship, going back to the Renaissance.[8]

Was Peiresc the antiquarian also Peiresc the thalassographer? If what we mean by this term is, as we have seen throughout this volume, a kind of historical scholarship permeated by the realities of maritime existence, but centered on the human experience of those realities, then the perspective of Peiresc of Marseille, not the usual one of Aix, gives us an early modern example. Just as "antiquarianism" is not exactly the same as "history" as we usually think about it today, Peiresc's "thalassographic" practice might not be the same as what we might think about as our own. And yet, these parallels show us connections between past and present that we may have missed, and reveal continuities which, once discovered, can change our sense of the norm against which both deviations and innovations are to be measured.

Because Peiresc's histories *of* the Mediterranean were conducted *in* the Mediterranean, the surviving documentation of his correspondence with those who facilitated his research—approximately 700 letters in two distinct dossiers—opens a window on to the everyday life of the Mediterranean in the 1620s and 1630s. The study of merchant letters is the key to this inside-outside perspective, both *in* and *of* the Mediterranean. Braudel the head of the VIe Section of the École Pratique des Hautes Études understood this by the time he was fifty (see chap. 1), even if he had not at thirty. Correspondence itself now appears as an agent of cultural exchange.[9] More particularly, London, Livorno, Greek, and Armenian merchants have been recently studied as subjects, not merely as vehicles of statistics.[10] There is no Geniza for the early modern Mediterranean republic of letters, but if there were, it would be the municipal library of Carpentras, where these letters are held.

Attending to the structure as well as content of Peiresc's Mediterranean correspondence, which Braudel did not, not only gives access to a Mediterranean that Braudel ignored (there is almost nothing on Marseille in *La Meditérranée*) but perhaps a different kind of sea-writing as well. Let us begin with the letters themselves. Jean Boutier, in a penetrating examination of the correspondence of the *érudit* Étienne Baluze, has used the term *constellation* to describe the letters to the lesser lights, defined either by smaller numbers or more occasional content. The Peiresc archive suggests that the more appropriate term might be *Cloud.* For, as in the astronomical imagery, we find clouds of correspondents clustered around, but also spreading out from, individual "stars."[11]

In these dossiers, letters to merchants cluster chronologically around the departure of ships to the Levant. Thus, for example, on May 14, 1629, Father Théophile Minuti departed for the Levant on Peiresc's bidding. A page from Peiresc's log of outgoing correspondence for that day (fig. 1) shows that in addition to the letters and memoranda carried by Minuti, and the twenty gold écus of Italy that Peiresc gave him, Peiresc also wrote letters to Estelle, vice consul in Sidon; Espannet, vice consul in Cyprus; Guez, a merchant in Constantinople; l'Empereur, in Jerusalem; Gabriel Farnoux and Cesar Lambert in Alexandria; and to the Marseille-based merchants de Gastines, Salicoffres and Messrs Mary, Douaille, and Fraise, for shipment to their unnamed contacts in Egypt. Those to the Marseille merchants would of course have stayed on this shore of the sea. But the others must all have been carried by Minuti, as if in his diplomatic pouch. The departure of a ship could, therefore, easily occasion the writing of upwards of ten to twelve letters. And all these were written by Peiresc within the space of a single day, after he learned exactly when the ship was to set sail.[12]

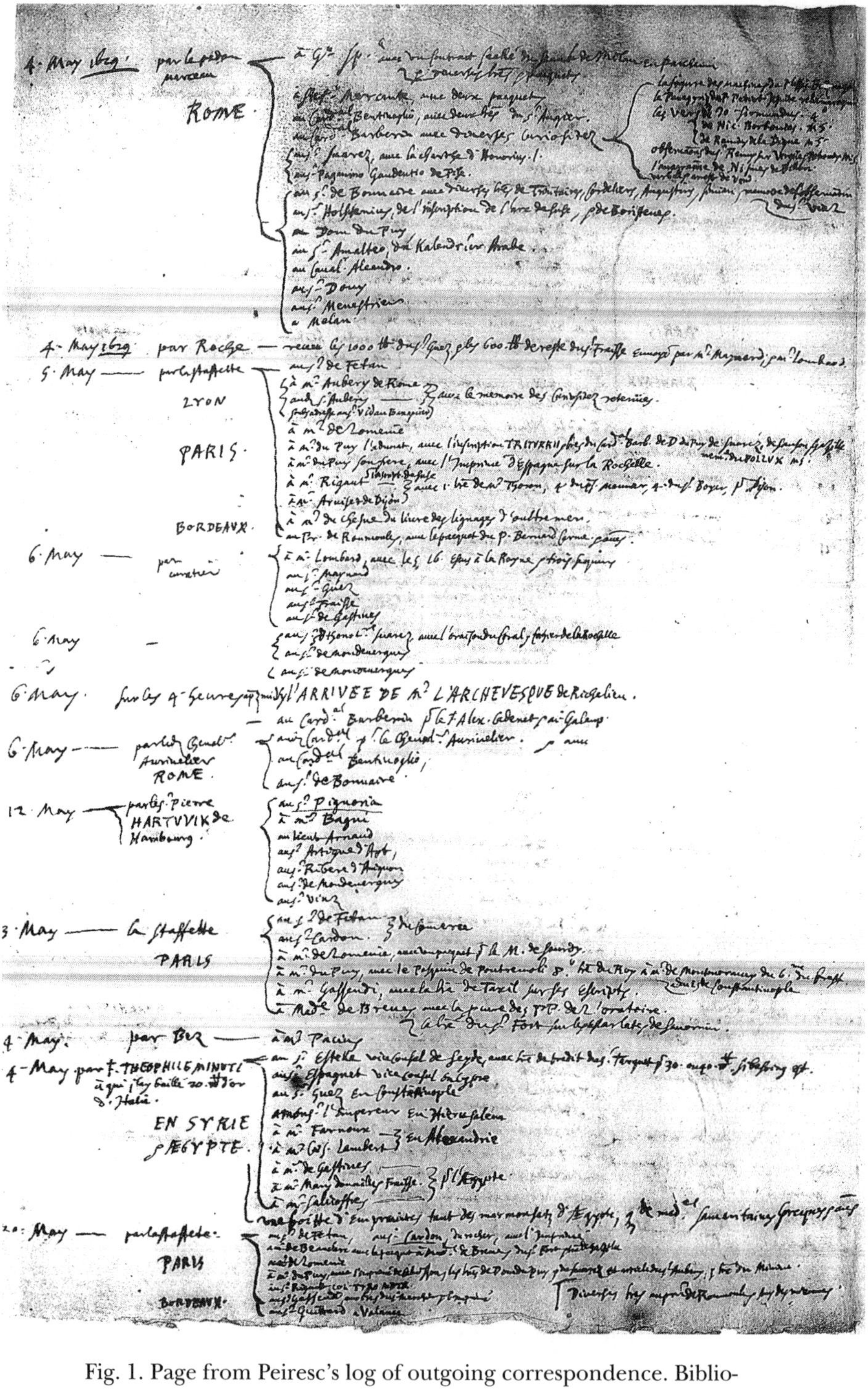

Fig. 1. Page from Peiresc's log of outgoing correspondence. Bibliothèque Nationale de France, Paris B.N. MS. N.a.f. 5169, fol.40v.

Let us now turn from the structure of Peiresc's letter-writing to the social structures we can tease out from the letters. This cloud of Marseille-based Mediterranean correspondence can be analyzed into several categories of persons (fig. 2). There are his contacts in the city of Marseille itself, mostly merchants, but also including some government officials and intellectuals. There are the shipowners and ships captains. There is also a much smaller circle in Toulon which must be included because during the years of plague in Marseille, roughly 1630–1632, Peiresc had to conduct most of his Mediterranean business through Toulon. Captains belong both to the near and distant shores of the sea. Diplomats were another category of correspondent who moved back and forth between Marseille and Ottoman North Africa and Levant. Then there are the Marseille merchants based in the Levantine ports—the *Échelles* of Aleppo, Sidon, Alexandria, and Cairo. Many of these represented family businesses with partners remaining back in Marseille. Finally, there were the travelers themselves. Some of these were personal friends such as Minuti, whom he sent on two expeditions, in 1629 and 1631, who was his confessor, and who closed Peiresc's eyes for eternity on June 24, 1637. But others were missionaries, including the Recollet Father Daniel Aymini, the Discalced Carmelite Celestin de Ste Lidwine and a group of Capuchins from the province of Tours who were staking out the French claim to Syria and Egypt. There were aristocrats such as the Parisian grandee François Auguste de Thou, and the Provençal François Galaup Sieur de Chasteuil. And, finally, there was a shadowy group of jeweler-goldsmiths who travelled east, to Ethiopia and India, in search of both raw materials and employment at Eastern courts.

What we learn from these letters is what a learned life by, in, on, and across the sea was like. One could write intellectual history from the Peiresc letters and come up with an account of the origins of oriental studies in Europe. One could write a cultural history from the Peiresc letters and come up with an account of the questions and practices that drove oriental studies.[13]

But one could also read these letters and come up with a material history of scholarship, one in which the categories could be those of the social or economic historian—numbers of ships travelling, shipping frequency, names of ships, names of captains, travel times across the sea, financing structures of overseas trade, profit margins and insurance premiums, quarantine practices, foreign merchant presence in the Mediterranean, etc. One could study these themes sequentially, but also synthetically, through the microhistories found in the archive,

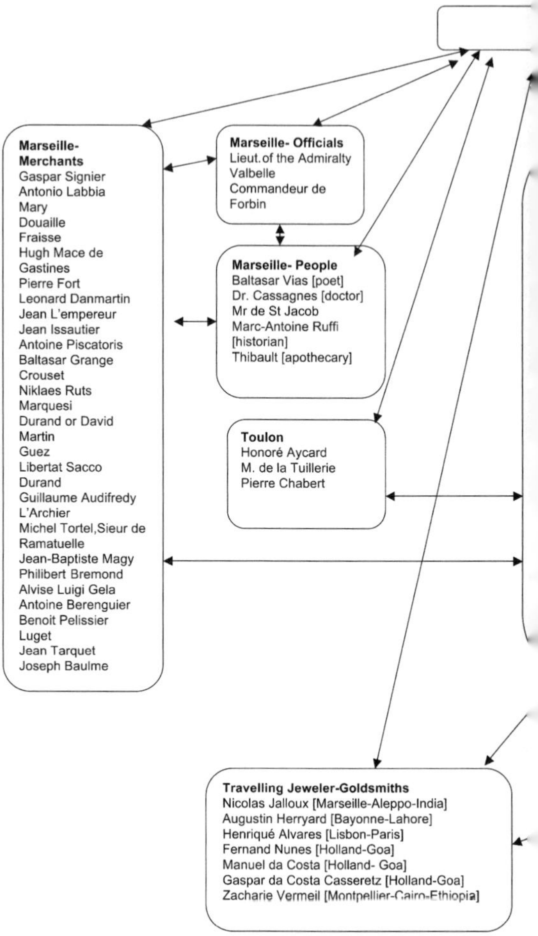

each of which demonstrates the varying impact of these material conditions. These might include the details of how to conserve books, metals, and a crocodile skin which had fallen into the sea, the naturalization of foreign-born Marseille merchants, the plan for ethnomusicological research at the Church of the Holy Sepulchre, the market for oriental manuscripts in Damascus, the depredations of Provencal corsairs on Provencal shipping, and the Mediterranean-wide hunt for a manuscript taken by corsairs.[14]

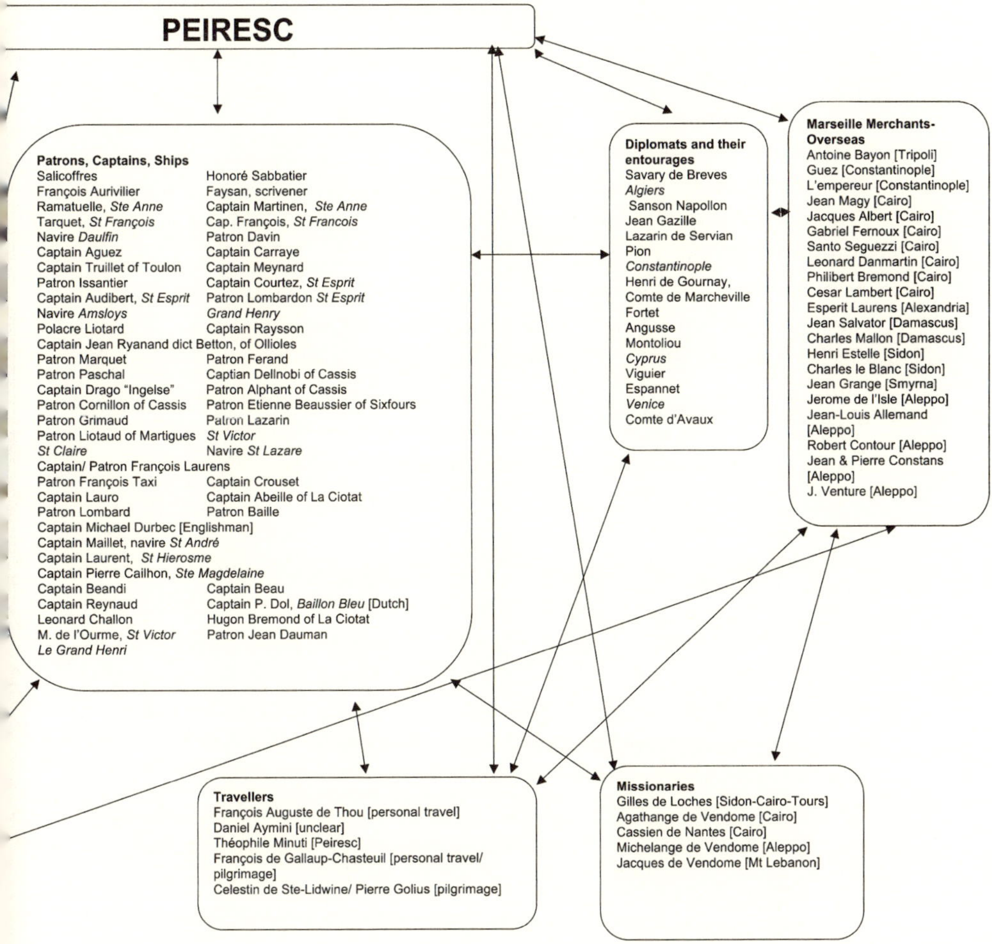

Fig. 2. Peiresc correspondence by category. Author's chart.

Lucien Febvre, famously, replying to Fernand Braudel's original idea for a study of Philip II, answered: "Philip II and the Mediterranean, a good subject. But why not the Mediterranean and Philip II? A subject far greater still."[15] "Peiresc and the Mediterranean," understood as Braudel did not—or at least did not in 1930—is a great subject for a historian because its central figure himself transcends dichotomization. He is an antiquarian but also historian; he studies natural and human events; he works with scholars and with merchants; he is fanatical about precision but unafraid to conjecture. He was both *in* and *of* the Mediterranean. Studying him in some sense means following him—and because he and his way of working has been so long forgotten that takes us, paradoxically, to a frontier of today's historical scholarship.[16]

The Mediterranean has become, since Pirenne's landmark book of 1937, but even more Braudel's of 1949, the arena for spectacular historiographical leaps forward. Braudel was followed by Goitein in the 1960s and 1970s. By the 1980s the first of the journals appeared (the *Mediterranean Historical Review,* in 1981), and then in the 2000s the monumental revisionist projects of Nicholas Purcell and Peregrine Horden, of Michael McCormick, and Chris Wickham.[17] But "Mediterranean" can, I think, be legitimately thought of not merely as a model of regional history. For its leading visionaries also used their Mediterranean scholarship to present models of how to do history that were quite self-consciously contrasted with practices elsewhere in the discipline. One crucial dimension so distinguished was the importance of material evidence. From Pirenne's "virtual material culture"—discussing objects only insofar as they are mentioned in texts—through Braudel's geography and social science, we have now come into an age in which archaeology, anthropology, and ecology drive the argument. For Michael McCormick, orienting economic history toward the movement of human actors and their agency led him to *Communications and Commerce* as the subtitle of a book with economic history in its title.[18] In fact, Peiresc described his own practice and priorities as "correspondance et communication" and "le commerce et la correspondence."[19] Peiresc's language suggests an affinity between his practice, and today's. Peiresc reminded Samuel Petit of Nîmes of his close relationship to commerce and yet of the ultimate difference between the goals of merchants and scholars. Discussing the implications of a possible breakdown in the Ottoman trade he wrote: "you know that I profess interests more sensitive than those of the merchants."[20]

The micro-ecological perspective of *The Corrupting Sea* is also echoed in the Peiresc correspondence. He was, as we have seen, attentive to the specific ecology of the Mediterranean, and always through the micrological lens afforded by correspondence. Writing to one person, in one place, his interest in the land and the sea was framed locally; writing to another, somewhere else, the question shifted to what that correspondence could know of and could answer.

In the following pages, I will try to sketch in outline the story of Peiresc and the sea, drawing out where possible its implications for a material culture of cultural history. This is an example of how thalassography offers opportunities for historiography (though not of course, to minimize the material turn elsewhere, in history of science, history of the book, and art history.) It also underlines the importance of the practical, and how the sea was one key contemporary junction of the learned and the commercial in the seventeenth century. Finally, looking back to the ambitions of Goitein and, especially, Braudel in the 1950s, we can see how a study of this sort could well have wound up in *Affaires et gens d'affaires* had it then been undertaken.

Marseille and the Mediterranean

Marseille, a city which does not figure at all in Braudel's Mediterranean, is at the heart of Peiresc's. Growing up and living in Aix would have put him only a day's travel away, and Peiresc was indeed often found in Marseille—sketches of ancient monuments and copies of inscriptions in his hand testify to a direct acquaintance from early in his life. His brother, Palamède, served as *viguier*—an honorary municipal chief executive—in 1633 and his financier, Pierre Fort, presented him and gave the *laudatio.*[21]

But more important is that Marseille was the main French gateway to the Mediterranean. The years spanned by Peiresc's life, 1580–1637, saw the rise of Marseille to a position of preeminence in the East, largely at the expense of Venice. By the end of the period, however, talk of decline was everywhere, including in the Peiresc archive, and spurred much discussion of reform.

As important as Marseille was, historians of the early seventeenth century remain reliant upon a single fifty-year-old history of Marseille's trade. Its age of Louis XIV and its eighteenth century are both now well studied. But no one has gone back to the first half of the seventeenth century in half a century.[22] Accounts of French Mediterranean trade are invariably centered on the later seventeenth century,[23] and wider histories of the Mediterranean for the most part persist in underrating its importance, and that of the French trade through it.[24]

The reason for this becomes clear when one actually goes to the archives: for the period in which Peiresc was active the public record is thin. It is thicker for the first two decades of the seventeenth century, and picks up again in the 1640s. For whatever reason, the 1620s and 1630s are very poorly documented.[25] Not only, then, does the Peiresc archive offer an unexpectedly rich vantage point on issues of economic history, but for the economic historian it is one of the few sources for learning about them.

As a magistrate and as a scholar, Peiresc showed a strategic interest in Mediterranean activity. This is attested in several different kinds of archival traces. There are documents which we could call "position papers," reflecting internal French discussions about whether and how much government support should be offered for maritime trade to the west and east. We also find a memo Peiresc drew up on the fisheries. Peiresc's commitment to these issues was such that he was thanked by the bishop of Marseille, Gabriel d'Aubespine, in 1627 for working to dispose the king toward "the re-establishment of trade in our seas." In this, Peiresc is presented as a crucial spokesman for a Mediterraneanist, commerce-oriented foreign policy, very much in the tradition of the high Gallican magistracy (Savary de Breves, Jacques Auguste de Thou).[26] The outbreak of open war between France and Spain in 1635 drew Peiresc still deeper into purely naval matters and made his news a valuable commodity for friends and contacts in metropolitan France.

One of Peiresc's friends was Alphonse de Richelieu, bishop of Aix. When he was elevated to the See at Lyon Peiresc kept him informed of Mediterranean affairs,[27] making the argument that the wealth of Lyon was connected to that of Marseille, thus simultaneously invoking the connectivity of commerce and casting himself as its defender.[28]

Since his friend's brother was *the* Cardinal Richelieu we can imagine that some of this information flowed on to Paris. Richelieu himself appointed Peiresc's brother-in-law, Henri de Seguiran, as his "lieutenant grand master of navigation" and gave him the task of drafting a report on military readiness on France's southern coast and of preparing a map. It is hard to believe that Seguiran did not consult Peiresc, since the first choice as cartographer was Pierre Gassendi—Peiresc's closest colleague and, later, biographer—and the second was Jacques Maretz, professor of mathematics at Aix. While the main part of the document was published in the nineteenth century, only in Peiresc's archive is an annex preserved ("Côte Maritime de Provence") which presents a detailed history of Marseille's trading relations in the Ottoman empire in the preceding decades.[29]

The author of the annex, whoever it was, explained the success of Marseille at displacing Venice in the first decades of the seventeenth century as directly related to the very nature of Marseille's maritime commerce: its more numerous, lighter, smaller, faster ships could make three trips in the time it took for the Venetians to make one. And, since the majority of the goods picked up in the East were textiles, paper, and other smaller items, speed and frequency mattered more than size. The luxury trades were smaller in size and smaller in overall volume, though they did account for a substantial amount of value.[30]

The economic historians who have tackled this subject, though, have remained within the familiar terrain of "big" versus "small," whether phrased in terms of subsistence versus luxury, *commerce de gros, commerce de détails,* or simply "the Northern invasion." In the end, the argument is always the same: because the northerners had the bigger ships they were inevitably going to be more successful and, implicitly, that carrying food for the masses always determines the course of history. Rethinking this in terms of the Mediterranean luxury trade follows from the general trend in recent decades toward consumption studies. And yet, in a striking indication of the current limits of even revisionist economic history, no economic historian has felt the need to make the obvious point that if the smaller, faster, more numerous Marseille shipping was well adapted for transporting small high-value goods, this same set of characteristics made it ideal for cultural communication. For letters, manuscript, memoranda, and books take up no more room than luxuries such as silk, coffee, spices, or even paper. Marseille, in short, offered an ideal base for someone with Peiresc's interests: a greater frequency, volume, and speed of shipping meant that his "commerce et correspondence" could reach further and faster than if he had been based anywhere else. It also explains why Braudel's favorite metric—tonnage —is not relevant to the cultural historian. The Mediterranean in the age of Peiresc, based on cultural practice, would measure connectivity; the Mediterranean of Phillip II, based on economic activity, would measure bulk.

Braudel's observation about the lasting importance of the invasion of the Mediterranean by Dutch and English shipping in the 1620s and 1630s finds corroboration in the Peiresc archive, but also takes on a different shape.[31] Braudel measures numbers. And we could, in fact, count mentions of Dutch ships or English shipping in the Peiresc correspondence. But it is more interesting—which is to say, the data is self-contextualized, and thus already interpreted for us—if we look at how the northern presence is found in Peiresc's practice.

He is himself less interested in their numbers than in their quality. Thus, he urges his merchant contacts in both Marseille and Egypt to use northern shipping whenever possible because the Dutch and especially the English were much tougher, hardier sailors. When confronted by corsairs the Provencaux fled, while the English fought back. This offered much better chances of survivability and of the safe return of cargo (and investment). The northerners, all Protestants, were also much less superstitious. They transported the Egyptian mummies that the Catholic Provençaux shunned.

The other clear and extraordinary demonstration of a northern, in this case Dutch, commercial role in Peiresc's Mediterranean has nothing at all to do with tonnage. It is the presence in Marseille of a branch of the Dutch trading firm of the Ruts family famous from Rembrandt's later portrait of the Russia trader, Niklaes Ruts, in The Frick Collection. (In fact, it may be this Niklaes, or another with this name, who was at Marseille.) "Ruts and Martin" was based in Marseille. And in the 1630s, especially from 1633, as Peiresc becomes very engaged with Aleppo, we find him turning to them to manage the communications and financing of his operations there. Hugh Mace de Gastines, his chief Marseille fixer, still plays a role, but one has the sense that the Dutch are better placed for the Aleppo trade. Even more surprising is that "Ruts and Martin," at least at one point, also had the more reliable connection to Bordeaux. When Peiresc needed to insure the transfer of funds from his abbey near St. Émilion to his bankers in Marseille, he turned to this Dutch firm to effectuate the transfer. Peiresc explained to Jean-Baptiste Magy, the Marseille-based brother of his Egyptian factor, that "there are only these English who traffic between Bordeaux and Marseille."[32]

Yet perhaps the most interesting commentary on the whole question of the Dutch penetration of the Mediterranean is the fact that when Jacob Golius in Leiden wanted to get in touch with his brother, Pierre, a discalced Carmelite who went by the name of Celestin de Ste Lidwine, he sent his letters via Paris to Peiresc, who then bundled them on board Marseille shipping to Aleppo, and then did the same in the opposite direction. No matter how mighty the Dutch fleet, no matter how well managed on the Syro-Lebanese coast, for speed, frequency, and reliability—what individuals sending precious goods, whether letters, manuscripts, or gems care about the most—Peiresc's Marseille was the destination.

Space and Time

From Marseille, the Mediterranean spread out. To Barcelona in the west, sailing time could be as short as a few days. But Peiresc had no regular correspondents in Spain. The iron curtain had already descended; Peiresc wrote to a French expatriate, Antoine Novel, in San Lucar de

Barrameda, to a noble Fleming, Lucas Torrius, in Madrid and to a visiting French cleric, Jean François of the Minims, in Barcelona. Beyond, to Lisbon, there were no Marseille connections at all.

To the south, it was around a fortnight's sail to Algiers. Peiresc's connection here was with Napollon and his entourage. More or less the same fortnight's sail connected Marseille to Tunis, where from 1631 Peiresc communicated with Thomas d'Arcos, a humanist secretary who apostatized and collected for Peiresc inscriptions, chameleons, and ethnographic data. There is even a small correspondence with Antoine Bayon in Tripoli in Lybia from whom Peiresc sought information about a lost manuscript (now in the Vatican Library as MS. Barb.-Or. 2).

To the east, letters reached Genoa in five to six days, and Genoa was a stop on the way to Civitavecchia, port of Rome. Rome, for the learned world of Oriental studies, had eclipsed Venice by the 1630s.[33]

Then there are the longer reaches, via Sicily or Malta to Rhodes or Chios for Constantinople and to Cyprus for Aleppo, Sidon, and Alexandria. How far was the long route? A Capuchin named Adrian de la Brosse made it in fifteen days from Marseille to Sidon, including a day-and-a half stop at Malta, and de Thou made it from Alexandria to Siracusa in Sicily in ten days. On the other hand, we learn of a forty-three-day trip from Sidon to Damiette instead of the usual six, because of a pirate-provoked flight to Cyprus.[34] More standard was a forty-day trip from Marseille to Egypt, including time spent in Malta and taking evasive action against corsaires.[35]

The letters map out the trunk routes, much as McCormick's prosopography helped him describe the roads of the sea in the early Middle Ages. There were, for example, "la routte ordinaire de Constantinople,"[36] and "la voye ordinaire de Marseille."[37] The route to Malta Peiresc actually called "the Malta caravan" (*la caravane de Malthe*). To the Toulon merchant de la Tuillerie, Peiresc described the route from Catalonia to Italy as "like a canal."[38] This meant direct, without stopping, which is how de Gastines described to Peiresc the route of de Thou from Malta to Barcelona, "ayant fait canal sans toucher a Rome."[39] In the shipping lanes, there were choke points. Islands, and by extension those who controlled them, Peiresc wrote to de Thou, were "Maitres de la Mer."[40]

For Peiresc, though, maps were not just metaphors. He possessed a map of the North African coast and another in which the portion around Tunisia had been rendered as if for a portolan, with rhumb lines and coastal detail.[41] D'Arcos in Tunis provided him with detailed line drawings of the area of Carthage, sometimes going into great detail and providing enough information to enable Peiresc to link ancient monuments referred to in texts, or the activities of ancients such as Scipio Africanus in 146 BCE, with extant physical remains.[42]

Peiresc, himself, from 1612 onwards—that is, as soon as he had mastered Galileo's observations of Jupiter's moons and realized that their movements could serve as a celestial clock for the purpose of regulating longitude at sea, all the way to the 1630s, when he switched to eclipse observation for its greater ease of record-keeping—was interested in clarifying the shape of the sea. He believed that the observation of 1635 would lead to a substantial correction in the shape and measurement of the Mediterranean and offered as proof of his approach the approval of the captains and seamen with whom he had shared his research.[43]

And then there are the distant reaches, the widest Mediterranean which Braudel bequeathed to us. From his Eastern correspondents Peiresc learned that it took forty days for the Aleppo caravan to reach Constantinople.[44] And, as he recorded:

> *D'Alep* en Bassora par le desert en 33 jours
> De *Bassorà* par mer à Muscat, en sept jours
> De *Muscat a Ormes* par mer en une jour à peu prez.
> De *Muscat à Goa* ou passe en 15 jours, a peu prez.
> De Goa on va à *Surate,* qui est un port du Mogor, ou à *Gaya,* qui est moings mercantile.
> De Gaya à Lahor y a 8 ou 10 jours de chemin. de Surate il y a moings.[45]

Merchants

Peiresc thought about the sea, and he thought about in terms of space and time. But above all, he thought about it in terms of people. In the Peiresc archive, we learn of a number of Arabs, North Africans, and Turks who were found in and around Marseille. Several of them we know well because they worked with Peiresc at different points on various Oriental-language projects, such as the Aleppo-born Mattouk Chiassan and the Berber Sayet of Taroudant in Marocco—the latter, together with the Greek Jew Salomon Casino, transcribed and translated Peiresc's collection of Islamic coins.[46] We also come across seasonal visitors, like the Ligurian citrus merchants—Domenico Majolo, Benedetto Gnieco, and Domencio del Monte—who passed through Provence every winter. There are also Maronite priests fleeing civil strife in the Lebanon in search of learned patronage, and maybe a cushy job as a professor or tutor, in Italy or France—three are mentioned in the Peiresc archive, Rabias, Moyse de Giacomo, and Sergio Gamario Reiskalla.

But most of the people in motion in the Mediterranean were merchants. Peiresc's age was one in which the Mediterranean was gradually ceasing to be a confrontation zone between Europe (what Peiresc generally called "Chrestienté") and the Ottomans. Corsairs, especially North Africans, but also rogue Christians, were a problem. Organized warfare was not. This made merchant life much more comfortable.

Of the great scholars and intellectuals of the seventeenth century it would be hard to imagine a closer set of working relationships with a merchant community. Peiresc entertained the merchants of Marseille at his home, and visited with them at theirs. They managed his financial affairs and he charged them with tasks both intellectual and practical, locally and overseas. He shared his research agendas with them, and they in turn helped him refine them. We can build up, prosopographically, a dense picture of the merchant community of Marseille, as perhaps from no other contemporary source, through Peiresc's correspondence.[47]

Peiresc also worked closely with the captains who sailed the ships fitted out by the merchants. Peiresc wrote to the captains directly, they visited him at Aix, and he shared with them his ideas, plans, and projects. They were in many cases his agents, and he was obliged to rely on their good judgment 2,000 miles from home. From his archive we can establish a list of the captains, seamen, and patrons who worked many of the Mediterranean routes: forty-nine are named and also thirteen ships.

Peiresc seems to have shared none of the contemporary prejudices about the inevitable superiority of the head to the hand, or theory to practice. He even articulated this in a letter of 1633. "There was," he wrote to the philosopher and mystic Jacques Gaffarel, "no evil but the defense of convention, which is no small 'impediment to acting well.'"[48]

This defense of "convention" was an attack on practical people and practical affairs. Peiresc had many friends in the highest of circles, but sometimes he preferred to work through a purely merchant channel. For instance, when he wanted a coin of Hadrian's from Belgium he wrote to the jeweler Henriqué Alvares in Paris and explained that he normally would have written to his friends Peter-Paul Rubens or Nicolaes Roccox for help on this, "But since the commerce of businessmen is freer than that of others, it would be easier for you to get this done than me."[49]

Among the intellectual projects Peiresc entrusted to merchants were

1. The measuring of ancient vases in Genoa.[50]
2. A Mediterranean-wide eclipse observation in August 1635—the key observation being made by Baltasar Claret, chancellor of the French consulate at Aleppo.[51]
3. Collecting of natural historical specimens, such as chameleons in Turkey—a task given to Baltasar Grange.[52]
4. Ethnographic reports on food ways in Egypt—from Jean Magi.[53]
5. Vases in Yemen—from the Turk from Aleppo, Mattouk Chiassan.[54]
6. Sub-Saharan caravan routes—also from Magi.[55]
7. Egyptian weights & measures—from a conversation with Cesar Lambert.[56]
8. Translation of Arabic documents—done by Philibert Bremond.[57]
9. Arabic epigraphy—the collaboration of Chiassan and Laurent Bremond of Marseille (the brother of Philibert).[58]

Peiresc gave—and took—with merchants in a way not so very different from how he operated with scholars. A long letter to the merchant Meynier in Damascus went into great detail about the language, literature, and script of the Samaritans so as to enable him to make somewhat informed purchases on Peiresc's behalf.[59] Another letter, to Santo Seghezzi in Egypt, took his understanding of canopic jars seriously enough to refute it.[60] Peiresc sharing his thoughts with merchant friends was part of the compensation they received for doing these favors for him: they were admitted to the world of erudition.[61]

Sometimes, indeed, it was their specific commercial training that was desired, as when Peiresc asked François Marchand in Rome to review evidence of tax revenues in Roman church registers.[62] Merchants were trained to pay attention to statistics, to revenue, and to expenses. Scholars usually were not. But Peiresc was different. His archive preserves three different analyses of Egyptian government administration, tax revenues and economic production, prepared by Cesar Lambert, Jacques Albert, and Santo Seguezzi.[63] These three merchants had been based in Egypt for decades and were closely involved with Marseille and Peiresc.[64] Their fascinating treatises, which blend ethnography, economic history, travel writing, and political science—they have since been published—belong to a genre that would be invented in Göttingen in the 1760s: *Statistik.*[65]

Some of the most interesting examples of the interpenetration of scholarly pursuits in the practical domain of merchants take us to the furthest reaches of Peiresc's Mediterranean: Ethiopia, India, and Yemen. A fascinating aspect of these is that in each of these three cases Peiresc's intellectual partner was a jeweler or goldsmith.

In Ethiopia, we find a jeweler from Montpellier, Zacharie Vermeil, probably a Huguenot, who left for Egypt—where he met some of Peiresc's friends—and then moved to Ethiopia where his talents caught the attention of the emperor. Once introduced into court, Vermeil became even more important as a military theorist, bringing modern Dutch notions of warfare to the Horn of Africa. He transmitted to Peiresc a request for books in 1633 via Cairo, and in a very long letter of February 1634, Peiresc replied.[66]

The Marseille jeweler Nicolas Jailloux had made two trips to the Levant and India before 1629, when he undertook his third and last. Peiresc had purchased gems and medals from him, and also solicited information about the Deccan.[67]

Augustin Herryard, of Bayonne and then Lahore, who worked as a jeweler for the Mughal ruler, provided information about jewels and precious stones, but also about court ceremonial. We know of his existence alongside of other French jewelers who worked at the Mughal courts.[68] But it is because of Peiresc that we possess his letters. Peiresc, in turn, asked him about the mountains in which gems were mined, and also about the fossils and shells, especially marine petrifications, found there.[69]

Peiresc may have learned of Herryard from Henriqué Alvarez in Paris. Alvarez had a brother named Fernand Nunes (though he was also called Guillaume Corner) and was from Hamfort in Holland. Alvarez, in turn, was married to the sister of Manuel da Costa Casseretz; Manuel had a brother named Gaspar da Costa Casseretz, and Gaspar da Costa and Fernand Nunes visited Peiresc in Aix on their way to India. Their background, profession—they were diamond merchants—and geographical dispersion identifies them as a family of Portuguese *Conversos*, or New Christians. There is no direct evidence of this in the Peiresc archive, but there is in that of the Lisbon Inquisition.[70] These family ties were summarized by Peiresc in an undated memo, probably prepared in 1630, when Nunes and da Costa visited him in the countryside near Toulon.[71] It was from them that Peiresc derived the information contained in his "Memoire pour les Indes." Letters, sent via Egypt, followed them to Goa.

The challenge of the Dutch and English East Indian ventures to Marseille's silk route was spelled out to Peiresc in a letter from one of his first

Marseille bankers, Gaspar Signier, already in March 1626. The English had begun to load silks at Ormuz, bypassing the Levantine emporia of Alexandria and Aleppo, and then shipping huge quantities back to England in their heavy merchant ships (600 bales of silk while no Marseille ship could carry more than 30).[72] The solution Signier proposed was for Marseille to found its own Indian ocean entrepôt.

We have seen that Peiresc took the protection of Marseille's Mediterranean commerce very seriously, as his own intellectual project of "correspondance et communication" depended upon it. He seems to have heard what Signier was saying because from then on we find Peiresc paying careful attention to Yemen. In the spring of 1629 Peiresc was soliciting news about Yemen, and already viewed it as a place of significance.[73] There are several memoranda[74] drawn up by Peiresc on the state of Yemen, its rebellion against the Ottomans, its governance, natural history, and arts and crafts. These seem to reflect face-to-face conversations with Jean Magy.[75]

It was, in the end, another "lapidaire" from Aix based in Marseille—Benoit Pelissier—who brought Peiresc to the mouth of the Indian Ocean. A memorandum composed by Peiresc in February 1635 and likely destined for de Thou at court, outlined a plan to send Pelissier to Moucal, or Mokka, on the Red Sea. From there he would be able to tap into the luxury trade not just of south Arabia (*Arabia Felix*), but of the Indies and Ethiopia, taking advantage of his contacts with Vermeil, Herryard, and Nunes.[76] The archive also explains something of why it never happened: we learn that Pelissier was murdered, on the road to Venice, sometime in the first half of 1636.[77]

Names

At this point an apology is in order. The preceding pages have been a fairly unremitting bombardment of names, almost all of them unfamiliar. I have done this on purpose, knowing all the while that it is bad technique, in order to convey something of the effect of reading through Peiresc's papers: the constant presence of names, almost always unidentified—and many unidentifiable outside of the correspondence itself. The gap between the strangeness of this to us and the obviousness of this to him outlines the contour of a question we need to ponder.

The attention to names was not accidental. Peiresc prepared whole memoranda that are lists of names with brief biographical information.

Three of them are from Egypt, which says something both about the density of foreigners there and about the value of the Egyptian trade to Marseille and Peiresc. One seems to have been drawn up before 1633, probably with information derived from Cesar Lambert.[78] A second was drawn up in July 1633 during conversations with Jean Magy, then visiting from Cairo.[79] A third seems to derive from the period before 1631.[80] Similar memoranda survive from Chios, which the Genoese held as a trading post,[81] and from Sicily.[82]

Names, in fact, may mark out the precise fault line between different historiographical regimes. Braudel, famously, omitted discussion of Philip II himself until the final pages of an 1,100–1,200 book. Goitein, sticking close to his documents, produced a picture of the Mediterranean and Indian Ocean traders that was full of names. If we want to understand Peiresc's thinking, we will find Goitein a much more helpful guide.

In reading through some of the letters published in Goitein's India book, we find this same thick nominative web. Yes, business was at a human scale. But it is also the case that names, especially far away, provided correspondents with hand-holds, people to whom they could turn for help. In a world where family constituted a ready-made network, names opened out on to whole prosopographies. Families, in turn, also shaped how people saw the past. Peiresc, for instance, built his history of Provence on the foundation of his genealogical studies of medieval Provencal families.[83] Names also allowed for the independent parallel reconstruction of projects if one or another part of it fell out, whether by accident, aggression, or ignorance. Names were shorthands for sets of associations, none of which needed to be spelled out. A name, in short, was worth a thousand words.

Names, then, take us into a whole world of belief structures and motivations. Names tell us what Peiresc thought was important for others, and for himself. Names remind us, even when they are only the names of long-forgotten merchants, customs officials, bankers, travelers, wanderers, sea captains, missionaries, and consuls, that all kinds of history, whether economic history or the history of scholarship, and however antiquarian, are, in the end, histories of people.

Peiresc's Mediterranean was four hundred years ago. But its reconstruction speaks directly to the concerns of historians today. As I explained in the introduction to this volume, the human subject has returned to the center in the best work being done today on the Mediterranean, and these—whether relating to corsairing, merchants, or mobility—are themselves creating models for historians outside of the Mediterranean. Once we get beyond the glare produced by Braudel's masterpiece, we see that from the beginning of modern historical research the Mediterranean has incubated scholarly practices which later spread widely, such as Peiresc's kind of antiquarianism. The same remains true today. But in an age of thalassography we can expect that evidence-rich archives, episodes, and microhistories from elsewhere ought also to begin enriching the historian's palette.

In the meantime, the close encounter of Braudel with the Peiresc archive, like that of Braudel with Goitein, offers a roadmap of sorts through the historiography of the twentieth century. These were less opportunities missed—that would be a very anachronistic approach, and would obscure the great force of what Braudel did accomplish—than markers of a whole series of affinities, blind spots, and accidents. Yet, from the perspective of the early twenty-first century, a more or less clear direction can be discerned. Peiresc's keywords—"commerce," "communication," and "correspondence," now mark our own interests. Peiresc and his archive may be as important for the future of historical scholarship as for its past.

NOTES

Earlier versions of this argument were presented to the Early Modern European Seminar at Princeton University, the Graduate Student Colloquium at the University of Chicago, the Centre Norbert Elias at the École des Hautes Études en Sciences Sociales in Marseille, and the Early Modern History Seminar at the Université de Paris-I (Sorbonne) in Paris. I am grateful especially to Tony Grafton, David Nirenberg, Jean Boutier, and Wolfgang Kaiser for their comments. A full-scale treatment is forthcoming with the title *The Mediterranean from Marseille: Merchants, Mariners, Missionaries and a Scholar.*

1. Jacques Ferrier exposed this story in "Une symphonie algérienne (letters inédites de Peiresc à Sanson Napollon)," in *L'Eté Peiresc Fioretti II, Nouveaux Mélanges,* ed. Jacques Ferrier, (Académie du Var: Aubanel, 1988), 215–60.
2. Braudel himself undertook the Peiresc-Napollon project for the Society ("Une Symphonie algérienne," 217). Ferrier's jumbling of other facts, such as placing the publication of Napollon's *Discorso* in 1932 rather than 1929, does not inspire confidence. And yet, the fact remains that Braudel was once active and remained always a member of this society—a series of facts not acknowledged in any of the now substantial autobiographical, biographical or hagiographical literature but which surely bears on Braudel's problematic relationship to North Africa in his *La Mediteranée* (see Colin Heywood, "Fernand Braudel and the Ottomans: The Emergence of an Involvement (1928–50)," *Mediterranean Historical Review* 23 (2008): 177).

3. David Abulafia's critical alternative to a Braudellian Mediterranean makes just this point in its subtitle: *The Great Sea: A Human History of the Mediterranean* (London: Allen Lane, 2011).
4. For all this see the essays in *Peiresc's Orient: Antiquarianism as Cultural History in the Seventeenth Century* (London: Ashgate/Variorum, 2012).
5. Peregrine Horden and Nicholas Purcell, *The Corrupting Sea: A Study of Mediterranean History* (Oxford: Blackwell, 2000), 43.
6. Miller, "Peiresc and the First Natural History of the Mediterranean," in *Sintflut und Gedächtnis,* eds. Jan Assman and Martin Mulsow (Paderborn: Wilhelm Fink Verlag, 2006), 167–98; and Miller, "Mapping Peiresc's Mediterranean: Geography and Astronomy, 1610–1636," in *Communicating Observations in Early Modern Letters, 1500–1575. Epistolography and Epistemology in the Age of the Scientific Revolution,* ed. Dirk van Miert (Oxford: Warburg Institute Colloquia, 2013).
7. Miller, *Peiresc's 'History of Provence' and the Discovery of a Medieval Mediterranean.* Transactions of the American Philosophical Society, volume 101 (Philadelphia, 2011).
8. This is, of course, a huge subject. I survey some of the relevant issues in "Writing Antiquarianism: Prolegomenon to a History," in *Antiquarianism and Intellectual Life in Early Modern Europe and China,* 1500–1800, eds. Miller and François Louis (Ann Arbor: University of Michigan Press, 2012), chapter 1.
9. For example, *Correspondence and Cultural Exchange in Europe, 1400–1700,* ed. Francisco Bethencourt and Florike Egmond (Cambridge: Cambridge University Press, 2007).
10. David Hancock, *Citizens of the World: London Merchants and the Integration of the British Atlantic Community, 1735–1785* (Cambridge: Cambridge University Press, 1997); Francesca Trivellato, *The Familiarity of Strangers: The Sephardic Diaspora, Livorno, and Cross-Cultural Trade in the Early Modern Period* (New Haven, CT: Yale University Press, 2009); Molly Greene, *Catholic Pirates and Greek Merchants: A Maritime History of the Mediterranean* (Princeton, NJ: Princeton University Press, 2010), esp. 131–37; Sebouh David Aslanian, *From the Indian Ocean to the Mediterranean: The Global Trade Networks of Armenian Merchants from New Julfa* (Berkeley and Los Angeles: University of California Press, 2011). Compare their treatments with an outstanding exemplar of the previous generation's perspective: Paul Bushkovitch, *The Merchants of Moscow 1580–1650* (Cambridge: Cambridge University Press, 1980).
11. Jean Boutier, "Étienne Baluze et l'Europe savante à l'age classique," in *Etienne Baluze, 1630–1718. Erudition et pouvoirs dans l'Europe Classique,* ed. Jean Boutier (Limoges: Presses Universitaires de Limoges, 2008), 291.
12. Paris B.N. MS. Nouvelles acquisitions français (henceforth N.a.f.) 5169, fol.40v.
13. These are the perspectives of, respectively, *Peiresc's Orient* and *Peiresc's 'History of Provence'.*
14. This paragraph describes the goal of my forthcoming *Mediterranean from Marseille.*

15. J. H. Hexter, "Fernand Braudel and the Monde Braudellien . . . ," *Journal of Modern History* 44 (1972): 510.
16. Putting intellectual history into manifold of maritime history precisely delineates the "next" frontier of sea studies as envisioned by Kären Wigen, "Introduction," in *Seascapes: Maritime Histories, Littoral Cultures, and Transoceanic Exchanges,* eds. Jerry H. Bentley, Renate Bridenthal, and Kären Wigen (Honolulu: University of Hawaii Press, 2007), 15–16.
17. Horden and Purcell, *The Corrupting Sea;* Michael McCormick, *Origins of the European Economy: Communications and Commerce AD 300–900* (Cambridge: Cambridge University Press, 2002); Chris Wickham, *Framing the Early Middle Ages: Europe and the Mediterranean, 400–800* (Oxford: Oxford University Press, 2005). For a review of the proliferating journals and smaller projects see Susan E. Alcock, "Alphabet Soup in the Mediterranean Basin: The Emergence of the Mediterranean Serial," in *Rethinking the Mediterranean,* ed. W. V. Harris (Oxford: Oxford University Press, 2005), 314–36.
18. The study of mobility has not only been his: see the essays in *Gens de passage en Méditerranée de l'Antiquité à l'époque moderne: Procédures de contrôle et d'identification,* eds. Claudia Moatti and Wolfgang Kaisar (Paris: Maisonneuve & Larose, 2007).
19. Carpentras, Bibliothèque Inguimbertine (henceforth Carp. Bib. Inguimb.) MS. 1777, fol.128r.
20. Peiresc to Petit, 7 April 1634, Carpentras, Bibliothèque Inguimbertine (henceforth Carp. Bib. Inguimb.) MS. 1875 fol. 268v: "Nous avons eu de grande apprehensions d'une rupture entiere du commerce avec le Turc ou vous scavez que je pretends des interests plus sensibles que les marchands sur les bruictz venuz du costé de Constantinople."
21. See Carp. Bib. Inguimb. MS. 1882, fols. 415r–419v.
22. This tendency continues up to the very minute; namely, Junko Thérèse Takeda, *Between Crown and Commerce: Marseille and the Early Modern Mediterranean* (Baltimore: Johns Hopkins University Press, 2011).
23. For example, Michel Fontenay, "Le Commerce des occidentaux dans les échelles du Levant vers la fin du XVIIe siècle," in *Chrétiens et musulmans à la Renaissance,* eds. Bartolomé Bennassar and Robert Sauzet (Paris: Honoré Champion, 1998), 337–70. The sole exception that I know of is over forty years old! Michel Morineau, "Flots de commerce et trafics français en mediterranée au XVIIe siècle (jusqu'en 1669)," XVIIe siècle, 86–87 (1970), 135–72.
24. Typical is the treatment in Faruk Tabak, *The Waning of the Mediterranean 1550–1870: A Geohistorical Approach* (Baltimore: Johns Hopkins University Press, 2008), 176, 224. An exception might be Romano Canosa, *Storia del Mediterraneo nel Seicento* (Roma: Sapere 2000, 1997). The recent boomlet in the study of the *corso* as itself a kind of trade has somewhat remedied this: see, for example, Wolfgang Kaisar, "Les 'hommes de crédit' dans les rachats provençaux (XVIe–XVIIe siècles)," in *Le Commerce des captifs. Les intermédiaires dans l'échange et le rachat des prisonniers en Méditerannée, XVe–XVIIIe siècle,* ed. Kasiar (Rome: Mélanges de l'École française de Rome, 2008), 291–318.
25. There is nothing for the later years of Peiresc's life at all like Wolfgang Kaisar's masterful *Marseille im Bürgerkrieg: Sozialgefüge, Religionskonflikt und Fak-*

tionskämpfe von 1559–1596 (Göttingen: Vandenhoeck & Ruprecht, 1991) for its earlier part.

26. Aubespine to Peiresc, 11 August 1627, Carp. Bib. Inguimb. MS. 1871, fol. 410r [*Les Correspondants de Peiresc,* 2 vols (Geneva: Slatikine Reprints, 1972), I, 250].
27. Peiresc to Richelieu, 24 July 1628, Carp. Bib. Inguimb. MS. 1875, fol. 486v.
28. Peiresc to Richelieu, 20 March 1633, Carp. Bib. Inguimb. MS. 1875, fol. 495r; Peiresc to Villeauxclercs, 20 March 1633, Carp. Bib. Inguimb. MS. 1876, fol. 623r.
29. The presentation copy of the report, Paris BN MS. F.fr. 24169, matches what is found in Carp. Bib. Inguimb. MS. 1775 fols. 1–75, adding only two more pages proposing additional seats of the admiralty for Provence (fols.81–2).
30. Fontenay, "Le Commerce des occidentaux dans les échelles du Levant," 348–49.
31. I will not dwell on this here, as it is a theme that has been touched on in several of my essays on Peiresc's Oriental studies. But it suffices as a place holder to note that when Peiresc was assembling books to send to Gedoin "le Turc" they included Greek books from England and Arabic ones from Holland (Peiresc to Aubery, 24 August 1629, Carp. Bib. Inguimb. MS. 1871, fol. 54v). On this more generally, see Molly Greene, "Beyond the Northern Invasion: The Mediterranean in the Seventeenth Century," *Past and Present* 174 (2000): 41–70, and now Colin Heywood, "The English in the Mediterranean, 1600–1630. A Post-Braudelian Perspective on the 'Northern Invasion," in *Trade and Cultural Exchange in the Early Modern Mediterranean: Braudel's Maritime Legacy,* eds. Maria Fusaro, Colin Heywood, Mohamed-Salah Omri (London: IB Tauris, 2010), 23–44.
32. Peiresc to Jean-Baptiste Magy, 5 April 1636, Carp. Bib. Inguimb. MS. 1874, fol. 413r.
33. For the latest detailed study of early seventeenth-century Roman erudition, see Federica Missere Fontana, *Testimoni parlanti. Le monete antiche a Roma tra Cinquecento e Seicento* (Rome: Edizioni Quasar, 2009).
34. Adrian de la Brosse to Raphael de Nantes, 25 November 1629 (from Beirut), Paris B.N. MS. N.a.f. 10220, fol. 95; Gilles de Losches to Raphael de Nantes, 15 December 1630 (from Cairo), fol. 103r.
35. Michelange de Nantes to Raphael de Nantes, 24 January 1633 (from Aleppo), Paris B.N. MS. N.a.f.) 10220, f.123r.
36. Peiresc to Guez (in Marseille), 6 September 1628, Carp. Bib. Inguimb. MS. 1876, fol. 372r.
37. Carp. Bib. Inguimb. MS. 1769, fol. 242v.
38. Peiresc to M. de la Tuillerie, undated, Carp. Bib. Inguimb. MS. 1873, fol. 425r: "il y a de l'apparence qu'il seroit canal pour aller aborder en Sardaigne & se joindre a ceux d'Italie avant que revenir en nos coste."
39. Gastines to Peiresc, 18 June 1629, Paris, Bibliothèque Nationale, MS. Fonds français (henceforth Paris B.N. MS. F.fr.) 9537, fol. 312.
40. Peiresc to de Thou, 31 July 1635, Carp. Bib. Inguimb. MS. 1877, fol. 463r. The context for this is the Spanish seizure of the Isles de Lerins.
41. Carp. Bib. Inguimb. MS. 1831, fols. 416v, 417v–418r.

42. Carp. Bib. Inguimb. MS. 1831, fols. 423–432.
43. "Mapping Peiresc's Mediterranean: Geography and Astronomy, 1610–1636," in *Communicating Observations in Early Modern Letters.*
44. Peiresc to Aycard, 5 June 1633, Carp. Bib. Inguimb. MS. 1871, fol. 31v.
45. Carp. Bib. Inguimb. MS. 1777, fol. 390r.
46. The story is told in "Peiresc and the Study of Islamic Coins in the Early Seventeenth Century," in *The Rebirth of Antiquity: Numismatics, Archaeology and Classical Studies in the Culture of the Renaissance,* ed. Alan G. Stahl (Princeton University Library Chronicle, Winter 2008), 315–85, reprinted as chapter 3 in *Peiresc's Orient.*
47. For example, while Boubaker Sadak is unusual in paying attention to the Marseille merchant community, he nevertheless feels the need to note that "Avant 1660, notre connaissance du commerce des Françias, essentiellement provençaux, est fragmentaire," *La Régence de Tunis au XVIIe siècle: ses relations commerciales avec les ports de l'Éurope méditerranéenne, Marseille et Livorne* (Zaghouan, 1987), 147. For Smyrna in 1851 Michel Morineau has produced a smaller scaled "map." "Naissance d'une domination. Marchands Européens, marchands et marchés du Levant aux XVIIIe et XIXe siècles," in *Commerce de Gros, Commerce de detail dans les pay méditérranéens (XVIe–XIXe siècles. Actes des journées d'études, Bendor 25–26 Avril 1975* (Nice, 1976), "Annexe," 184
48. Peiresc to Gaffarel, 4 July 1633, Carp. Bib. Inguimb. MS. 1873, fol. 403r: "Il n'y a aucun mal que la deffense de la prattica, qui n'est pas un petit Impedimentum rerum bene agendarum, car les bonnes lettres ont de besoing d'une correspondance et communication plus libre et sans entremise de tant de truchementz et tierces personnes." I thank my friend Jérôme Delatour for discussion of this phrase.
49. Peiersc to Alvares, 1 August 1634, Carp. Bib. Inguimb. MS. 1871, fol. 334v. This letter is omitted from those to Alvares printed by Tamizey de Larroque in volume VII of *Lettres de Peiresc.*
50. Peiresc to Pallavicino, 7 September 1634, Carp. Bib. Inguimb. MS. 1872, fol. 389r.
51. Peiresc to Constans, 13 May 1635, Carp. Bib. Inguimb. MS. 1873, fol. 46v. The best example of this is the work of Baltasar Claret in Aleppo: Peiresc to Claret, 21 May 1636, Paris B.N. MS. N.a.f. 5172, fol. 73r. See also Peiresc to Dupuy, 12 August 1636, *Lettres de Peiresc,* III, pp. 542–3 and discussed in Miller "Mapping Peiresc's Mediterranean.
52. Peiresc for Grange, 5 January 1635, Carp. Bib. Inguimb. MS. 1874, fol. 363r: "Memoire au sieur Jehan Grange allant à Smyrne, et au sieur François Grange son cousin. Neveux du sieur Baltasar Grange."
53. Paris B.N. MS. F.fr. 9530, fol.179r.
54. Paris B.N. MS. F.fr. 9532, fol. 43r.
55. Peiresc to Magy, 27 July 1636, Carp. Bib. Inguimb. MS. 1874, fol.422r.
56. Paris B.N MS. F.fr. 9532, fol. 38r.
57. Peiresc to Petit, 2 November 1633, Carp. Bib. Inguimb. MS. 1875, fol. 263r.
58. Paris, Bibliothèque Nationale, MS. Lat. 9340, fol. 226v.
59. Peiresc to Meynier, 16 November 1629, Carp. Bib. Inguimb. MS. 1876, fol. 363r–v.

60. Peiresc to Seghezzi, 25 April 1636, Carp. Bib. Inguimb. MS. 1874, fol. 423r.
61. Peiresc to Gela, 13 January 1636, Carp. Bib. Inguimb. MS. 1874, fol. 409v.
62. Peiresc to Marchand, early March 1633, Carp. Bib. Inguimb. MS. 1874, fol. 570r.
63. Lambert's is now in Paris, Bibliothèque Nationale, MS. Dupuy 669, fols. 219–36; Albert's in V Cents Colbert 483 (formerly Dupuy 475), fols.554–56 with a copy at MS. Dupuy 669, fols. 239–51v; Seghezzi's in Carp. MS. 1777, fols. 157–61 with copies at MS.V Cents Colbert 483 (Dupuy 475), fols. 554–564 and MS. Dupuy 669, fols.253–58.
64. Albert's "Memoire ample de l'estate de l'AEgypte" was written in 1634, Cesar Lambert's relation of Egypt around 1633 (see Lambert to Peiresc, 10 March 1633, Carp. Bib. Inguimb. MS. 1777, fol.342r), and Santo Seghezzi's detailed examination of the "Revenues d'Egypte" sometime in the early to mid-1630s.
65. They are published and discussed in Oleg V. Volkoff, *À la recherce de manuscrits en Égypte.* Recherches d'archéologie, de philologie et d'histoire, vol. 30 (Institut francais d'archéologie orientale) (Cairo, 1974); see also Sydney Aufrère, "Nicolas-Claude Fabri de Peiresc et ses correspondants de la nation du Caire: Santo Seghezzi, Jacques Albert et César Lambert," *Annales Islamologiques* 25 (1991): 311–19.
66. For more on this story, see Miller, "Peiresc's Ethiopia: How? and Why?," *Lias* 37 (2010): 55–88, reprinted as chapter 10 in *Peiresc's Orient.*
67. For Jailloux's activities, see Miller, "Peiresc and the Study of Islamic Coins in the Early Seventeenth Century."
68. Nuno Vassallo e Silva, "Precious Stones, Jewels and Cameos: Jacques de Coutre's Journey to Goa and Agra" in *Goa and the Great Mughal,* eds. Jorge Flores and Nuno Vassallo e Silva (Lisbon: Calouste Gulbenkian Foundation, 2004), 116–33. The throne made by Herryard for Jahangir is described on page 132.
69. Sneyders de Vogel, "Une lettre de Herryard, joaillier du Grand Mogol," *Neophilologus* 39 (1955): 1–8.
70. James Boyajian, *Portuguese Trade in Asia under the Habsburgs, 1580–1640* (Baltimore, MD: Johns Hopkins University Press, 1993), 138–41. I am grateful to Sanjay Subrahmanyam for drawing my attention to this parallel account.
71. Carp. Bib. Inguimb. MS. 1777, fol. 390r: A Gaspar da Costa Casseretz | ausente a Fran.co Tinoco de Carvallo E em ausensia d'ambos a Ruy. Lopes da Silva que de Sr g.de [?] Em Goa" This is all in another's writing. Peiresc adds, beneath, "India Oriental," and continues in his own hand. "Fernand Nunes ou Guill. Corner de Hamfort en Hollande qu'a un frere a Paris nommé Mr Alvarez Flamand qui se tient rüe Michel le Comte. Manuel de la Costa Casseretz, qui a une soeur marie audit Alvarez et qui est frere dudit Gaspar de Costa."
72. Signier to Peiresc, 20 March 1626, Carp. Bib. Inguimb. MS. 1879, fol. 636r.
73. Peiresc to de Thou, 25 April 1629, Carp. Bib. Inguimb. MS. 1876, fol. 354v: "Ceste affaire de l'Hyemen est de de trez grande importance si on l'eust mesnager mais on l'aura sans doubte laissé ruiner comme celle d'Erzeron." The context here suggests what we learn later of the province's rebellion against its Ottoman suzerains.

74. Peiresc to Magy, 17 May 1635, Carp. Bib. Inguimb. MS. 1874, fol. 368v. Peiresc added that Magy was not to tell the Venetian consul anything of this. Peiresc to Magy, 10 August 1635, Carp. Bib. Inguimb. MS. 1874, fol. 382bis v, thanking him for news of Suaquin and Moucal.
75. For example, Carp. Bib. Inguimb. MS.1777, fol. 374r: "1633. 7 Juill. IEAN MAGI avec le memoire des livres & graines / pour le ROY D'AETHIOPIE" [Aix, Bib. Mejanes MS. 207 (1025), 3].
76. Carp. Bib. Inguimb. MS.1777, fol.128r–v.
77. Peiresc to Gela, 24 Feb 1637, Carp. Bib. Inguimb. MS. 1874, fol. 442r.
78. Paris B.N. MS. N.a.f. 5174, fols. 25r–v.
79. Carp. Bib. Inguimb. MS. 1864, fol. 256r.
80. Carp. Bib. Inguimb. MS. 1864, fol. 257r.
81. Carp. Bib. Inguimb. MS. 1777, fol. 365r. In addition to the memo's index, we know that Stefano Giustiniani of Scio was in Aix with the Capuchins for a legal case. Peiresc to Aycard, 31 October 1636, Carp. Bib. Inguimb. MS. 1871, fol. 67v.
82. Carp. Bib. Inguimb. MS. 1821, fol. 140r.
83. This is elaborated in my *Peiresc's "History of Provence" and the Discovery of a Medieval Mediterranean.*

Afterthoughts: Histories in Bottles

Sanjay Subrahmanyam

> "God! he said quietly. Isn't the sea what Algy calls it: a great sweet mother? The snotgreen sea. The scrotumtightening sea. *Epi oinopa pon-ton*. Ah, Dedalus, the Greeks! I must teach you. You must read them in the original. *Thalatta*! *Thalatta*! She is our great sweet mother. Come and look."
>
> —James Joyce, *Ulysses*

We had all heard of the proverbial message-in-a-bottle long before it was immortalized by Sting and The Police in a song from their album *Reggatta de blanc* in 1979. It symbolizes the dilemma of the Robinson Crusoe figure, the castaway on an "island lost at sea" sending out a message not to anyone in particular but to everyone in general, any person that is who cares to open the bottle that has washed up on a beach somewhere. But what if one were a sailor from Brittany or Normandy passing through the Malvinas in 1700, who picked up a bottle launched in Batticaloa with scribblings inside in Sinhala or Tamil? "Oceans Connect" was no doubt the evocative slogan of an ambitious program launched to great fanfare at Duke University in 1997–1998 (with generous support from the Ford Foundation), but do these connections produce meaning and intelligibility or merely noise?

This was at the center of the challenge that the participants in the meeting of which this volume is a product set themselves when they met in Manhattan at the Bard Graduate Center in October 2009. The challenge itself was threefold. The first aspect—not new by any means—was that of writing histories that were conceptually oriented towards the sea rather than the land, thus thalassographies, and not the histories of camel-drivers and other landlubbers. Not surprisingly, a name that was frequently evoked in this context was that of Fernand Braudel and his work on the Mediterranean in the age of Philip II (first edition 1949; second edition 1966); but there were also other competing names and visions, to which I will return presently, even if in brief. The second aspect was that of historiographical models, and their transportability. Could a model of thalassography that was born, let us say, in the Mediterranean be transported with little or no modification or friction to another body of water—say the Indian Ocean, the China Sea, or the Baltic? Third, and finally, there was the problem of refining the appropriate units of analysis, and of whether one should always stick to those named and renowned seas and oceans that one finds today and one has found over the centuries in maps and books as the most meaningful units for reflection. Here one returns to the problem of oceans that connect, for a bottle that is launched in the Atlantic may respect the logic of currents and be subject to its share of vagaries and accidents, but is not necessarily aware that Atlantic studies is not supposed to be the same as Indian Ocean studies according to some normative convention laid down by wise people in universities.

Writing histories of seas, or where the sea is at the center of affairs, is certainly not new, as many of the preceding essays have been at some pains to point out. But equally, the tendency has always been a minority one within the larger sphere of historiography and is likely to remain so. This is for obvious reasons. Historiography all across the world has been dominated over the past two millennia by two central objects; namely, accounts (or biographies, if one likes that terminology) of states and empires on the one hand, and of ethnicities and communities on the other (the two objects have at times naturally overlapped, especially in the "national histories" of the last two centuries). In the nineteenth century, these two trends came to be described in shorthand as Hegelian and Romantic, respectively, but they are in fact far older and far more entrenched than that simple opposition would suggest. The relationship of the sea to both these objects exists, but is somewhat tenuous. States and empires that are organized using the sea as their central point of reference are certainly not unknown historically from the early Greeks and Phoenicians, but they are the exception rather than the rule. In the 1960s, a series of historical works in English was launched by the influential British historian J. H. Plumb (within his larger History of Human Society project) which used as its rubric the seaborne empire, and treated the Spanish, the Portuguese, the Dutch, and other cases from the early modern period. However, the reader of these volumes—

written by such major historians of the time as Charles Boxer and John Parry—often found that beyond a chapter (or at best, two) dealing with maritime transport and communications and seaborne trade patterns, the works themselves soon abandoned the sea for the land. Imperial administration, after all, was very largely a land-bound affair, and organized using territories divided into provinces. Viceroys and governors lived in and built ports and interior cities, and treated the oceanic moments in their lives largely as painful episodes to be hurried over in their reports and writings. The sea for them was more hiatus than lived reality. The main genre of histories with an overt maritime orientation that the early modern Iberians produced was after all the *História Trágico-Marítima*—a rather hair-raising collection of shipwreck narratives that ran over not one, but several oceans; in the seventeenth century, Dutch printers also made a great of deal of money on the back of the colorful horrors of their versions of shipwrecks and storms. It was almost as if thalassophobia, rather than any affection for the sea, ran like a red thread through the thalassography of these seaborne empires, and it certainly permeates even such epic works as Luís Vaz de Camões's *Lusíadas*.

Histories that were primarily oriented around communities or ethnicities also used the sea as a central reference only rarely. It is a well-known paradox that the early modern empires of the Iberians did not mobilize littoral populations for the most part, but men from the interior who often did not even know how to swim. The Spanish Extremadura, for example, is vastly overrepresented when one looks to participants in the seaborne empire of the Catholic monarchs to 1550. On the other hand, there are regions in the world where ethnohistories have been attempted based largely on the oral traditions of populations of "sea nomads"; historians of maritime Southeast Asia and the South Pacific have been at the forefront of this endeavor which concerns what in Malay are sometimes termed *orang laut* (or "people of the sea"). A particularly interesting variant of this work is that which deals with "ship-shape societies"—societies where the social and political imaginary is centrally shaped by maritime and seafaring metaphors. To be sure, many of these histories do have a markedly Romantic flavor, often the case in projects that celebrate the small voice of history in an aggressively anti-Hegelian mode (from Ranajit Guha to James Scott). More to the point, they often turn out to be remarkably structuralist and immobilist in orientation, which is perhaps inevitable given their dependence on oral material. It is thus some relief when one reads an account such as that of James Warren in this volume, which insists on the one hand on processes of ethnogenesis as a crucial phenomenon, and also shows on the other hand that such groups were not merely the victims of expanding states but also the perpetrators of violence themselves.

Fernand Braudel, in his study of the Mediterranean, was not particularly concerned with such groups, though he mentions them as he does so much else. While drawing on physical and even cultural geography to a large extent, he apparently did not take ethnography very seriously (a fact that possibly exacerbated his celebrated quarrel with Claude Lévi-Strauss). So much attention has been devoted to Braudel's work since the publication of its first edition in 1949 that it seems that little that is new can be said about it. But it is worth noting that the reactions to *The Mediterranean* themselves are in fact completely contradictory. Where some saw order, coherence, and high abstraction in the work, allowing it to be a model for maritime history in general, others (such as the Atlanticist Bernard Bailyn, writing as a graduate student in 1951) saw disorder, fragmentation, and incoherence, papered over with large doses of purple prose. The last sections of Braudel's work, which largely comprised a very conventional politico-diplomatic history of negotiations between the Western powers and the Ottomans, have been puzzling even to other members of the *Ecole des Annales.* How exactly do these passages of a *Chartiste* orientation fit in with the long sections replete with reflections on physical geography and its impact, which are used to launch a swingeing attack on event-oriented history? As I have suggested elsewhere, a clue may be found in the overly long gestation of the work itself, which began in the 1920s as a modest dissertation on the place of renegades in the society of sixteenth- and seventeenth-century Istanbul, and thus carried traces of several distinct projects in it. Furthermore, even if this reiterates the obvious, Braudel's Mediterranean remained all-too asymmetric, and oriented toward materials and perspectives from Spain, France, and Italy to the considerable detriment of the southern and eastern Mediterranean. The works of his disciples and younger collaborators such as Pierre Chaunu and Frédéric Mauro somewhat improved on this, but only because a single, quite coherent set of archives and materials could be legitimately deployed by them to deal with the Spanish or Portuguese Atlantic, where nothing like the archives of the Topkapı Saray could be found "on the other side."

However, could Braudel's Mediterranean be transported as a model of thalassography? We are aware that Braudel used as an ironic epigraph a phrase from the Jesuit José de Acosta, where Acosta pointed to the fact that no Mediterranean had been found in America. It would appear that Atlanticists for the most part were never particularly attracted to Braudel's vision, as Wim Klooster's essay here confirms. Amongst other reasons, this was because Atlantic historiography always tended to be organized nationally or imperially: units were the Anglo-Atlantic, the Hispano-Atlantic, the Luso-Atlantic, the Franco-Atlantic, and so on. Of late, some attempts at comparison have been made, notably between the Spanish and British empires in the Atlantic, as well as some timid sorties suggesting that we might read the histories of the two as somewhat entangled—especially in the context of the wider Caribbean world. But, with the possible exception of projects intended to establish a comprehensive quantitative base for the slave trade, there are few views that take on board the whole of the Atlantic, both north and south, and across its European and indigenous diversity.

A well-known sea shanty that was once taught to English schoolchildren began with how "there were three sailors of Bristol city/ who took a boat and went to sea/ there was Gorging Jack and Guzzling Jimmy/ and the youngest he was Little Billy." Now in terms of thalassography, the Mediterranean and the Atlantic have certainly been Gorging Jack and Guzzling Jimmy, with the Baltic Sea, or the Pacific and Indian Oceans being habitually reduced to Little Billies pretty much starved of oxygen. This is a curious and unfortunate fact which is only recently being rectified, and it has meant that some rather obvious parallels have not been sought out. For example, it is clear that the most obvious space to which a Mediterranean model could be transposed would be the Baltic, with the Danish Sound closely approximating the Straits of Gibraltar, and the Gulfs of Bothnia, Finland, and Riga corresponding rather obviously to the interior seas in the Mediterranean such as the Aegean and the Adriatic. Yet the 1960s and 1970s saw no German historian of the Baltic slavishly (or otherwise) producing a work entitled *Die Ostsee und die baltische Welt in der Epoche des Königs Gustav II. Adolf von Schweden*—a measure, perhaps, of the distance between French and German historiographies of the time. Rather, the Braudelian influence came to felt most strongly in the field of Indian Ocean studies, despite the fact that the very loosely bounded Indian Ocean as a space had no resemblance whatsoever to the Mediterranean. This concrete spatial question, while it was met with profound indifference by authors like K. N. Chaudhuri and Anthony Reid, who by 1980 were wholly under the Braudelian spell, was reflected on at some length by Braudel's younger Parisian colleague, Denys Lombard. Lombard proposed the existence not of a general parallel between the Indian Ocean and the Mediterranean, but between the seas of Southeast

Asia and the Mediterranean. To this former spatial entity, he gave the name *laut tengah* (or "sea in-between"), and he defined it as including not merely the waters between mainland Southeast Asia and Java and Sumatra, but also a good part of the South China Sea. Lombard's intervention in this matter has been debated somewhat, though perhaps with less vigor than one might have expected. It is therefore heartening to see a version of this discussion emerge once more here, with a focus on the South China Sea.

It is surely necessary now, though, to go beyond models of maritime spatial organization that can travel or be transposed to think more flexibly about the appropriate units for study. Jorge Luis Borges once wrote that it was his destiny "to have sailed across the diverse seas of the world/ or across that single and solitary sea of diverse names." This should give us reason for pause. Were one to write a history of the lascars—the Indian maritime laborers who manned so many ships in the late nineteenth and early twentieth centuries—would the distinction between the Indian and Atlantic Oceans be so crucial? Perhaps, but only because different forms of labor legislation might prevail in different spaces and these might be defined in terms other than the names of oceans. To take another example, a proper consideration of Dutch maritime hegemony in the seventeenth century would embrace the Atlantic, the Baltic, and the Mediterranean—to say nothing of the Indian Ocean. We would be very foolish then to allow the mere names of these diverse seas to oblige us to separate them as distinct objects, rather than interconnected spaces—especially when few of them had such well-defined gates as Gibraltar and the Danish Sound. Or, to follow the track proposed by Roxani Margariti, we might wish to consider in a comparative perspective other spaces entirely such as islands, which at times were "crossroads" (to borrow another favored term of Denys Lombard), and at others very large spaces in themselves of the dimensions of Madagascar, Sumatra or Vancouver Island.

We might also want to reflect on the convention that excludes certain sizeable water bodies from consideration when we engage in thalassography. The Caspian Sea, or Darya-yi Khazar, has attracted the attention of hydrographers and most recently of natural resource specialists (interested both in its oil and natural gas, and its caviar), but it is very poorly represented in historiographical terms despite the fact that it has a deep and complex history. In comparison, the neighboring Black Sea—discussed here by Nicola di Cosmo—has traditionally fared far better, whether when discussed on its own or when conjugated (as Di Cosmo does) with the study of large territorial empires. Furthermore, if the purpose of thalassographers is at least in some measure to be gadflies, challenging the smug received wisdom of national histories and their manifest destinies, we must equally wonder at the ongoing neglect of the Great Lakes region in North America as a coherent object of reflection. It is only the odd Scandinavian savant who has wondered whether Baltic and Great Lakes history could be studied together or in a comparative framework.

Though we often attempt in retrospect to render it coherent, history-writing in reality is often a fragmented and somewhat solitary affair. Even the most obvious of scholarly conversations—as between Braudel and Goitein—do not really take place, despite quite propitious circumstances. Historians often see themselves too as castaways, and it is not by chance that I began this brief reflection by evoking the message-in-a-bottle, which one supposes could also be linked to another familiar image, that of the genie—benevolent or malevolent—imprisoned in a bottle. As historians of the maritime world, many of us have no doubt felt quite another sentiment: that of the patient persons who piece by piece assemble ships in bottles. But I would like to end with a fourth such *mise en bouteille:* of those large pears that are grown gradually inside bottles (the technical term is to be held "captive"), so that they may eventually be pickled by and lend more flavor to the heady doses of *Poire Williams* that will eventually surround them. Maritime historians and their models could do much worse in terms of their relationship to their own intellectual surroundings.

Contributors

Nicola Di Cosmo is the Henry Luce Foundation Professor of East Asian Studies at the Institute for Advanced Study since 2003. Previously he taught at Harvard University and at the University of Canterbury, New Zealand. His main research fields are the archaeology and history of ancient China, Chinese military history, and Mongol and Manchu history. His current research is on the political culture, economic conditions, and social development of the Manchus in the half-century prior to their conquest of China (1590–1640). He occasionally teaches at Princeton University and other institutions. His publications include *Diary of a Manchu Soldier in Seventeenth-Century China* (2006), *A Documentary History of Manchu-Mongol Relations (1616–1626)* (coauthored, 2003), *Ancient China and Its Enemies: The Rise of Nomadic Power in East Asian History* (2002, Chinese and Korean translations), *The Cambridge History of Inner Asia: The Chinggisid Age* (coedited, 2009), and *Military Culture in Imperial China* (edited, 2009).

David Kirby is Emeritus Professor of Modern History at University College London. He has specialized for many years in the history of the Baltic region, and is the coauthor with Dr. Merja-Liisa Hinkkanen of *The History of the Baltic and North Seas* (Routledge, 2000). His most recent book is *A Concise History of Finland* (Cambridge University Press, 2006).

Wim Klooster is Professor at Clark University, in Worcester, Massachusetts, where he has taught since 2003. He earned his PhD at the University of Leiden in 1995, and was the Inter-Americas Mellon Fellow at the John Carter Brown Library, a Charles Warren Fellow at Harvard University, a Postdoctoral Fellow in Atlantic History at the National University of Ireland, Galway, a Fulbright Fellow, and the recipient of the NWO-Talent (Dutch government) stipend. Since 2001, he has been coeditor of Brill's Atlantic World series. He has authored or (co)edited eight books, including *Revolutions in the Atlantic World: A Comparative History* (2009) and *Illicit Riches: Dutch Trade in the Caribbean, 1648–1795* (1998), and is currently working on a book entitled *The Dutch Moment in Atlantic History*.

Roxani Eleni Margariti is Associate Professor of Middle Eastern and South Asian Studies at Emory University. Born and raised in Athens, Greece, she received her BA in Western Asiatic Archaeology from University College, London, her MA in Nautical Archaeology from Texas A&M University, and her PhD in Near Eastern Studies from Princeton University. She is the author of *Aden and the Indian Ocean Trade: 150 Years in the Life of a Medieval Arabian Port* (University of North Carolina Press, 2007), and coeditor of *Histories of the Middle East: Studies in Middle Eastern Society, Economy, and Law in Honor of A.L. Udovitch* (Brill, 2011).

Peter N. Miller is Professor and Dean of the Bard Graduate Center in New York City. He is the general editor of *Cultural Histories of the Material World* and writes widely on the history of historical research.

Nicholas Purcell FBA is Camden Professor of Ancient History and a Fellow of Brasenose College, Oxford. He was University Lecturer in Ancient History at Oxford University and held posts at All Souls College and St John's College, Oxford. Professor Purcell has written widely on Mediterranean history and the ancient city of Rome.

Angela Schottenhammer is professor of Chinese Studies at the Department of South and East Asian Languages and Cultures at Ghent University, Belgium, and adjunct professor of East Asian History at the History Department, McGill University, Montreal. She obtained her PhD in 1993 from Würzburg University, Germany, with a thesis entitled "Song Period Tomb Inscriptions" (MA 1989 on Liao Mosha and the Cultural Revolution) and her Habilitation degree (postdoctoral university professor teaching qualification) 2000 from Munich University with a thesis entitled "Song Time Quanzhou in a Conflict Situation Between Central Government and Maritime Trade: Unexpected Consequences of the Central

Government's Grasp for the Wealth of a Coastal Region." She is the editor of the online journal *Crossroads—Studies on the History of Exchange Relations in the East Asian World,* and of the book series *East Asian Maritime History,* part of *East Asian Economic and Socio-cultural Studies* (*Dongya jingji yu shehui wenhua luncong* 東亞經濟與社會文化論叢), and has widely published on China's and East Asian economic, social, and cultural history.

Sanjay Subrahmanyam is professor of history and the Navin and Pratima Doshi Chair of Indian History at UCLA. He received his PhD in economics from the Delhi School of Economics. Previously he was Professor of Economic History at the Delhi School of Economics, he has taught at Paris as Directeur d'études in the Ecole des Hautes Etudes en Sciences Sociales, and was appointed as the first holder of the Chair in Indian History and Culture at the University of Oxford. He is Joint Managing Editor of the *Indian Economic and Social History Review,* and on the editorial board of the multivolume *Cambridge History of the World,* and will jointly edit volume VI (in two parts). His publications include *The Political Economy of Commerce: Southern India, 1500–1650* (1990), *Improvising Empire: Portuguese Trade and Settlement in the Bay of Bengal, 1500–1700* (1990), *The Portuguese Empire in Asia, 1500–1700: A Political and Economic History* (1993), *The Career and Legend of Vasco da Gama,* (1997), *Penumbral Visions: Making Polities in Early Modern South India,* (2001), and *Explorations in Connected History: From the Tagus to the Ganges* (2004).

James Francis Warren is Professor of Southeast Asian Modern History at Murdoch University. He has held teaching and research positions at the Australian National University, Yale University, McGill University, National University of Singapore, and Kyoto University. His current research focuses on slavery and the creation of transcultural identities and aspects of the environmental history of Southeast Asia. James Warren's more recent publications include *The Sulu Zone, the World Capitalist Economy and the Historical Imagination* (1998) and *Pirates, Prostitutes and Pullers: Explorations in the Ethno—and Social History of Southeast Asia* (2008). In 2003, he was awarded the Centenary Medal of Australia for service to Australian society and the humanities in the study of ethnohistory. He lives in Perth, Western Australia, with his wife Carol, an anthropologist, and daughter Kristin, a wildlife medicine veterinarian and conservationist.

Index